Agricultural Water Conservation: Tools, Strategies, and Practices

Agricultural Water Conservation: Tools, Strategies, and Practices

Editor

Aliasghar Montazar

MDPI • Basel • Beijing • Wuhan • Barcelona • Belgrade • Manchester • Tokyo • Cluj • Tianjin

Editor
Aliasghar Montazar
University of California Agriculture
and Natural Resources
USA

Editorial Office
MDPI
St. Alban-Anlage 66
4052 Basel, Switzerland

This is a reprint of articles from the Special Issue published online in the open access journal *Agronomy* (ISSN 2073-4395) (available at: https://www.mdpi.com/journal/agronomy/special_issues/Water_Conservation_Agronomy).

For citation purposes, cite each article independently as indicated on the article page online and as indicated below:

LastName, A.A.; LastName, B.B.; LastName, C.C. Article Title. *Journal Name* **Year**, *Volume Number*, Page Range.

ISBN 978-3-0365-2646-1 (Hbk)
ISBN 978-3-0365-2647-8 (PDF)

Cover image courtesy of Aliasghar Montazar

Contents

About the Editor

Aliasghar Montazar is currently Irrigation and Water Management Advisor with the University of California Cooperative Extension in Southern California. He has a PhD in Irrigation and Drainage and more than 20 years of research, extension, teaching, and technical consulting experience and has served in several leadership positions in agricultural water management and irrigation engineering in California and abroad. Before joining University of California in 2011, he was associate professor at the Department of Irrigation Engineering, University of Tehran. He has a well-developed applied research and training program on irrigation and soil–water-related issues. His focus is sensor-based irrigation management, water conservation, evapotranspiration, and the best irrigation and nutrient management practices.

 *agronomy*

Editorial

Irrigation Tools and Strategies to Conserve Water and Ensure a Balance of Sustainability and Profitability

Aliasghar Montazar

UC Cooperative Extension Imperial County, University of California Division of Agriculture and Natural Resources, 1050 East Holton Road, Holtville, CA 92250, USA; amontazar@ucanr.edu; Tel.: +1-442-265-7707

1. Introduction

Efficient management and conservation practices for agricultural water use are essential for adapting to and mitigating the impacts of the current and future discrepancy between water supplies and water demands. The importance of water conservation in agriculture is fundamental to ensuring a balance of sustainability and profitability. While profitability is a primary concern in any sustainable enterprise, farmers typically adopt new tools and practices that result in higher profits or reduced risks. Irrigation management practices that reduce water use with acceptable impacts on production would be viable strategies and cost-effective tools to cope with diminished water supplies and generate new sources of water to transfer for other agricultural uses and urban and environmental demands.

The water conservation measures may reduce crop water use and/or improve efficiency, and consist of advanced irrigation scheduling, deficit irrigation, on-farm irrigation system conversion and improvement, tailwater recovery systems, precision irrigation, and crop rotation and alternative low water use cropping systems. This Special Issue focuses on "Agricultural Water Conservation: Tools, Strategies, and Practices", which aims to bring together a collection of recent cutting-edge research and advancements in agricultural water conservation.

2. Deficit Irrigation Strategies

Several studies investigated the effect of mid-summer irrigation cut-off (no irrigation after June until the following spring) on alfalfa yield [1–3]. Moderate deficit irrigation strategies applying 12.5–33% less irrigation water than farmers' normal irrigation practices during the summer period were evaluated in the low desert of California [4]. This strategy demonstrated a promising and decent amount of water conservation and simultaneously generated desirable hay yields and quality. Implementation of the proposed summer deficit irrigation strategies on alfalfa could provide a reliable source of seasonally available water as well as sustain the economic viability of agriculture in the region. These strategies might be sustainable as an effective water conservation tool if such measures provide adequate economic incentives to the participating farmers. Incentive programs to farmers must offset the risk of implementing the proposed practices, as a tool for adopting water conservation practices.

The effect of drip irrigation integrated with partial root drying (PRD) and soil mulching was studied on squash plants [5]. The PRD strategy improved both the squash yield and water use efficiency (WUE). Soil mulching enhanced the physiological properties of the squash plants, fruit quality, squash yield, and WUE. This study reported that sowing squash plants in the winter season, applying water to compensate 50% of evapotranspiration, and using plastic mulch as water-saving strategies could be used as a water-saving strategy without reducing yields.

Citation: Montazar, A. Irrigation Tools and Strategies to Conserve Water and Ensure a Balance of Sustainability and Profitability. *Agronomy* **2021**, *11*, 2037. https://doi.org/10.3390/agronomy11102037

Received: 18 September 2021
Accepted: 8 October 2021
Published: 11 October 2021

Publisher's Note: MDPI stays neutral with regard to jurisdictional claims in published maps and institutional affiliations.

3. Water-Conserving Irrigation Practices in Row-Crops

A survey conducted among row-crop farmers in Mississippi [6] showed that the amount of irrigated area, years of education, perception of a groundwater problem, and participation in conservation programs are positively associated with practice adoption, while the number of years farming, growing rice, and pumping cost are negatively associated with adoption. However, not all factors are statistically significant for all water conservation practices. Survey results indicated that only a third of growers are aware of groundwater problems at the farm or state level; this lack of awareness is related to whether farmers noticed a change in the depth to water distance in their irrigation wells. The study recommended adopting more efficient irrigation technology and practices and precision agriculture technologies, such as soil moisture sensors and irrigation automation.

4. Accurate Estimates of Evapotranspiration for Irrigation Scheduling

Various methods of evapotranspiration measurement have been discussed by several researchers [7–9]. The most common methods for evapotranspiration (ET) measurement may be categorized as hydrological approaches (soil water balances and lysimeter measurements), micrometeorological approaches (eddy covariance, surface renewal, and Bowen ratio energy balances), and plant physiology approaches (chamber systems and sap flow measurements) [7]. In recent decades, satellite-based ET estimates using vegetation indices and scintillometer systems were developed as a result of rapid advances in instrumentation, data acquisition, and remote data access. The surface renewal method may estimate the surface fluxes at a relatively low cost, ultimately improving calculations of evapotranspiration and providing an economical tool for improving crop water management [10].

Remote sensing-based ET estimation is considered a promising tool for irrigation water management; however, uncertainties associated with satellite-based ET estimation still exist, especially with various remotely sensed platforms due to variations in spatial and temporal resolution [11]. In a study, satellite-based ET was evaluated using Landsat under semi-arid conditions in Texas under irrigated and dryland conditions. The Landsat-based ET overestimated the measured ET early and late in the growing season and underestimated ET during the peak of the growing season. More satellite-based ET assessment under arid and semi-arid conditions is required, where the magnitude and frequency of precipitation are erratic, and irrigation is the only source under arid conditions to replenish crop water needs [11].

5. Water Efficient Crop Management

A study was conducted to determine the optimum irrigation levels, row spacing, and tillage to maximize WUE while maintaining stable forage production [12] in pearl millet (a warm season C4 grass well adapted to semiarid climates) in the semi-arid region of the Southern Great Plains, USA. The greatest average forage production was achieved with the highest irrigation level; however, the greatest WUE was attained in tilled soil due to greater LAI, light interception, and plant growth than in no-till [12]. While the application of water increases the forage production, low LAI values increase E (soil evaporation) and reduce WUE, especially without adequate nutrient application.

The adoption of integrated management strategies will be useful for growing tolerant genotypes under saline water conditions and increasing water use efficiency. For the sustainable management of crop growth in saline environments, soil–crop–water management interventions consistent with site-specific conditions need to be adopted [13]. A study conducted in Tunisia [14] recommended that farmers with higher salinity water for irrigation should grow tolerant barley genotypes, allowing them to reduce the cost, on average by 30%. Changing cropping patterns is also regarded as a useful strategy for the rehabilitation and management of saline soils, especially when only saline water is available for irrigation [15,16].

Postharvest drip irrigation of asparagus cultivated in very light sandy soil significantly contributed to an increase in crop productivity. A significant increase in the height, number, and diameter of summer stalks, as well an increase in the marketable yield, weight, and number of green spears were observed for several American asparagus cultivars [17].

Ground-based remote sensing data of NDVI and canopy temperature along with soil moisture and ET-based smart controllers were assessed for efficient irrigation management of hybrid bermudagrass and tall fescue turfgrass in central California [18,19].

6. Irrigation with Wastewater

Irrigation with wastewater may contribute to the reduction of water abstraction in agriculture with an especial interest in arid and semiarid areas. The results of a study conducted in Italy [20] suggested that low-diluted hydrocyclone filtered digestate liquid fractions could directly be injected in a drip irrigation system with few drawbacks for the system. It may significantly contribute to water conservation since such wastewater are available from the late spring to the early fall when water requirements are high in the region [20].

7. Surface Irrigation Improvements

The results of a study conducted in Mexico showed that an efficient design and operation of surface irrigation considering initial irrigation tests and evaluation, characterization of irrigation plots, and calculation of the optimal flow rate using an analytical formula may considerably reduce irrigation-applied water [21]. The study demonstrated that irrigation application efficiencies increased more than 100% in some cases, while the WUE increased by 27, 38, and 47% for sorghum, barley, and corn, respectively [21].

8. Improvements in Small-Scale Irrigation Systems

Solar-powered drip irrigation using solar MajiPump along with conservation agriculture (CA) farming systems were found to be efficient to expand small-scale irrigation and improve productivity and livelihoods of smallholder farmers in Ethiopia [22]. Compared to the farmers' practices, water productivity was significantly improved under the CA farming and drip irrigation systems for both irrigated vegetables (garlic, onion, cabbage, potato) and rainfed maize production.

Funding: This research received no external funding.

Conflicts of Interest: The author declare no conflict of interest.

References

1. Ottman, M.J.; Tickes, B.R.; Roth, R.L. Alfalfa yield and stand response to irrigation termination in an arid environment. *Agron. J.* **1996**, *88*, 44–48. [CrossRef]
2. Cabot, P.; Brummer, J.; Hansen, N. Benefits and impacts of partial season irrigation on alfalfa production. In Proceedings of the 2017 Western Alfalfa and Forage Conference, Reno, NV, USA, 28–30 November 2017.
3. Guitjens, J.C. Alfalfa irrigation during drought. *J. Irrig. Drain. Eng.* **1993**, *119*, 1092–1098. [CrossRef]
4. Montazar, A.; Bachie, O.; Corwin, D.; Putnam, D. Feasibility of Moderate Deficit Irrigation as a Water Conservation Tool in California's Low Desert Alfalfa. *Agronomy* **2020**, *10*, 1640. [CrossRef]
5. Farah, A.H.; Al-Ghobari, H.M.; Zin El-Abedin, T.K.; Alrasasimah, M.S.; El-Shafei, A.A. Impact of Partial Root Drying and Soil Mulching on Squash Yield and Water Use Efficiency in Arid. *Agronomy* **2021**, *11*, 706. [CrossRef]
6. Quintana-Ashwell, N.; Gholson, D.M.; Krutz, L.J.; Henry, C.G.; Cooke, T. Adoption of Water-Conserving Irrigation Practices among Row-Crop Growers in Mississippi, USA. *Agronomy* **2020**, *10*, 1083. [CrossRef]
7. Montazar, A.; Krueger, R.; Corwin, D.; Pourreza, A.; Little, C.; Rios, S.; Snyder, R.L. Determination of Actual Evapotranspiration and Crop Coefficients of California Date Palms Using the Residual of Energy Balance Approach. *Water* **2020**, *12*, 2253. [CrossRef]
8. Snyder, R.L.; Spano, D.; Pawu, K.T. Surface Renewal analysis for sensible and latent heat flux density. *Bound. Layer Meteorol.* **1996**, *77*, 249–266. [CrossRef]
9. Paw, U.K.T.; Snyder, R.L.; Spano, D.; Su, H.-B. Surface renewal estimates of scalar exchange. In *Micrometeorology in Agricultural Systems*; Hatfield, J.L., Baker, J.M., Eds.; Agronomy Society of America: Madison, WI, USA, 2005; pp. 455–483.
10. Wang, J.; Buttar, N.A.; Hu, Y.; Lakhiar, I.A.; Javed, Q.; Shabbir, A. Estimation of Sensible and Latent Heat Fluxes Using Surface Renewal Method: Case Study of a Tea Plantation. *Agronomy* **2021**, *11*, 179. [CrossRef]
11. Hashem, A.A.; Engel, B.A.; Bralts, V.F.; Marek, G.W.; Moorhead, J.E.; Radwan, S.A.; Gowda, P.H. Assessment of Landsat-Based Evapotranspiration Using Weighing Lysimeters in the Texas High Plains. *Agronomy* **2020**, *10*, 1688. [CrossRef]

12. Crookston, B.; Blaser, B.; Darapuneni, M.; Rhoades, M. Pearl Millet Forage Water Use Efficiency. *Agronomy* **2020**, *10*, 1672. [CrossRef]

13. Wiegmanna, M.; William, T.B.; Thomasb, H.J.; Bullb, I.; Andrew, J.; Flavellc, J.; Annette, Z.; Edgar, P.; Klaus, P.; Andreas, M. Wild barley serves as a source for biofortification of barley grains. *Plant Sci.* **2019**, *283*, 83–94. [CrossRef] [PubMed]

14. Hammami, Z.; Qureshi, A.S.; Sahli, A.; Gauffreteau, A.; Chamekh, Z.; Ben Azaiez, F.E.; Ayadi, S.; Trifa, Y. Modeling the Effects of Irrigation Water Salinity on Growth, Yield and Water Productivity of Barley in Three Contrasted Environments. *Agronomy* **2020**, *10*, 1459. [CrossRef]

15. FAO. The Irrigation Challenge. In *Increasing Irrigation Contribution to Food Security through Higher Water Productivity from Canal Irrigation Systems*; Issue Paper; FAO: Rome, Italy, 2003.

16. Barnston, A. Correspondence among the Correlation [root mean square error] and Heidke Verification Measures; Refinement of the Heidke Score. *Notes Corresp. Clim. Anal. Cent.* **1992**, *7*, 699–709.

17. Rolbiecki, R.; Rolbiecki, S.; Figas, A.; Jagosz, B.; Prus, P.; Stachowski, P.; Kazula, M.J.; Szczepanek, M.; Ptach, W.; Pal-Fam, F.; et al. Response of Chosen American Asparagus officinalis L. Cultivars to Drip Irrigation on the Sandy Soil in Central Europe: Growth, Yield, and Water Productivity. *Agronomy* **2021**, *11*, 864. [CrossRef]

18. Haghverdi, A.; Reiter, M.; Sapkota, A.; Singh, A. Hybrid Bermudagrass and Tall Fescue Turfgrass Irrigation in Central California: I. Assessment of Visual Quality, Soil Moisture and Performance of an ET-Based Smart Controller. *Agronomy* **2021**, *11*, 1666. [CrossRef]

19. Haghverdi, A.; Reiter, M.; Singh, A.; Sapkota, A. Hybrid Bermudagrass and Tall Fescue Turfgrass Irrigation in Central California: II. Assessment of NDVI, CWSI, and Canopy Temperature Dynamics. *Agronomy* **2021**, *11*, 1733. [CrossRef]

20. Bergonzoli, S.; Brambilla, M.; Romano, E.; Saia, S.; Cetera, P.; Cutini, M.; Toscano, P.; Bisaglia, C.; Pari, L. Feeding Emitters for Microirrigation with a Digestate Liquid Fraction up to 25% Dilution Did Not Reduce Their Performance. *Agronomy* **2020**, *10*, 1150. [CrossRef]

21. Chávez, C.; Limón-Jiménez, I.; Espinoza-Alcántara, B.; López-Hernández, J.A.; Bárcenas-Ferruzca, E.; Trejo-Alonso, J. Water-Use Efficiency and Productivity Improvements in Surface Irrigation Systems. *Agronomy* **2020**, *10*, 1759. [CrossRef]

22. Assefa, T.T.; Adametie, T.F.; Yimam, A.Y.; Belay, S.A.; Degu, Y.M.; Hailemeskel, S.T.; Tilahun, S.A.; Reyes, M.R.; Prasad, P.V.V. Evaluating Irrigation and Farming Systems with Solar MajiPump in Ethiopia. *Agronomy* **2021**, *11*, 17. [CrossRef]

 *agronomy*

Article

Feasibility of Moderate Deficit Irrigation as a Water Conservation Tool in California's Low Desert Alfalfa

Ali Montazar [1,*], Oli Bachie [1], Dennis Corwin [2] and Daniel Putnam [3]

[1] University of California Division of Agriculture and Natural Resources,
     UC Cooperative Extension Imperial County, 1050 East Holton Road, Holtville, CA 92250, USA;
     obachie@ucanr.edu
[2] USDA-ARS, United States Salinity Laboratory, 450 West Big Springs Road, Riverside, CA 92507, USA;
     Dennis.Corwin@usda.gov
[3] Department of Plant Sciences, University of California Davis, One Shields Ave., Davis, CA 95616, USA;
     dhputnam@ucdavis.edu
* Correspondence: amontazar@ucanr.edu; Tel.: +1-442-265-7707

Received: 16 September 2020; Accepted: 22 October 2020; Published: 24 October 2020

Abstract: Irrigation management practices that reduce water use with acceptable impacts on yield are important strategies to cope with diminished water supplies and generate new sources of water to transfer for other agricultural uses, and urban and environmental demands. This study was intended to assess the effects of moderate water deficits, with the goal of maintaining robust alfalfa (*Medicago sativa* L.) yields, while conserving on-farm water. Data collection and analysis were conducted at four commercial fields over an 18-month period in the Palo Verde Valley, California, from 2018–2020. A range of deficit irrigation strategies, applying 12.5–33% less irrigation water than farmers' normal irrigation practices was evaluated, by eliminating one to three irrigation events during selected summer periods. The cumulative actual evapotranspiration measured using the residual of energy balance method across the experimental sites, ranged between 2,031 mm and 2.202 mm, over a 517-day period. An average of 1.7 and 1.0 Mg ha^{-1} dry matter yield reduction was observed under 33% and 22% less applied water, respectively, when compared to the farmers' normal irrigation practice in silty loam soils. The mean dry matter yield decline varied from 0.4 to 0.9 Mg ha^{-1} in a clay soil and from 0.3 to 1.0 Mg ha^{-1} in a sandy loam soil, when irrigation water supply was reduced to 12.5% and 25% of normal irrigation levels, respectively. A wide range of conserved water (83 to 314 mm) was achieved following the deficit irrigation strategies. Salinity assessment indicated that salt buildup could be managed with subsequent normal irrigation practices, following deficit irrigations. Continuous soil moisture sensing verified that soil moisture was moderately depleted under deficit irrigation regimes, suggesting that farmers might confidently refill the soil profile following normal practices. Stand density was not affected by these moderate water deficits. The proposed deficit irrigation strategies could provide a reliable amount of water and sustain the economic viability of alfalfa production. However, data from multiple seasons are required to fully understand the effectiveness as a water conservation tool and the long-term impacts on the resilience of agricultural systems.

Keywords: Colorado River Basin; drought; irrigation management strategy; water deficit; water productivity

---

1. Introduction

Due to recurring droughts and altered weather patterns, the Colorado River Basin is facing increasing uncertainty concerning water supplies. Hence, implementing impactful agricultural water conservation tools and strategies might have a significant value to the resiliency and profitability of agricultural systems in the low desert region of California and Arizona. It is likely that water deficits will be the reality for agriculture in the future, mostly affecting agronomic crops such as alfalfa (*Medicago sativa* L.), which accounts for about 28% of the crops grown in the area [1] and is the dominant water user in the region.

Currently, efficient use of irrigation water and improved irrigation management strategies are the most cost-effective tools to address water conservation issues. While more than 95% of California's low desert alfalfa (nearly 80,000 hectares) is currently irrigated by surface irrigation systems [2], one strategy to enhance water-use efficiency and on-farm water conservation in alfalfa fields is through improved technology of water delivery. Improved systems, such as subsurface drip irrigation, overhead linear move sprinkler irrigation, automated surface irrigation, and tailwater recovery systems might enable more precise control of irrigation water. Many of these technologies were already adopted by local farmers, although the process of adoption is continuing.

Another strategy is deficit irrigation of alfalfa, applying less water than the full crop water requirements for a season. Deficit irrigation was investigated as a valuable and sustainable crop production strategy over a wide variety of crops, including alfalfa, to maximize water productivity and to stabilize—rather than maximize—yields while conserving irrigation water [3–7].

The overall effect of deficit irrigation highly depends on the type of crop and adopted irrigation strategy. Although alfalfa is frequently criticized for its high seasonal water requirements, it has positive biological features, environmental benefits, and greater yield potential than many other crops under water stressed conditions [8], such as deep-rootedness, high yield and harvest index, contribution to wildlife habitat, and ability to survive a drought. If water allotments to alfalfa are significantly curtailed, it will result in reduced evapotranspiration (ET), CO_2 exchange, symbiotic N_2 fixation, and dry matter yield [9]. There is a positive relationship between alfalfa yield and its ET [10,11], but alfalfa yields are not always reduced in direct proportion to the reduction in applied water during droughts [12,13]. Non-stressed total season ET values are greater than most crops because of long periods of effective ground cover. Alfalfa, being a herbaceous crop, exhibits rapid growth characteristics and its yield is linearly related to ET [3,14], under optimum growing conditions. Dry matter per unit of water used in alfalfa is compared favorably with other C_3 plants.

Several studies investigated the effect of mid-summer irrigation cut-off (no irrigation after June until the following spring) on alfalfa yield. Ottman et al. [15] found that alfalfa yield under mid-summer deficit irrigation in Arizona was very low and did not recover in sandy soil, but summer irrigation termination had less effect on sandy-loam soils. At a site in the San Joaquin Valley of California, yields of a mid-summer irrigation treatment were 65% to 71% of that of a fully irrigated alfalfa, over a 2-year period [16]. Mid-summer deficit irrigation in the Imperial and Palo Verde Valleys of California reduced yields to 53–64% [17] and 46% [18], relative to a fully irrigated alfalfa. Studies in multiple environments of California showed reductions in alfalfa yield, significant conservation of water, and the consistent ability of the crop to recover after drought [3]. In long-term studies conducted on the western slope of the Rocky Mountains of Colorado, researchers determined that late-season deficits impacted yields in some cases, but not all, and that full recovery in the year following re-watering occurred in most cases [19]. In a study conducted in Nevada, significant yield reductions occurred from deficit irrigation over a three-year period, but yields were recovered under adequate irrigation during the fourth year [20]. Another study conducted in Kansas [21] showed no yield reductions in alfalfa hay yields, under 20% and 30% sustained deficit subsurface drip irrigation over the season. The researchers suggested that rainfall had a major contribution to the crop water use, with precipitation contributing to about 25% of seasonal crop water use between June and October, for the Kansas experiment.

The main purpose of this study was to identify and optimize moderate summer deficit irrigation strategies with profitable and sustainable alfalfa forage production, while conserving water under limited water conditions. The study intended to develop a dataset that could serve as a reference for further studies and an opportunity to better understand this underutilized water conservation strategy in the desert environment.

2. Materials and Methods

2.1. Field Experiments

Studies were conducted at four commercial alfalfa fields (designated "sites A1" through "A4") in the desert environment of the Palo Verde Valley, CA, USA. The study area had a true desert climate with an annual average air temperature, total annual precipitation, and ET_0 of 21.4 °C, 78.2 mm, and 1782 mm, respectively (Table 1). Soil characteristics for all experimental sites pertaining to four generic horizons are provided in Table 2. Dominant soil type ranged from loams at sites A1 and A2, to clay at site A3, and sandy loam at site A4. Soil cation exchange capacity (CEC) ranged between 7.9 and 12.7 meq/100 g at site A4 to between 9.2 and 22.4 meq/100 g at site A3. The Colorado River was the source of irrigation water with an average pH of 8.1 and an average electrical conductivity (EC_w) of 1.1 dS m^{-1} for all fields.

Table 1. Monthly mean long-term (10-year, 2009–2018) climate data for Blythe, CA, USA, data from CIMIS (California Irrigation Management Information System) station NE #135.

Month	Rain (mm)	Solar Radiation (W m^{-2})	Average Air Temp. (°C)	Average Dew Point Temp. (°C)	Average Wind Speed (m s^{-1})	Total ET_0 (mm)
Jan	13.4	126.4	11.1	2.8	2.1	65.2
Feb	10.2	171.9	13.5	4.2	2.2	86.2
Mar	6.4	228.8	17.5	5.4	2.4	144.0
Apr	1.3	282.3	21.0	5.8	2.8	188.7
May	2.3	322.3	24.3	8.8	2.7	227.5
Jun	0.0	331.2	29.4	13.6	2.3	232.4
Jul	8.3	295.0	32.3	19.5	2.3	221.5
Aug	9.0	275.7	32.1	20.0	2.2	202.4
Sep	5.6	232.0	28.4	17.1	2.0	157.8
Oct	6.0	186.9	22.0	11.2	1.9	122.9
Nov	3.2	140.9	14.9	5.4	1.9	76.3
Dec	12.5	111.7	10.5	2.7	2.0	58.0

Table 2. Physical and chemical properties of the soil of the four experimental sites. CEC represents the cation exchange capacity (meq/100 g). A1, A2, A3, and A4 represent alfalfa experimental sites.

Experimental Site	Generic Horizon (m)	Soil Texture			Organic Matter (%)	CEC (meq/100 g)	pH
		Sand (%)	Clay (%)	Silt (%)			
A1	0–0.3	44.2	11.1	44.7	1.7	20.6	8.0
	0.3–0.6	46.9	8.1	45.0	0.8	17.4	8.1
	0.6–0.9	41.7	7.7	50.5	0.9	18.6	8.2
	0.9–1.2	47.8	5.9	46.3	0.7	16.3	8.2
A2	0–0.3	39.7	20.6	39.7	2.3	22.4	8.1
	0.3–0.6	50.1	11.9	37.9	0.7	14.1	8.0
	0.6–0.9	75.0	5.1	19.9	0.9	8.0	8.2
	0.9–1.2	83.7	4.1	12.2	0.8	7.2	8.2
A3	0–0.3	31.5	13.2	55.3	1.3	18.1	8.0
	0.3–0.6	26.1	18.7	55.2	0.8	22.4	8.1
	0.6–0.9	88.6	2.8	8.6	0.5	10.1	8.2
	0.9–1.2	94.8	1.4	3.8	0.5	9.2	8.4
A4	0–0.3	69.3	13.1	17.6	1.7	12.7	7.9
	0.3–0.6	91.9	2.8	5.3	0.8	8.1	8.2
	0.6–0.9	82.0	8.3	9.7	1.2	8.7	8.1
	0.9–1.2	85.9	5.7	8.4	0.9	6.4	8.3

All four fields were planted in October 2018. Low desert alfalfa fields are typically harvested in a 28-day to 33-day cycle during spring and summer, while a total of 8–10 harvests per year is common in the region. Eleven harvest cycles were investigated in this study.

The experimental fields represent soil types (a wide range from sandy loam to clay) and irrigation management practices in the low desert region of California. The surface irrigation practices consisted of straight border irrigation (sites A3 and A4) and graded furrow irrigation (sites A1 and A2). An average of 75% may be assumed as the irrigation efficiency of the irrigation systems [22]. Sites A2 and A4 were divided into three individual sections (large plots) and sites A1 and A3 were divided into six individual plots. Five irrigation strategies were studied within the experimental plots, including normal farmer irrigation practice (NI, as control strategy) and four summer deficit irrigations (DI1, DI2, DI3, DI4; Table 3). The deficit irrigation strategies were implemented by eliminating irrigation events during summer harvest cycles. For the harvest cycles of July–September, three irrigation events per cycle is common irrigation practice in the Palo Verde Valley. Three plots were accommodated for each of the deficit irrigation strategies.

Table 3. Description of irrigation management strategies imposed at the experimental sites over the study period (18-month).

Site	Number of Irrigation Events					Description of Irrigation Management Strategies
	NI	DI1	DI2	DI3	DI4	
A1	30	27	28	-	-	DI1: irrigation events were eliminated 20 July 2019, 23 August 2019, and 26 September 2019 DI2: irrigation events were eliminated 23 August 2019 and 26 September 2019
A2	31	28	29	-	-	DI1: irrigation events were eliminated 21 July 2019, 24 August 2019, and 26 September 2019 DI2: irrigation events were eliminated 23 August 2019 and 26 September 2019
A3	24	-	-	22	23	DI3: irrigation events were eliminated 31 July 2019 and 2 September 2019 DI4: irrigation event was eliminated 2 September 2019
A4	25	-	-	23	24	DI3: irrigation events were eliminated 19 July 2019 and 24 August 2019 DI4: irrigation event was eliminated 24 August 2019

NI: following normal farmer irrigation practice over the study period
DI1: 33% less applied water than corresponding NI strategy during selected summer period
DI2: 22% less applied water than corresponding NI strategy during selected summer period
DI3: 25% less applied water than corresponding NI strategy during selected summer period
DI4: 12.5% less applied water than corresponding NI strategy during selected summer period

2.2. ET Monitoring and Data Processing

The actual evapotranspiration (ET_a) was measured using the residual of energy balance (REB) method with a combination of surface renewal (SR) and eddy covariance (ECov) techniques. The SR and ECov are well-recognized methods to estimate the sensible heat flux density (H) and to calculate the latent heat flux density (LE), using the REB approach [23–28].

A full flux density tower was set up in the plots under normal farmers' irrigation practice at each of the experimental site, totaling four towers (Figure 1). In each tower, several sensors were set up. An NR LITE 2 net radiometer (Kipp & Zonen, Ltd., Delft, The Netherlands) was used to measure net radiation (Rn). Two 76.2 μm diameter, type-E, chromel-constantan thermocouples model FW3 (Campbell Scientific, Inc., Logan, UT, USA) were used to measure high frequency temperature data for computing uncalibrated sensible heat flux (H0), using the SR technique. An RM Young Model 81000RE sonic anemometer (RM Young Inc., Traverse City, MI, USA) was used to collect high frequency wind velocities in three orthogonal directions at 10 Hz, to estimate H for the latent heat flux density calculations using the ECov technique.

Each tower also consisted of three HFT3 heat flux plates (REBS Inc., Bellevue, WA, USA) inserted at a 0.05 m depth below the soil surface, to measure soil heat storage at three different locations; three 107 thermistor probes (Campbell Scientific, Inc., Logan, UT, USA) to measure soil temperature at three depths in the soil layer above the heat flux plates; three EC5 soil moisture sensors (METER Groups Inc., Pullman, WA, USA) to measure soil volumetric water content at soil depths and locations near the heat flux plate and the thermistor probes; EE181 temperature and RH sensor (Campbell Scientific, Inc., Logan, UT, USA) to measure air temperature and relative humidity; an SP LITE 2 Pyranometer (Kipp & Zonen, Ltd., Delft, The Netherlands) to measure solar radiation; and a TE525MM tipping-bucket rain gauge with magnetic reed switch to measure precipitation.

Figure 1. A fully automated surface renewal and eddy covariance evapotranspiration (ET) tower in the plot under normal farmer irrigation practice (NI) at site A3.

Except for the soil sensors, all other sensors were set up at 1.8 m above the ground surface. The data were recorded using a combination of a Campbell Scientific CR1000X data logger and a CDM-A116 analog input module. Direct two-way communication with each monitoring flux tower was possible using a cellular phone modem model CELL210 (Campbell Scientific, Inc., Logan, UT, USA). The data of the sonic anemometer and fine wire thermocouples were collected at a 10 Hz sampling rate and the data of the other sensors were sampled once per minute. Half-hourly data were archived for later analysis.

Both the EC and SR techniques were individually employed to determine sensible heat flux density. The available energy components, R_n and G (ground heat flux density) were also measured throughout the study period. After acquiring the half-hourly H_0 data, a calibration factor (α) was established by determining the slope through the origin H values from the ECov technique versus H_0 from the SR technique, separately for the positive and negative values of H_0. The calibrated SR H value was finally estimated as H = $\alpha \cdot H_0$. The SR H and the ECov H were used to determine the LE values. The advantage from using both the ECov and SR methods was that they are independent and similar results provide a high level of confidence in the data used [26,28,29]. Latent heat flux density was calculated using the Residual of Energy Balance equation, as follows:

$$LE = Rn - G - H \tag{1}$$

where LE, G, H are positive away from the surface, and Rn is positive towards the surface. G is the ground heat flux density at the soil surface. It is assumed that Rn, G, and H are measured accurately. While use of the full eddy covariance method often does not demonstrate closure, Twine et al. [30] recommended that ECov results could be forced to have closure by holding the measured Bowen ratio (H/LE) constant, and increasing the H and LE values until Rn − G = H + LE. Twine et al. [30] also reported that using the REB method provides nearly the same accuracy as using the Bowen ratio correction. After determining LE, ET_a in mm d^{-1} was calculated by dividing the LE in MJ m^{-2} d^{-1} by 2.45 MJ kg^{-1}, to obtain the ET values in kg m^{-2} d^{-1}, which was equivalent to mm d^{-1}.

A Tule sensor (Tule technologies, Inc., Oakland, CA, USA) was used in each of the deficit irrigation plots at the experimental sites, to estimate ET_a using a surface renewal technique. The estimated daily ET_a from the Tule sensors at each site was verified by comparing with the ET_a measured from the full flux density towers.

Using the daily ET_a determined in each experimental site and the daily reference ET (ET_o) retrieved from the spatial CIMIS (California Irrigation Management Information System) data [31] for the coordinates of the monitoring station, the daily actual crop coefficient K_a (=$K_s \times K_c$) was calculated using Equation (2):

$$K_a = ET_a/ET_o \tag{2}$$

The daily stress coefficient (K_s) represents water and salt stresses, management, and environmental multipliers. To obtain the actual ET, K_s is needed to adjust crop coefficient (K_c). Spatial CIMIS combines remotely sensed satellite data with traditional CIMIS stations data, to produce site-specific ET_o on a 2-km grid, which provides a better estimate of ET_o for the individual sites.

2.3. Canopy Temperature and Soil Moisture Monitoring

Two SI-411 fixed view-angle infrared thermometers (IRTs, Apogee Instruments, Logan, UT, USA) were used to measure canopy temperature in each experimental plot. The IRTs were installed on a pole with a 47.5° angle below horizon, in opposite direction, viewing north and south to match for consistency. The IRTs were installed 1.8 m from the ground surface. The average temperatures of IRTs viewing north and south were considered to be the canopy temperatures. Canopy temperature was scanned with the IRTs units every minute and readings were averaged over a 30-min interval, using ZL6 cellular data logger (METER Groups Inc., Pullman, WA, USA).

Crop Water Stress Index (*CWSI*) was estimated using the difference between measured canopy and air temperatures (dT_m), using Equation (3):

$$CWSI = \frac{(dT_m - dT_{LL})}{(dT_{UL} - dT_{LL})} \tag{3}$$

The dT_m was compared against lower (dT_{LL}) and upper (dT_{UL}) limits of the canopy–air temperature differential, which could be reached under non-water-stressed and non-transpiring crop conditions. The Idso et al. approach [32] was used for estimating dT_{LL} and dT_{UL}.

Watermark Granular Matrix Sensor (Irrometer company, Inc., Riverside, CA, USA) was used to measure soil water tension at multiple depths of 15, 30, 45, 60, 90, and 120 cm, on a continuous basis. The data of Watermark sensors were recorded by a 900M Monitor data logger (Irrometer company, Inc., Riverside, CA, USA), on a 30-min basis.

2.4. Soil Salinity Assessment

Soil properties were surveyed and characterized within an approximate footprint area of 200 m × 200 m, around the ET monitoring stations in each plot, to assess soil salinity following deficit irrigation regimes. Surveys of apparent soil electrical conductivity (EC_a) were conducted in October 2019 (right after the alfalfa harvest), using mobile electromagnetic induction (EMI) equipment, following the guidelines developed by the U.S. Salinity Laboratory of the United States Department of Agriculture, for field-scale salinity assessment [33–36]. EC_a measurements were taken with a dual-dipole EM38 sensor (Geonics Ltd., Mississauga, ON, Canada), in horizontal (EM_h) and vertical (EM_v) dipole modes, to provide shallow (0 to 0.75 m) and deep (0 to 1.5 m) measurements of EC_a, respectively. At each of the plots, soil cores at four distinct depth ranges (0–0.3, 0.3–0.6, 0.6–0.9, and 0.9–1.2 m) were taken from 6 sampling locations, which were selected using the ESAP software to reflect the spatial variability of root zone soil salinity. A comprehensive laboratory analysis was conducted on all soil samples.

2.5. Yield Measurements and Plant Stand Evaluation

Yield sampling from the sub-plots was conducted on the same day or the day before, when the participating growers scheduled to harvest the entire experimental fields. In each irrigation plot, yield samples were taken from 12 sub-plots with a dimension of 1.5 m wide and 2.0 m long (Figure 1). The sub-plots were harvested using a hand cutter. A portable PVC quadrate was used to accurately sample uniform sub-plot sizes. Plant cutting height was 6–8 cm. Fresh weights of plants harvested within the quadrate was recorded, after which samples were dried for three days in conventional oven at 60 °C and recorded for alfalfa dry matter (DM). The significance of deficit irrigation strategies on mean dry matter yields was evaluated using a *t*-test.

Forage quality test for each collected sample was conducted using the Near Infrared Reflectance Spectroscopy (NIRS) method [37], to determine Crude Protein (CP), Acid Detergent Fiber (ADF), and Lignin percentage.

All irrigation plots were evaluated for plant stand density in February 2020, four months after switching the deficit irrigated plots back to normal irrigation practices. A portable PVC quadrate of 0.6 m wide and 0.6 m long was used to count the number of plants from the center of the 12 sub-plots that were used for yield measurements. Mean plant numbers per hectare were compared to the plots under normal farmers' irrigation practices.

2.6. Water Productivity

Two water productivity indices were calculated to compare the efficiency of irrigation water use and actual ET, using Equations (4) and (5):

$$\text{Irrigation water productivity (IWP)} = \frac{Alfalfa\ dry\ matter}{Irrigation\ water\ applied} \tag{4}$$

$$\text{Evapotranspiration water productivity (ETWP)} = \frac{Alfalfa\ dry\ matter}{ETa} \tag{5}$$

where the unit of alfalfa dry matter is kg ha^{-1}, and the units of irrigation water applied and ET_a are mm.

2.7. Statistical Analysis

The statistical significances of mean dry matter alfalfa yield and mean forage quality indices were performed using *t*-tests.

3. Results

3.1. Weather Parameters

There was higher than normal rainfall between November 2019 and March 2020 (Figure 2 and Table 1). Total precipitation was 105 mm over these five months. This precipitation amount was more than a long-term annual rainfall of the study area. The average daily air temperature, dew point temperature, solar radiation, and wind speed for the 2019 season (January 2019–December 2019) were approximately 21.0 °C, 234.8 W m^{-2}, 7.5 °C, and 2.3 m s^{-1}, respectively. However, peak daily temperatures of 30 °C to 42 °C during the late summer months when deficit treatments were imposed were common for this desert environment (Figure 2). More windy days were observed during the first half of the 2019 season, when compared with the 2020 season. The average daily wind speed was nearly 9% higher from January–June 2019 than the same season in 2020, although maximum average wind speed for the 2020 season, as recorded on 4th of February 2020 was 6.7 m s^{-1}. During the 2019 season, the month of June had the highest average daily solar radiation of 334.1 W m^{-2}, followed by May (324 W m^{-2}) and July (310 W m^{-2}).

3.2. Applied Water

The cumulative applied water (irrigation water + rainfall) for the different irrigation strategies at each site is shown in Figure 3. The total applied water was 3006, 2933, 2783, and 3125 mm in irrigation strategy NI for sites A1, A2, A3, and A4, respectively, over the 2019 season. These amounts were 788, 897, 716, and 1007 for sites A1, A2, A3, and A4, respectively, during the study period of the 2020 season. Although, most water amounts were provided by the irrigation events, the number of irrigation events were slightly different among the study sites. A total of 23 irrigation events occurred at sites A1 and A2 during the 2019 season, in irrigation strategy NI. The number of irrigation events was 22 at site A3 and 21 in at site A4.

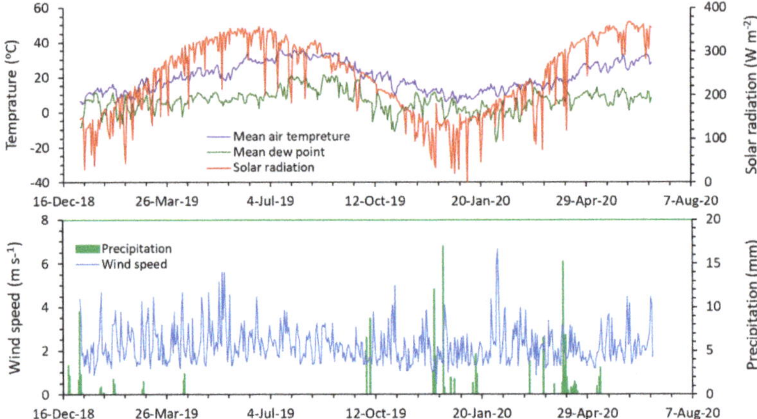

Figure 2. Daily weather data of the study area over the 18-month period of this experiment (January 2019 to June 2020).

The total reduction in the water used in the deficit irrigation strategies (DI1 to DI4) when compared with strategy NI was different at each of the experimental fields. For example, it was 314 mm in strategy DI1 and was 203 mm in strategy DI2 at site A1. The reduction in applied water was 310 mm in strategy DI1 at site A2, 185 mm in strategy DI3 at site A3, and 239 mm in strategy DI3 at site A4. The results demonstrated that the greatest amount of water conservation with the proposed deficit irrigation strategies was achieved under irrigation strategy DI1 at site A1, and the least amount was conserved in strategy DI4 at site A3.

3.3. Soil Moisture Status and Crop Water Stress

The soil moisture sensors placed within the effective root zone (15–120 cm) provided a representative condition of the soil water status. Figure 3 depicts the half-hourly soil water tension for irrigation strategy NI at sites A2 and A4, from March 2019 to June 2020. Due to the necessity of drying soils during and after alfalfa harvest, and to allow equipment to be used on alfalfa fields and for drying of hay, irrigation events typically stopped approximately 4–5 days before cuttings and resumed 4–5 days after cuttings. Therefore, changes in soil water tension could be observed before and after these dry-down periods (Figure 4). Soil water tension responded most within the top 60 cm depth, while some responses were also observed at deeper soil depths over time. For example, water tension values at site A4 sharply increased to more than 200 kPa at the topsoil (15–45 cm), before the first irrigation events, right after the alfalfa harvests. Soil water tension readings below 60 cm indicated that soil moisture was effectively maintained at a relatively uniform and desirable level during the study period, even at site A4, with a sandy loam soil where soil water tension values were less than 25 kPa.

Soil moisture data indicated that the soil at site A2, which had a silty loam soil, was generally not within the stressed range. However, alfalfa at site A4 might occasionally experience moderate water stress around cuttings. The recommended average soil water tension levels within the effective alfalfa root zone at which irrigation was triggered on loamy and sandy loam soil was at 60–90 kPa and 40–50 kPa ranges, respectively [12]. The insufficient soil moisture levels at site A4 during summer harvest cycles, from July through September (Figure 4) might have caused some mild alfalfa water stress. Soil moisture was clearly impacted by irrigation strategy DI3 at this site (Figure 5). However, additional potential mild water stress could have occurred in the middle of the harvest cycles due to halted irrigation water. For instance, soil water tension values increased to 134 and 93 kPa on 24 July 2019, at 30 and 60 cm of soil depth, as a result of eliminating the irrigation event on 19 July 2019 (Figure 5). Similarly, soil moisture readings for the same soil depths and dates at this site was less than 28 kPa for irrigation strategy NI.

Alfalfa CWSI was estimated for different irrigation strategies at each site, based on canopy temperature and air temperature measurements for three consecutive hourly periods of 1100–1200, 1200–1300, and 1300–1400 PST. The seasonal trend of midday CWSI estimated for the period of early-June 2019 through mid-October 2019 for different irrigation strategies at site A4, is shown in Figure 6. Average midday CWSI at site A4 for the period was estimated to be 0.13, 0.15, and 0.14 for irrigation strategies NI, DI3, and DI4, respectively, suggesting a similar, but relatively lower CWSI responses within the normal farmers irrigation practices. The CWSI values illustrated that moderate short-term midday water stress occurred around the alfalfa cuttings (and mid-harvest cycles) of July and August, thus, there was a good match between the findings obtained from the soil moisture data and the CWSI analysis.

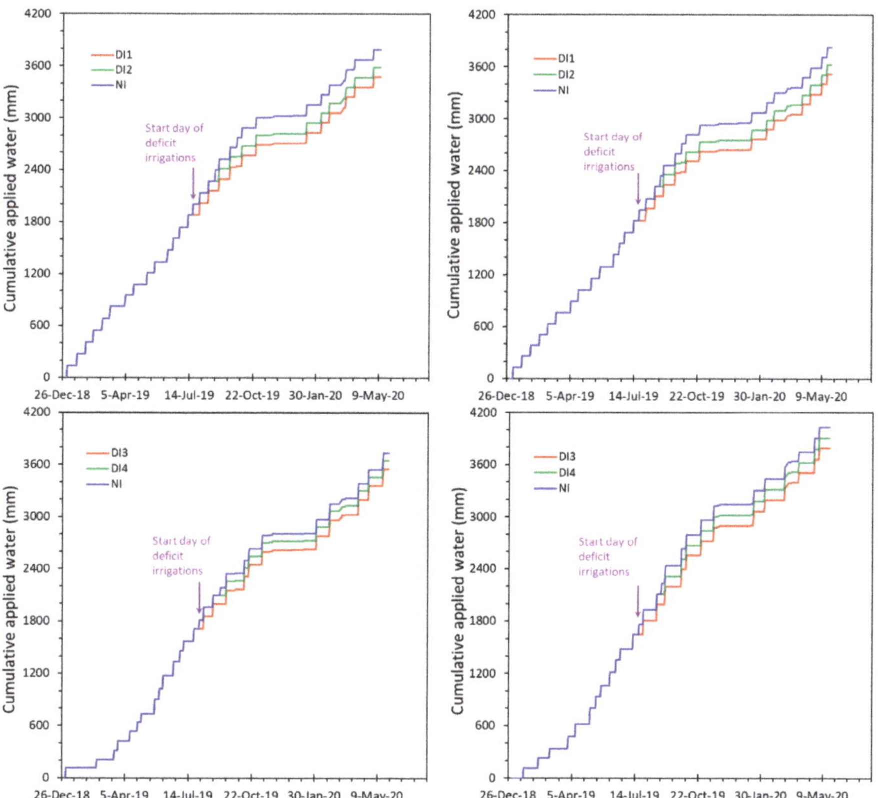

Figure 3. Cumulative applied water in each of the experimental irrigation strategies (NI, DI1–DI4) and sites (A1–A4) over the study period.

3.4. Actual Evapotranspiration and Crop Coefficients

ET_a was determined by calculating half-hourly latent heat flux density, using the REB approach with the SR and EC techniques. The daily ET_o, ET_a from the SR calculations, and K_a values for the irrigation strategy NI, at sites A1 and A4 are shown in Figure 7. The ET_a varied widely for each crop harvest cycle and throughout the experimentation seasons at both sites. For example, the ET_a

at site A4 ranged between 3.4 mm d^{-1} after alfalfa cutting and 9.3 mm d^{-1} at midseason full crop canopy, from June through August. The maximum and minimum ET$_a$ at site A1 were 2.6 mm d^{-1} and 0.3 mm d^{-1} during three-months of the study period, November 2019 to January 2020.

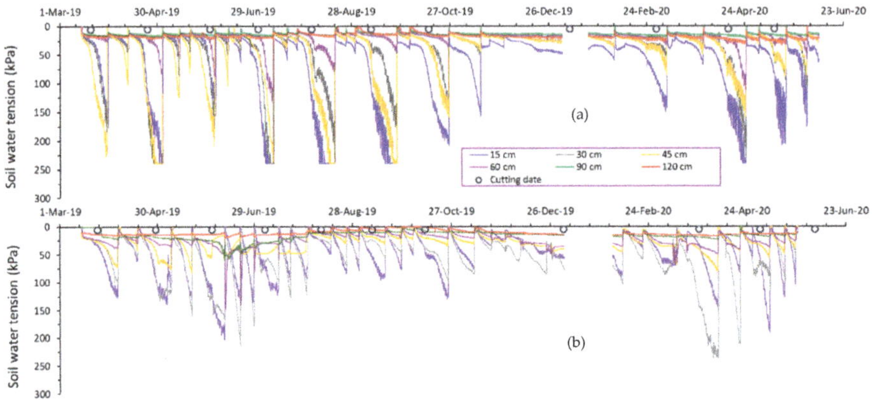

Figure 4. Half-hourly soil water tension (kPa) measured at multiple depths of 15 cm, 30 cm, 45 cm, 60 cm, 90 cm, and 120 cm in plots under normal farmer irrigation practice (NI) at—(**a**) site A4 and (**b**) site A2, from March 2019 to June 2020. Cutting dates are demonstrated with circles on the x-axes.

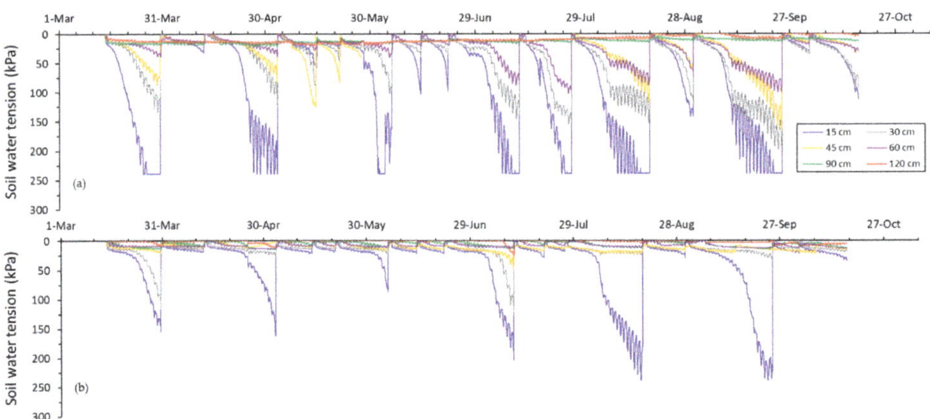

Figure 5. Half-hourly soil water tension (kPa) measured at multiple depths of 15 cm, 30 cm, 45 cm, 60 cm, 90 cm, 120 cm in plots under (**a**) irrigation strategy DI3 at site A4 and (**b**) irrigation strategy DI4 at site A4, from March 2019 to mid-October 2019.

The cumulative ET$_a$ (CET$_a$) in irrigation strategy NI at sites A1–A4, for a 517-day period (1 January 2019 to 31 May 2020) was 2202, 2187, 2031, and 2175 mm, respectively (Figure 8). For a one-season 12-month irrigation period (2019) for the same irrigation strategy, the cumulative ETa was 1596 mm at site A1, 1582 mm at site A2, 1423 mm at site A3, and 1558 mm at site A4. Comparing the total applied water and the cumulative ET$_a$ under normal farmer irrigation strategies indicated that the plots under NI irrigation strategy remained over-irrigated during the whole study period. However, moderate water stress was occasionally observed for the NI irrigation strategy, particularly at sites A3 and A4, because of the dry-down around alfalfa harvests or the unprecedented delays in irrigation schedules.

The daily K$_a$ values for the sites A1 and A4 (irrigation strategy NI) over the study period are shown in Figure 7. The K$_a$ value depended on alfalfa growth stages, ranging from smallest during initial growth stage, just after each harvest, and reaching the maximum when the crop height was at mid and full canopy development stages, attained prior to each harvest cycle. Large K_a values were attained at both sites during March and April (ranging from 0.47 after harvest to 1.24 at full canopy). Growth of alfalfa in early season (January) and late season (November to December) was slower due to cooler weather and lower solar radiation, in which lower K_a values were observed. The average K_a values of alfalfa sites over the study period varied from 0.8 at site A3 to 0.87 at site A1 (Figure 8 and Table 4).

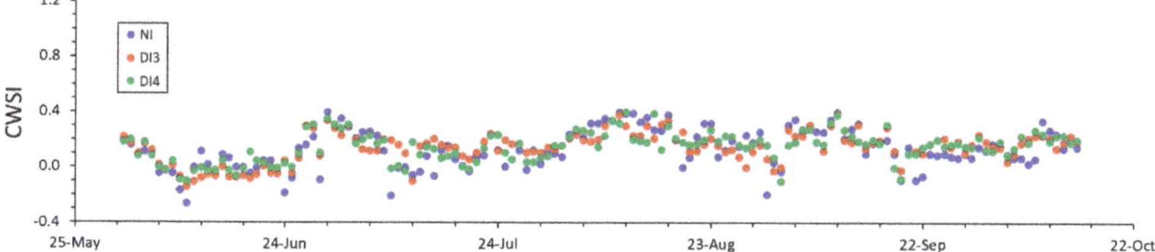

Figure 6. Daily crop water stress index (CWSI) values for the plots under different irrigation strategies (NI, DI3, and DI4) at site A4. NI, DI3, and DI4 represent normal farmer irrigation practice, applying 25% less water than NI, and applying 12.5% less water than NI, respectively.

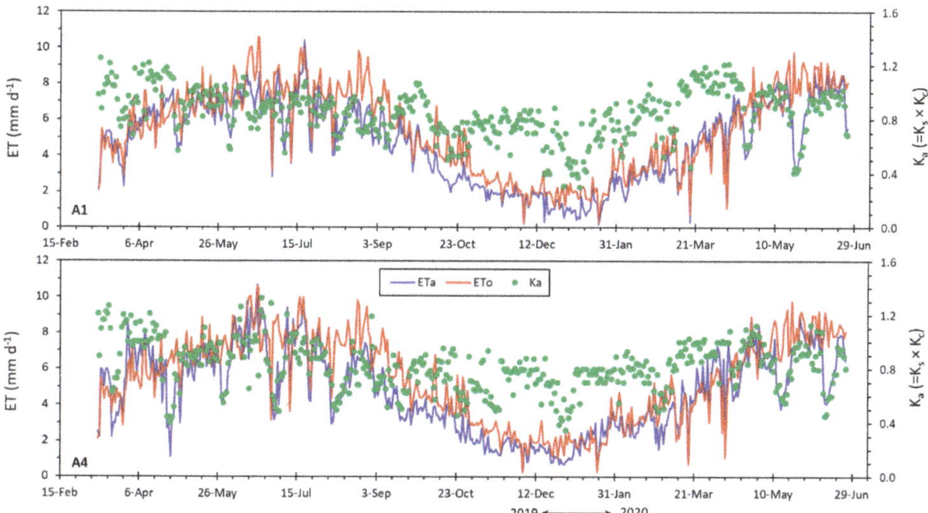

Figure 7. Daily reference evapotranspiration (ET_o), daily actual evapotranspiration (ET_a), and daily actual crop coefficient (K_a) at sites (**A1**) and (**A4**), from March 2019 to June 2020.

Average K_a values at harvest cycles (eight cuttings in 2019 and three cuttings in 2020) for each experimental site are provided in Table 4. The results demonstrated seasonal variabilities in the harvest cycle K_a values. With an average seasonal K_a value of 0.87 for the 2019 season at the site A1, the average cutting cycle K_a values varied from 0.72 (cutting cycle 8) to 1.0 (cutting cycle 2). Average seasonal K_a values for the 2019 season at sites A2, A3, and A4 were 0.86, 0.79, and 0.84, respectively. There was a considerable increase in average K_a value (12.5%) at site A3 from the 2020 cropping season compared to 2019. Lower K_a values in the first three harvest cycles of the 2019 season might have been due to poor irrigation management. Site A3 received 424 mm water during the first three months of 2019, which was the lowest amount of water applied amongst the experimental sites (Figure 3). While trivial differences (an average of 5%) were found in the average K_a values at sites A1 and A2 during harvest cycles of June–August (cutting cycles 4 to 6) at site A1 and A2, substantial differences (average of 20%) were obtained in the mean K_a values of cutting cycles at sites A3 and A4.

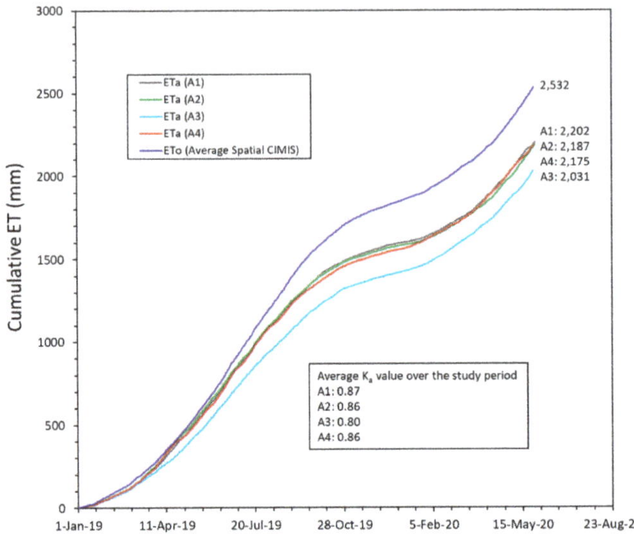

Figure 8. Cumulative reference evapotranspiration (Spatial CIMIS ET$_o$) and cumulative actual evapotranspiration (ET$_a$) at each of the experimental sites (A1–A4). The cumulative ET$_a$ are provided for the plots under normal farmer irrigation practices at each site. The average spatial CIMIS ET$_o$ of the four sites is provided as ET$_o$.

Table 4. Mean (±standard deviation) K$_a$ values of harvest cycles for each experimental site (A1–A4) determined from surface renewal measurements. The values are reported for the normal farmer irrigation practices over eight cuts in the season 2019 (Year 1) and three cuts in the season 2020 (Year 2).

Cut—Year Number	Harvest Time	K$_a$			
		A1	A2	A3	A4
Cut 1—Year 1	23 Mar–4 Apr	0.81 (±0.13)	0.81 (±0.14)	0.79± (0.16)	0.80 (±0.14)
Cut 2—Year 1	24 Apr–8 May	1.00 (±0.14)	1.03 (±0.15)	0.83 (±0.16)	1.01 (±0.11)
Cut 3—Year 1	1 Jun–12 Jun	0.94 (±0.13)	0.92 (±0.13)	0.78 (±0.14)	0.83 (±0.14)
Cut 4—Year 1	3 Jul–12 Jul	0.89 (±0.12)	0.87 (±0.13)	0.85 (±0.13)	0.98 (±0.12)
Cut 5—Year 1	5 Aug–16 Aug	0.88 (±0.14)	0.86 (±0.11)	0.83 (±0.10)	0.86 (±0.15)
Cut 6—Year 1	5 Sep–19 Sep	0.85 (±0.10)	0.82 (±0.11)	0.75 (±0.12)	0.77 (±0.11)
Cut 7—Year 1	9 Oct–29 Oct	0.78 (±0.12)	0.74 (±0.13)	0.74 (±0.11)	0.73 (±0.12)
Cut 8—Year 1	31 Dec–15 Jan	0.72 (±0.14)	0.73 (±0.14)	0.71 (±0.13)	0.70 (±0.15)
Cut 1—Year 2	4 Mar–30 Mar	0.78 (±0.16)	0.79 (±0.15)	0.83 (±0.17)	0.87 (±0.14)
Cut 2—Year 2	20 Apr–5 May	1.02 (±0.13)	0.93 (±0.15)	0.97 (±0.15)	0.94 (±0.15)
Cut 3—Year 2	22 May–14 Jun	0.90 (±0.14)	0.89 (±0.13)	0.90 (±0.12)	0.90 (±0.12)
Average	-	0.87 (±0.10)	0.86 (±0.09)	0.82 (±0.07)	0.86 (±0.10)

The observed average K$_a$ value from this study was lower than the value (0.95) suggested by Doorenbos and Kassam [38] for dry climate and 0.99 reported by Hanson et al. [3] for the Sacramento Valley. However, the average K$_a$ value was about the same as what was found by Kuslu et al. [39].

3.5. Soil Salinity and Water Availability Features

It is well-known that salinity associated problems are a major challenge for global food production, with particularly critical impact in the low desert region. Applications of excess water to control soil root zone salinity is an important agricultural practice for these regions and considered a 'beneficial use' of water, since soils and crop production can only be sustained by controlling salinity. Buildup of salinity might be considered a serious concern and likely a key limitation for any reduced water demand strategies in the region. Therefore, it is important to understand the impact of deficit irrigation on potential soil salinity buildup and soil water balances, vis-à-vis evapotranspiration and devising optimal irrigation management.

Spatially interpolated map of EM_v at sites A1, A2, and A4 in late October of 2019, just before all deficit irrigated plots were switched to normal farmer irrigation practice after the 1.5-year study, is shown in Figure 9. The entire surveyed areas that were affected by the different irrigation strategies at these sites exhibited small EM_v measurements (11.13–174.57 mS m^{-1} or 0.11–1.17 dS m^{-1}). These measurements approximate the differences in EM values (which is affected by salinity, texture, and moisture) in approximately the top 1.2 m of soil. Inconsequential differences were observed among EM_v values of the plots treated with different irrigation strategies. However, we should point out that these were moderate deficit treatments, and more severe deficits might cause greater excess salinity buildup.

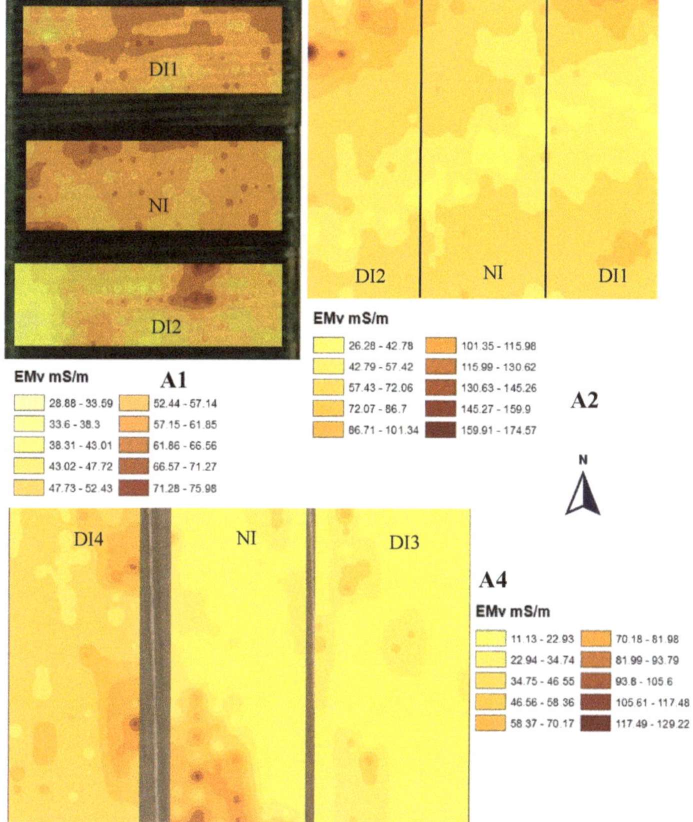

Figure 9. Spatial distribution of ancillary variable EM_v (electromagnetic induction measurement in the vertical coil orientation) at sites (**A1**), (**A2**), and (**A4**). 100 m S m^{-1} is equal 1 dS m^{-1}.

In this study, the mean EC_e at the effective crop root zone (30–120 cm), across the experimental sites in late October 2019, demonstrated that deficit irrigation strategies had some impacts on the soil salinity (Figure 10), however, these values were always in the 'acceptable' range for alfalfa. A higher level of ECe values were observed in plots under irrigation strategy DI1, in comparison to plots under irrigation strategies NI and DI2, at sites A1 and A2 (furrow irrigated alfalfa fields). For instance, the mean ECe at site A2 was 2.2 dS m^{-1} and 4.2 dS m^{-1} at 60 and 90 cm soil depths, respectively, in irrigation strategy DI1. However, the values were 1.47 and 1.76 dS m^{-1} in irrigation strategy NI, and were 1.45 and 2.2 dS m^{-1} in irrigation strategy DI2, at the corresponding depths, respectively.

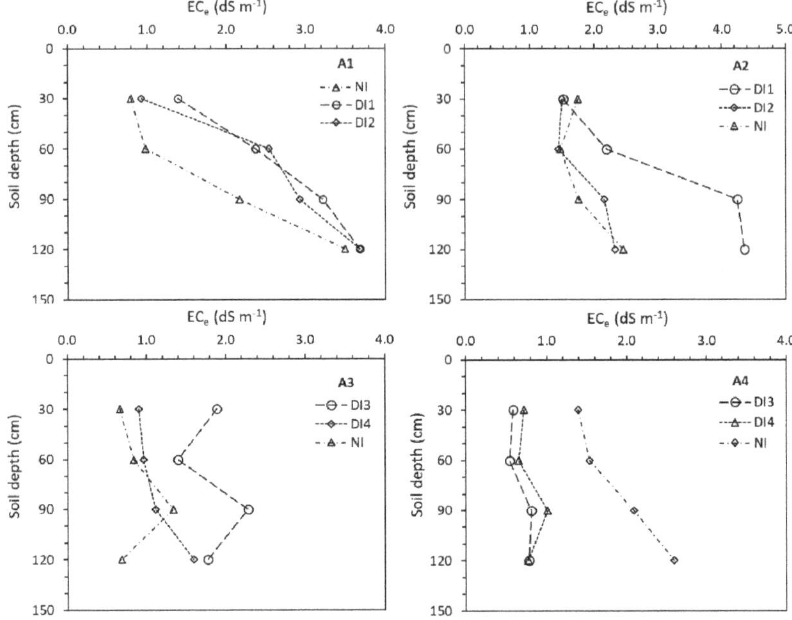

Figure 10. Whole-soil profile representations of mean EC$_e$ (electrical conductivity of the saturation extract) distribution of observed values in different irrigation strategies at four experimental sites. The complete data (collected in late October 2019) from the six soil core sampling locations in plots under different irrigation strategies at each site were used to plot EC$_e$.

3.6. Alfalfa Yield Responses

Alfalfa dry matter values for eight seasonal cuttings 2019 (Y1) and the first three cuttings for 2020 (Y2) under each irrigation strategies of all experimental sites are illustrated in Figure 11. The results indicated that the first four cuttings were the most productive for all sites in 2019. For instance, mean DM yield at site A2 from irrigation strategy NI were 4.7, 4.0, 4.7, and 3.9 Mg ha^{-1} for first, second, third, and fourth cuttings, respectively. For sites A1, A2, and A4, mean DM values for the first three cuttings in 2020 were lower than the corresponding DM values in 2019. At site A3, alfalfa DM yields were relatively lower than the other sites over the first three cuttings of the 2019 season.

Alfalfa mean DM yields from the 5th to 7th cuttings in 2019 were much lower than the first four cuttings even under full irrigation practices. Yields in late summer were moderately affected by deficit irrigation strategies. For example, yield reduction at site A1 from cuttings 5, 6, and 7 in irrigation strategy DI1, compared to irrigation strategy NI were 0.3, 0.8, and 0.6 Mg ha^{-1}, respectively (Figure 11). The reduction in DM yield was 0.5 Mg ha^{-1} for cutting 6 and was 0.4 Mg ha^{-1} for cutting 7 in irrigation strategy DI2. Yields summed over the 11 cuttings, over 1.5 years, were reduced from 0.7% to 4.6% (0.3 to 1.7 Mg ha^{-1}) by the deficit irrigation practices at the four sites.

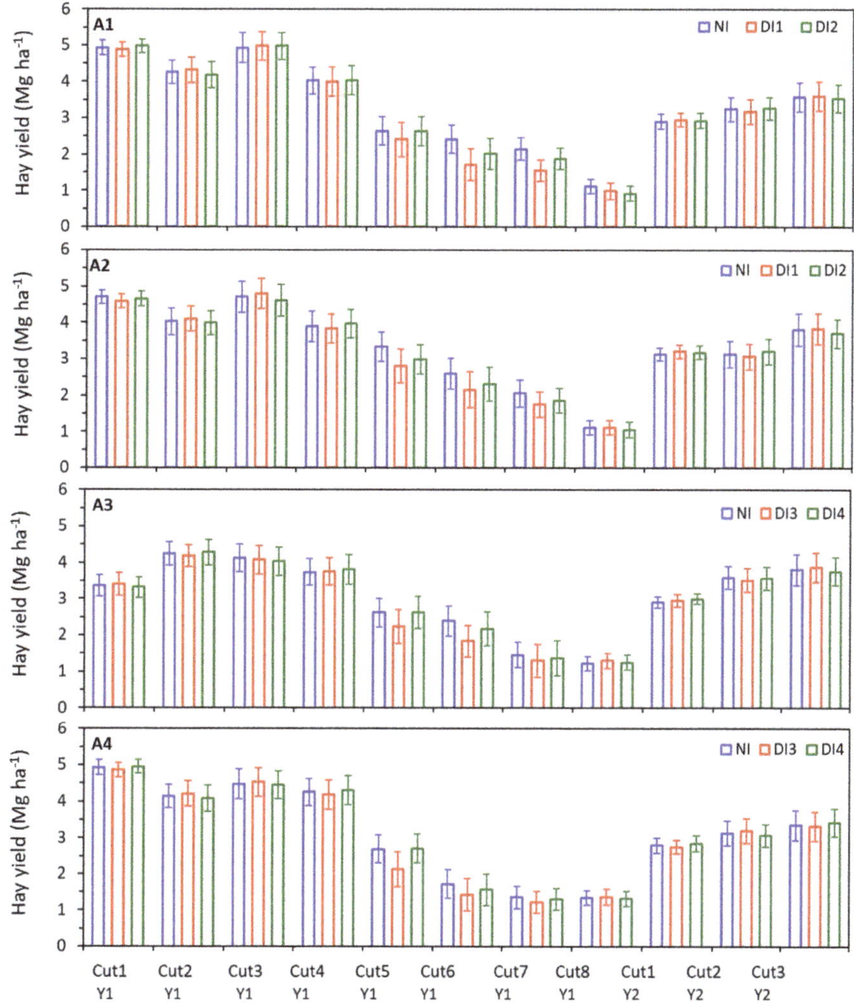

Figure 11. Mean dry matter (DM) yields of each irrigation strategy (NI, DI1–DI4) at the experimental sites (**A1**–**A4**) for eight cuttings in the season 2019 (Cu1-Y1 to Cut8-Y1) and three cuttings in the season 2020 (Cut1-Y2 to Cut3-Y2). The bars demonstrate the standard deviation of DM values.

3.7. Forage Quality

There was a trend for a small improvement in forage quality due to the deficit irrigation strategies, but not at all sites (Table 5). A significant reduction in acid detergent fiber was observed in deficit irrigation strategies DI1 at site A2 and DI2 at site A1, compared to normal irrigation practice (p values of 0.001 for DI1 at site A2, 0.02 for DI2 at site A1). Significant crude protein increase was also found from implementing deficit irrigation regimes DI1 at site A2 (p value of 0.04) and DI3 at site A4 (p value of 0.01), but not at other sites. No significant impact was observed on lignin percentage. The improved forage quality might be attributed to a reduction in stem growth (increase % leaf) under such irrigation practices. Small improvements in alfalfa forage quality under deficit irrigation regimes was also reported by other researchers [5,40,41].

Table 5. Mean forage quality indices (Acid Detergent Fiber (ADF), Crude Protein (CP), and Lignin) of normal farmer irrigation practices against deficit irrigation strategies. The forage quality data of June through September 2019 was used for this analysis (*t*-test).

Site	ADF (%)					CP (%)					Lignin (%)				
	NI	DI1	DI2	DI3	DI4	NI	DI1	DI2	DI3	DI4	NI	DI1	DI2	DI3	DI4
A1	29.1	28.9 ns	27.8 *	-	-	21.0	20.7 ns	20.8 ns	-	-	5.2	5.1 ns	5.2 ns	-	-
A2	31.2	26.4 *	28.2 ns	-	-	19.6	22.5 *	21.2 ns	-	-	5.2	5.1 ns	5.1 ns	-	-
A3	27.6	-	-	25.7 ns	27.5 ns	18.3	-	-	19.4 ns	18.4 ns	4.5	-	-	4.6 ns	4.7 ns
A4	31.2	-	-	26.5 ns	30.0 ns	17.6	-	-	20.6 *	18.3 ns	5.1	-	-	4.9 ns	5.0 ns

ns Non-significant. * Significant at the 5% level of probability.

3.8. Water Conservation Versus Yield Reduction

Deficit strategies with alfalfa are primarily feasible due to the seasonal yield patterns of the crop, with heavy yields during early season, and very light yields in late summer. Approximately 73–74% of total alfalfa seasonal yield productivity at the experimental sites occurred by mid-July (20 July 2019), right before starting summer deficit irrigation strategies (Figure 12). This finding is similar to results from research reported for the Sacramento Valley of California [5,39]. The deficit irrigation strategies could affect the 5th through 7th cuttings, while only 21–22% of the annual DM yield was produced during this period.

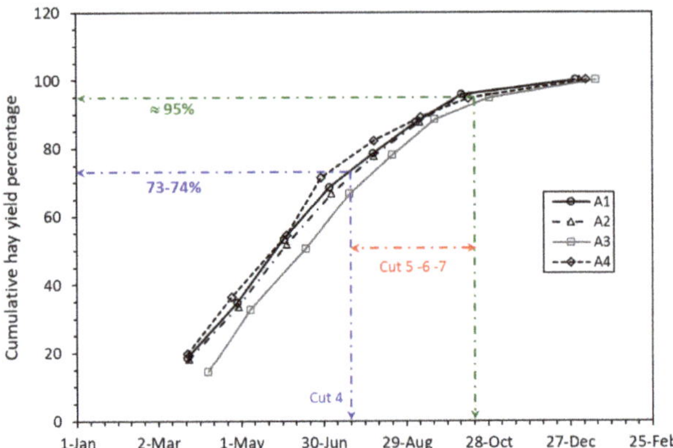

Figure 12. Cumulative alfalfa yield percentage over the growing season 2019 at the experimental sites (A1–A4). Results are provided for the normal irrigation practices at each site.

A significant DM reduction was observed in deficit irrigation strategies DI1 and DI2, compared to normal irrigation practice at site A2 (*p* value of 0.0004 for DI1 and 0.002 for DI2). There was also a significant yield reduction in deficit irrigation DI1 at site A1 (*p* value of 0.0005) (Table 6). No significant DM reduction was affected by deficit irrigation regimes DI3 and DI4 at sites A3 and A4; and deficit irrigation DI2 at site A2. The findings suggest an average of 1.7 Mg ha^{-1} and 1.0 Mg ha^{-1} dry matter yield reduction in deficit irrigation strategies DI1 and DI2 at sites A1 and A2, respectively (Table 6). The average DM yield reduction at sites A3 and A4 was nearly 1.0 Mg ha^{-1} in deficit irrigation strategy DI3 and 0.4 Mg ha^{-1} in deficit irrigation strategy DI4.

The total amount of conserved water across the experimental sites varied from 83 mm (3.0%) at site A3 to 314 mm (10.5%) at site A1, relative to what was used under seasonal water applied in normal irrigation practice. Summer deficit irrigation strategies enhanced the IWP values, but not the ETWP values (Table 7). For instance, irrigation strategies DI1 and DI2 at site A2 improved the IWP value by 0.5 and 0.4 kg ha^{-1} mm^{-1}, respectively, compared to the irrigation strategy NI (with an IWP of 8.9 kg ha^{-1} mm^{-1}).

Table 6. The total mean dry matter yield (Mg ha^{-1}) for different irrigation management strategies and total mean dry matter yield in normal irrigation practice at the experimental sites (18-month period). The significance of independent *t*-tests is provided.

Site	Normal Irrigation Practice (NI)	Deficit Irrigation Strategy			
		DI1	DI2	DI3	DI4
A1	36.2	34.6 *	35.4 [ns]	-	-
A2	36.6	35.0 *	35.5 *	-	-
A3	33.5	-	-	32.6 [ns]	33.1 [ns]
A4	34.3	-	-	33.2 [ns]	34.0 [ns]

[ns] Non-significant. * Significant at the 5% level of probability.

Table 7. The irrigation water productivity (IWP) and actual evapotranspiration (ET) water productivity (ETWP) values in different irrigation strategies (NI, DI1–DI4) at each of the experimental sites (A1–A4).

Site	Irrigation Strategy	IWP (kg ha^{-1} mm^{-1})	ETWP (kg ha^{-1} mm^{-1})
A1	NI	8.9	16.6
	DI1	9.3	16.1
	DI2	9.3	16.4
A2	NI	8.9	16.4
	DI1	9.4	16.1
	DI2	9.3	16.3
A3	NI	8.4	16.1
	DI3	8.6	15.8
	DI4	8.6	16.1
A4	NI	8.0	15.9
	DI3	8.3	15.6
	DI4	8.2	15.9

4. Discussion

Alfalfa was historically reported to be a moderately sensitive crop to salinity, with estimated yield declines above a saturated soil extract (EC$_e$) of 2.0 dS m^{-1} [42]. However, more recent reports and experiments in California confirmed that alfalfa has a much higher tolerance of salinity. Field and greenhouse experiments estimated tolerance of alfalfa varieties up to approximately 6.0 EC$_e$ or higher [43]. The findings from this study suggest that the proposed deficit irrigation strategies might cause salt accumulation at the crop root zone (particularly for furrow irrigated alfalfa fields) and this practice might elevate soil salinity class from a non-saline soil (0–2 dS m^{-1}) to slightly saline (2–4 dS m^{-1}) condition. Soil salinity can be managed by switching to farmer irrigation practices in early fall (mid- to late-October), with no need for the excessive water to leach salts.

The deliberate re-filling of the soil profile with irrigation water after implementing summer deficits should be considered, both in terms of water availability, salinity management, and water-use policy. The continuous soil moisture readings in this study verified that there was insignificant soil moisture depletion from deficit irrigations to require some recharge (Figures 4 and 5). As can be seen from the soil water tension data plots, the average soil water tension was kept constant at about 88 kPa, at the top 30 cm, and maintained at about 15 kPa, for the 45–120 cm soil depth. Consequently, farmers might confidently refill the soil profile after implementing summer deficit irrigation strategies and switching to their normal irrigation practices with little excessive water needs in the fall.

At sites A1–A4, the total mean annual DM yield in irrigation strategy NI for 2019 was 26.2, 26.5, 23.2, and 25.0 Mg ha^{-1}, respectively (Figure 11). These alfalfa yield values from the long-seasoned low desert sites are generally higher than the average alfalfa yield in California, which ranges from 14.6 to 16.1 Mg ha^{-1}, and a thirty-year (1984–2013) statewide average yield of 15.3 Mg ha^{-1} [44]. Dry matter yields of 11.2 Mg ha^{-1} was reported in an earlier deficit study that imposed severe water deficits, restricting applications to 1,249 mm seasonal water use in the Imperial Valley [3].

Alfalfa plant stand evaluation conducted on 18 February 2020 showed no significant differences in the plant population between the deficit irrigation strategies and normal irrigation practices, suggesting that there is no evidence of losing the alfalfa plant density from the implemented deficit irrigation strategies. For instance, plant population per hectare at site A1 was estimated to be 103×10^6, 105×10^6, and 102×10^6 in the plots under irrigation strategies NI, DI1, and DI2, respectively. Additionally, no yield reduction was observed from the summer deficit irrigation strategies within the first three harvest cuttings of the 2020 season (Figure 10), indicating a full recovery of the crop upon re-watering.

None of the deficit irrigation strategies produced severe water or salinity stress at the experimental sites. Soil water availability in non-sandy soils (sites A1–A3) was retained at a desired level during and after summer irrigation strategies. At site A4 (sandy loam soil), the residual soil moisture below the depth of 45 cm was consistent at a non-stressed level, and supplied enough water for alfalfa, suggesting that this might be a primary reason why there was only a small reduction in total ET$_a$ of deficit irrigated plots. The maximum ET$_a$ reduction (54 mm) was observed in irrigation strategy DI1 at site A2, when compared with strategy NI. The findings revealed that ET$_a$ was not directly reduced in relation to the level of reduction in applied water. This might be part of the reason why no improvement was gained in the ETWP values from moderate deficit irrigation strategies. Further measurement is necessary to provide a more solid conclusion on the impact of the deficit irrigation strategies on ET$_a$. Overall, the ETWP values computed for the normal farmer irrigation practices and deficit irrigation trials at the experimental sites (an average of 16.1 kg ha^{-1} mm^{-1}) were as high as the values predicted by Montazar et al. [45] for alfalfa fields, under subsurface drip irrigation in the low desert of California.

Imposition of summer water deficits is likely to result in yield reductions, a finding that was similar to other researches [3–7,17]. The findings suggest that conserving 0.083–0.314 (ha.m) ha^{-1} water through summer deficit irrigation strategies might result in 0.3–1.7 Mg ha^{-1} yield loss in California's low desert alfalfa production system. Therefore, while insignificant yield reduction is unpreventable, the proposed management strategies could serve as an effective water conservation tool. The hay yield reduction might be a consequence of reduced water distribution uniformity caused by the deficit irrigation regimes. Large-scale farming systems in the low desert, along with the use of surface irrigation methods, resulted in lower water distribution uniformity values (over time and space) with deficit irrigation strategies. Results expected from these deficit irrigation strategies would likely be more favorable in more efficient irrigation systems such as subsurface drip irrigation systems or advanced overhead sprinkler systems.

The practice of filling the soil profile so that it holds as much water as possible would be an effective early-season alfalfa irrigation strategy. Such practice might allow alfalfa to take full advantage of the available water and promote its rapid, early season growth, when the yield potential was highest, and when soil and water temperatures were not likely to be high enough to stress the crop and limit crop productivity. Consequently, combining full irrigation in winter-spring with moderate deficit irrigation during summer could be an efficient approach in conserving water than continuously irrigating (or over irrigating) for the entire season.

5. Conclusions

This study aimed at assessing the effectiveness of moderate deficit irrigation strategies during summer harvest cycles (less applied water than normal farmers' irrigation practices) on conserving water and maintaining a robust hay production.

The proposed deficit irrigation strategies conducted showed a promising and decent amount of water conservation and simultaneously generated desirable hay yields and quality. However, yield penalties of this practice must be considered. These moderate deficit irrigation practices resulted in an average of 1.47 Mg ha^{-1} and 0.31 Mg ha^{-1} hay yield reduction, but used

33% ($\approx$0.31 (ha.m) ha^{-1}) and 12.5% less applied-water than normal farmer practice over the summer ($\approx$0.10 (ha.m) ha^{-1}), respectively.

Several cautionary notes need to be considered with the data reported and the analysis provided in this study:

- The ET and crop coefficient values reported herein are referred to as observed or actual values, which are limited by water deficits and salinity, and the 'dry down' required for frequent harvests. The maximum crop ET (ET_c) is limited only by energy availability to vaporize water and not soil hydrology or salinity, or droughts imposed by harvest scheduling. Imposed stress, such as this, is common to almost all alfalfa growing regions.

- Although stand density under desert conditions decays more rapidly than other locations, there were negligible differences between different deficit irrigation strategies and normal farmers' irrigation practices in this study after one year of water deficits. However, it is uncertain whether multiple years of summer irrigation strategies might threaten the long-term viability of the crop stand and yields.

- Although, it might be unlikely to prevent salinity buildup due to summer water deficits, salinity issues are likely to be managed through irrigation practices that flush salts in the months after implementing deficit irrigations. Continuous monitoring of soil salinity is recommended to ensure flushing/leaching salts out of root zone over multiple deficit irrigation seasons.

- The importance of re-filling of the soil profile with water, in the year after implementing summer irrigation strategies, need to be considered both in terms of water availability, crop production, and policy. Such practices might enable shifting water demand to water-rich time periods in early spring. This practice would benefit both early season growth and salinity management in subsequent years. In this study, continuous soil moisture readings verified that soil moisture was insignificantly depleted in the deficit irrigation fields. However, data from multiple irrigation seasons are required to fully certify this conclusion.

- Implementation of the proposed summer deficit irrigation strategies on alfalfa could provide a reliable source of seasonally available water as well as sustain the economic viability of agriculture in the region. These strategies might be sustainable as an effective water conservation tool if such measures provide adequate economic incentives to the participating farmers. Incentive programs to farmers must offset the risk of implementing the proposed practices (even trivial production loss), as a tool for adopting water conservation practices.

Author Contributions: Conceptualization, A.M. and D.P.; Data curation, A.M.; Formal analysis, A.M.; Funding acquisition, A.M.; Investigation, A.M., O.B., and D.C.; Methodology, A.M., D.C., and D.P.; Project administration, A.M.; Supervision, A.M.; Writing—original draft, A.M.; Writing—review & editing, A.M., O.B., D.C., and D.P. All authors have read and agreed to the published version of the manuscript.

Funding: Funding for this study was made possible by the U.S. Department of Agriculture (USDA)—Natural Resources Conservation Service (NRCS), through Palo Verde Water Conservation RCPP (Regional Conservation Partnership Program).

Acknowledgments: The authors would like to thank the local NRCS office in Blythe, Palo Verde Resource Conservation District, and Palo Verde Irrigation District for providing their support and input. The authors gratefully acknowledge the Seiler Farms and Chaffin Farms for their sincere collaboration during this study, and for allowing us to implement the project in their agricultural operations. The authors wish to acknowledge the technical support of Tayebeh Hosseini and Tait Rounsaville, whose conscientious works in the field and in the laboratory were crucial to the success of this study.

Conflicts of Interest: The authors declare no conflict of interest.

References

1. Montazar, A. Update alfalfa crop water use information: An estimation for spring and summer harvest cycles in California low desert. *Agric. Briefs-Imp. Cty.* **2019**, *22*, 202–205.
2. Montazar, A. Deficit Irrigation Program Study. In Proceedings of the Law of Colorado River Conference, Scottsdale, AZ, USA, 12–13 March 2020.

3. Hanson, B.; Putnam, D.; Snyder, R. Deficit irrigation of alfalfa as a strategy for providing water for water-short areas. *Agric. Water Manag.* **2007**, *93*, 73–80. [CrossRef]

4. Orloff, S.; Hanson, B. Conserving water through deficit irrigation of alfalfa in the intermountain area of California. *Forage Grazing Lands* **2008**. [CrossRef]

5. Orloff, S.; Putnam, D.; Bali, K. *Drought Strategies for Alfalfa*; ANR Publication no. 8522; University of California: Oakland, CA, USA, 2015.

6. Montazar, A.; Radawich, J.; Zaccaria, D.; Bali, K.; Putnam, D. Increasing water use efficiency in alfalfa production through deficit irrigation strategies under subsurface drip irrigation. Water shortage and drought: From challenges to solution. In Proceedings of the USCID Water Management Conference, San Diego, CA, USA, 17–20 May 2016.

7. Montazar, A. Partial—Season Irrigation as an Effective Water Management Strategy in Alfalfa. *Agric. Briefs* **2017**, *10*, 3–7.

8. Peterson, P.R.; Sheaffer, C.C.; Hall, M.H. Drought Effects on Perennial Forage Legume Yield and Quality. *Agron. J.* **1992**, *84*, 774–779. [CrossRef]

9. Kirkham, M.B.; Johnson, D.E.; Kanemasu, E.T.; Stone, L.R. Canopy temperature and growth of differentially irrigated alfalfa. *Agric. Meteorol.* **1983**, *19*, 235–246. [CrossRef]

10. Grimes, D.W.; Wiley, P.L.; Sheesley, W.R. Alfalfa yield and plant water relations with variable irrigation. *Crop Sci.* **1992**, *32*, 1381–1387. [CrossRef]

11. Asseng, S.; Hsiao, T. Canopy CO_2 assimilation, energy balance, and water use efficiency of an alfalfa crop before and after cutting. *Field Crops Res.* **2000**, *67*, 191–206. [CrossRef]

12. Orloff, S.; Putnam, D.; Bali, K.; Hanson, B.; Carlson, H. Implications of deficit irrigation management on alfalfa. In Proceedings of the California Alfalfa and Forage Symposium, Visalia, CA, USA, 12–14 December 2005.

13. Ottman, M. Irrigation cutoffs with alfalfa-what are the implications? In Proceedings of the 2011 Western Alfalfa and Forage Conference, Las Vegas, NV, USA, 11–13 December 2011.

14. Sheaffer, C.C.; Tanner, C.B.; Kirkham, M.B. Alfalfa water relations and irrigation. In *Alfalfa and Alfalfa Improvement*; Hanson, A.A., Barnes, D.K., Hill, R.R., Eds.; Agronomy Monographs, ASA-CSSA-SSSA; The American Society of Agronomy: Madison, WI, USA, 1988; Volume 29, pp. 373–409.

15. Ottman, M.J.; Tickes, B.R.; Roth, R.L. Alfalfa yield and stand response to irrigation termination in an arid environment. *Agron. J.* **1996**, *88*, 44–48. [CrossRef]

16. Frate, C.A.; Roberts, B.A.; Marble, V.L. Imposed drought stress has no long-term effect on established alfalfa. *Calif. Agric.* **1991**, *45*, 33–36. [CrossRef]

17. Robinson, F.E.; Teuber, L.R.; Gibbs, L.K. *Alfalfa Water Stress Management During Summer in Imperial Valley for Water Conservation*; Research Report; Desert Research and Extension Center: El Centro, CA, USA, 1994.

18. Putnam, D.; Takele, E.; Kallenback, R.; Graves, W. *Irrigating Alfalfa in the Low Desert: Can Summer Dry-Down be Effective for Saving Water in Alfalfa?* United States Bureau of Reclamation: Washington, DC, USA, 2000.

19. Cabot, P.; Brummer, J.; Hansen, N. Benefits and impacts of partial season irrigation on alfalfa production. In Proceedings of the 2017 Western Alfalfa and Forage Conference, Reno, NV, USA, 28–30 November 2017.

20. Guitjens, J.C. Alfalfa irrigation during drought. *J. Irrig. Drain. Eng.* **1993**, *119*, 1092–1098. [CrossRef]

21. Lamm, F.L.; Harmoney, K.R.; Aboukheira, A.A.; Johnson, S.K. Alfalfa production with subsurface drip irrigation in the Central Great Plains. *Trans. ASABE* **2012**, *55*, 1203–1212. [CrossRef]

22. Irmak, S.; Odhiambo, L.O.; Kranz, W.L.; Eisenhauer, D.E. *Irrigation Efficiency and Uniformity, and Crop Water Use Efficiency*; Extension Circular EC732; University of Nebraska-Lincoln: Lincoln, NE, USA, 2011.

23. Snyder, R.L.; Spano, D.; Pawu, K.T. Surface Renewal analysis for sensible and latent heat flux density. *Bound. Layer Meteorol.* **1996**, *77*, 249–266. [CrossRef]

24. Shaw, R.H.; Snyder, R.L. Evaporation and eddy covariance. In *Encyclopedia of Water Science*; Stewart, B.A., Howell, T., Eds.; Marcel Dekker: New York, NY, USA, 2003.

25. Paw, U.K.T.; Snyder, R.L.; Spano, D.; Su, H.-B. Surface renewal estimates of scalar exchange. In *Micrometeorology in Agricultural Systems*; Hatfield, J.L., Baker, J.M., Eds.; Agronomy Society of America: Madison, WI, USA, 2005; pp. 455–483.

26. Montazar, A.; Rejmanek, H.; Tindula, G.N.; Little, C.; Shapland, T.M.; Anderson, F.E.; Inglese, G.; Mutters, R.G.; Linquist, B.; Greer, C.A.; et al. Crop coefficient curve for paddy rice from residual of the energy balance calculations. *J. Irrig. Drain. Eng.* **2017**, *143*, 04016076. [CrossRef]

27. Shapland, T.M.; McElrone, A.J.; Paw, U.K.T.; Snyder, R.L. A turnkey data logger program for field-scale energy flux density measurements using eddy covariance and surface renewal. *Ital. J. Agrometeorol.* **2013**, *1*, 1–9.

28. Snyder, R.L.; Pedras, C.; Montazar, A.; Henry, J.M.; Ackley, D.A. Advances in ET-based landscape irrigation management. *Agric. Water Manag.* **2015**, *147*, 187–197. [CrossRef]

29. Montazar, A.; Krueger, R.; Corwin, D.; Pourreza, A.; Little, C.; Rios, S.; Snyder, R.L. Determination of Actual Evapotranspiration and Crop Coefficients of California Date Palms Using the Residual of Energy Balance Approach. *Water* **2020**, *12*, 2253. [CrossRef]

30. Twine, T.E.; Kustas, W.P.; Norman, J.M.; Cook, D.R.; Houser, P.R.; Meyers, T.P.; Prueger, J.H.; Starks, P.J.; Wesely, M.L. Correcting eddy-covariance flux underestimates over a grassland. *Agric. Forest Meteorol.* **2000**, *103*, 279–300. [CrossRef]

31. CIMIS. Available online: http://wwwcimis.water.ca.gov/SpatialData.aspx (accessed on 7 September 2020).

32. Idso, S.B.; Jackson, R.D.; Pinter, P.J., Jr.; Reginato, R.J.; Hatfield, J.L. Normalizing the stress-degree-day parameter for environmental variability. *Agric. Meteorol.* **1981**, *24*, 45–55. [CrossRef]

33. Corwin, D.L.; Lesch, S.M. Application of soil electrical conductivity to precision agriculture: Theory, principles, and guidelines. *Agron. J.* **2003**, *95*, 455–471. [CrossRef]

34. Corwin, D.L.; Lesch, S.M. Characterizing soil spatial variability with apparent soil electrical conductivity: I. Survey protocols. *Comput. Electron. Agric.* **2005**, *46*, 103–133. [CrossRef]

35. Corwin, D.L.; Lesch, S.M. Apparent soil electrical conductivity measurements in agriculture. *Comput. Electron. Agric.* **2005**, *46*, 11–43. [CrossRef]

36. Corwin, D.L.; Scudiero, E. Field-scale apparent soil electrical conductivity. In *Methods of Soil Analysis*; Logsdon, S., Ed.; Soil Science Society of America: Madison, WI, USA, 2016; Volume 1.

37. Norris, K.H.; Barnes, R.F.; Moore, J.E.; Shenk, J.S. Predicting forage quality by near infrared reflectance spectroscopy. *J. Anim. Sci.* **1976**, *43*, 889–897. [CrossRef]

38. Doorenbos, J.; Kassam, A.H. Yield response to water. In *FAO Irrigation and Drainage Paper 66*; FAO: Rome, Italy, 1979; p. 193.

39. Kuslu, Y.; Sahin, U.; Tunc, T.; Kiziloglu, F.M. Determining water-yield relationship, water use efficiency, seasonal crop and pan coefficient for alfalfa in as semiarid region with high altitude. *Bulg. J. Agric. Sci.* **2010**, *16*, 482–492.

40. Jones, L.P. Agronomic Responses of Grass and Alfalfa Hayfields to No and Partial Season Irrigation as Part of a Western Slope waTer Bank. Master's Thesis, Colorado State University, Fort Collins, CO, USA, 2015.

41. Putnam, D.H.; Ottman, M.J. *Your Alfalfa Crop Got the Blues? It May be 'Summer Slump'*; Cooperative Extension; University of California: Half Moon Bay, CA, USA, 2013.

42. Ayers, R.S.; Westcott, D.W. Water Quality for Agriculture. In *FAO Irrigation and Drainage Paper No. 29*; FAO: Rome, Italy, 1985; p. 174.

43. Putnam, D.; Benes, S.; Galdi, G.; Hutmacher, B.; Grattan, S. Alfalfa (*Medicago sativa* L.) is tolerant to higher levels of salinity than previous guidelines indicated: Implications of field and greenhouse studies. In Proceedings of the 19th EGU General Assembly, Vienna, Austria, 23–28 April 2017.

44. U.S. Department of Agriculture. Farm and Ranch Irrigation Survey, 2012. In *Census of Agriculture*; USDA: Washington, DC, USA, 2014; Volume 3.

45. Montazar, A.; Putnam, D.; Bali, K.; Zaccaria, D. A model to assess the economic viability of alfalfa production under subsurface Drip Irrigation in California. *Irrig. Drain.* **2017**. [CrossRef]

Publisher's Note: MDPI stays neutral with regard to jurisdictional claims in published maps and institutional affiliations.

 *agronomy*

Article

Adoption of Water-Conserving Irrigation Practices among Row-Crop Growers in Mississippi, USA

Nicolas Quintana-Ashwell [1,*], Drew M. Gholson [1], L. Jason Krutz [2], Christopher G. Henry [3] and Trey Cooke [4]

[1] National Center for Alluvial Aquifer Research, Mississippi State University, 4006 Old Leland Rd, Leland, MS 38756, USA; drew.gholson@msstate.edu
[2] Mississippi Water Resources Research Institute, 885 Stone Blvd, Ballew Hall, Mississippi State University, Leland, MS 38756, USA; j.krutz@msstate.edu
[3] Rice Research and Extension Center, University of Arkansas Cooperative Extension Service, Stuttgart, AR 72160, USA; cghenry@uark.edu
[4] Delta Farmers Advocating Resource Management, Delta F.A.R.M., Stoneville, MS 38776, USA; trey@deltawildlife.org
*   Correspondence: n.quintana@msstate.edu; Tel.: +1-662-390-8508

Received: 16 June 2020; Accepted: 17 July 2020; Published: 27 July 2020

Abstract: This article identifies irrigated row-crop farmer factors associated with the adoption of water-conserving practices. The analysis is performed on data from a survey of irrigators in Mississippi. Regression results show that the amount of irrigated area, years of education, perception of a groundwater problem, and participation in conservation programs are positively associated with practice adoption; while number of years farming, growing rice, and pumping cost are negatively associated with adoption. However, not all factors are statistically significant for all practices. Survey results indicate that only a third of growers are aware of groundwater problems at the farm or state level; and this lack of awareness is related to whether farmers noticed a change in the depth to water distance in their irrigation wells. This evidence is consistent with a report to Congress from the Government Accountability Office (GAO) that recommends policies promoting the use of: (1) more efficient irrigation technology and practices and (2) precision agriculture technologies, such as soil moisture sensors and irrigation automation.

Keywords: irrigation; groundwater; alluvial aquifer; water conservation adoption; row crops; Mississippi Delta; precision agriculture; Lower Mississippi River Valley

1. Introduction

The Mississippi River Valley Alluvial Aquifer (MRVAA) sustains irrigated agriculture in the Mississippi Delta. Almost 22,000 permitted wells [1] withdrawing more than 370 million m³ of water per year [2] continue to reduce the stock of groundwater available in the MRVAA at an unsustainable rate [3]. A shortage of irrigation water would be a critical challenge to agricultural production in the region [4]. To address this threat, researchers, regulators, and conservation agencies promote the adoption of water-conserving practices in irrigated agriculture. However, little is known about what drives growers in Mississippi to adopt water conservation practices that improve irrigation efficiency.

Profitability is a primary concern in any sustainable enterprise. Hence, farmers would adopt new practices that result in higher profits or reduced risks. However, profitable practices are not universally adopted; which suggests there are other factors related to the farmers and their ecosystem that influence their choice of agricultural practices. This implies that the combination of practices adopted and the factors that influence their adoption vary by practice and location [4]. In some cases, factors such as farmer age or the practicality of the technology are more important than monetary factors [5]. Recent literature identifies factors likely to be associated with the adoption of certain water management and conservation practices at the state (for example, Nian et al. [4] for Arkansas) and national level [6]. However, no recent study examines the factors driving conservation practice adoption in Mississippi.

This article describes water conservation practices that have the potential to profitably reduce the rate of depletion of the MRVAA and identifies social, economic, and environmental factors associated with the adoption of those practices among irrigators in Mississippi. The adoption of conservation practices and farmer characteristics are identified from a comprehensive survey of irrigators in Mississippi that achieved 148 valid responses. A choice model estimated using probit regression on the dataset identifies which factors have a statistically significant association with each of the practices considered.

The predominant irrigation method in the Delta region of Mississippi is continuous flow furrow irrigation [7] on row-crops. This is a modified gravity irrigation system that employs pipes with holes aligned to deliver the flow of water on the furrows. The system is better suited to the relatively flat Delta area than it would be for other regions. Elevation goes from 62.5 m above sea level in the northern tip just south of Memphis, TN, to 24.4 m above sea level at the southern tip in Vicksburg, MS, while the center of the Delta averages an elevation of 38 m between Greenville and Greenwood. Furthermore, the fields are often precision leveled, which results in less "pooling of water" on the fields and more uniform irrigation. Consequently, the baseline case is a relatively inefficient gravity system in terms of costs and irrigation performance. The conservation practices assessed in this article are modifications, "add-ons", or substitutes to this baseline prevalent system.

There is compelling evidence that even minor modifications to existing irrigation and agronomic practices in the Mid-South USA region can result in noticeable water savings while achieving similar yields at harvest [7–10]. As anticipated, the practices evaluated in those studies and considered here are not universally adopted in the area. Despite the expected profitability of adoption, the costs of these practices occur at the time of adoption while their benefits accrue over time. Consequently, producers may require generous incentives or returns to the cost of investing in conservation practices to adopt them [11] or the assurance of witnessing several years of neighboring farmers employing them.

The adoption of conservation practices in the Delta area of Mississippi is partially driven by regulatory mandate. All wells drilled in the area with a casing diameter of 15 cm or greater are required to have a permit. The permits must be renewed every five years and require crop producers to file an Acceptable Agricultural Water Efficiency Practices (AAWEP) form. Irrigators must claim to employ a high efficiency irrigation system, such as a sprinkler irrigation system; or claim the use or proposed use of three water conservation practices; see [12] for the permit application and list of acceptable practices.

1.1. Conservation Practices

This article considers practices that show potential to profitably conserve irrigation water, are accepted in the AAWEP form, and have been adopted by multiple respondents in the 2016 Survey of Mississippi Irrigators. Although many irrigators may decide on adopting several practices simultaneously, this article analyzes adoption practice by practice in order to maximize the number of valid observations for each case.

Soil Moisture Sensors (SMS) are used in irrigation event decision scheduling to prevent yield-limiting water deficit stress on a given crop. SMS gives producers the knowledge of the moisture within the soil profile to make informed and confident irrigation initiation and termination decisions [8] that typically result in increased irrigation efficiency [7,13].

The simplest and most inexpensive (free) upgrade to the baseline irrigation system is Computerized Hole Selection (CHS). Instead of punching uniform holes in layflat poly-tubing, CHS calculates relatively larger holes for parts of the field with long irrigation runs, while shorter parts of the field receive smaller holes to allow water to uniformly reach the end of the field and minimize water runoff [8]. On-farm studies in the Delta area found that CHS achieves water savings of 20 to 25 percent in most situations [14].

"If it can be measured, it can be managed" goes the adage. Pumping flow-meters (meters) are are an important irrigation water management tool compatible with all irrigation systems. Although they do not provide an intrinsic ability to conserve water, they are crucial components, for example, to calculate the optimal size of the holes with CHS [15]. They are also required for cooperator farmers participating in NRCS conservation contracts.

Surge irrigation (surge) allows fields to be divided in two in order to deliver a higher flow rate of water to each half. Water surges down one half of the field until the surge valve flips to deliver water to the other side of the field [8]. The wetting and soaking cycles reduce surface runoff and deep percolation loss while improving water application efficiency by up to 25 percent on the baseline systems with improved infiltration rates documented in sealing silt loam soils [16].

On-Farm Water Storage systems (OFWS) are irrigation water storage structures that are typically designed by NRCS in the Mississippi Delta with the capacity to apply 77 mm of water per hectare per season and meet irrigation requirements for eight out of 10 years [17]. Depending on seasonal conditions, storage capacity, and farmed area, these systems can completely substitute groundwater pumping in some years. NRCS provides technical and financial assistance to producers interested in building OFWS, but many producers face a high opportunity cost to retire productive land to be used for water storage.

The Tailwater Recovery System (TWS) collects irrigation and storm water runoff on the farm in a reservoir (OFWS). TWS increases the amount of water available to irrigation compared to an OFWS filled only by precipitation. This allows OFWS to occupy a smaller surface area and more hectares to be farmed. NRCS estimates that TWS by itself can reduce groundwater pumping by 25 percent.

Micro-irrigation (micro) is a low pressure, low volume, frequent application of water directly to the plant's root zone [18]. It can increase yields and decrease water use by drastically reducing non-beneficial evaporation and virtually eliminating irrigation water runoff. Micro-irrigation is rarely found in the Delta where the soil types and the water quality make the emitters (i.e., nozzles) prone to clogging.

Center pivot-irrigation (pivot) is a type of irrigation that delivers water through sprinklers that create artificial precipitation and are attached to a wheel-driven frame that rotates radially (the arm pivoting on the center). They are highly configurable for a variety of field and crop requirements. Most center pivot systems in the Delta were installed in the 1980s and designed for cotton [19]. However, the original designs are inappropriate to meet the maximum water demands of corn and soybeans. Consequently, there is both adoption of and migration away from center pivot irrigation in the Delta. Similar to micro-irrigation, the soil types and water quality in the Delta present challenges in the form of nozzle clogging and wheels getting stuck in mud.

Pump timers (timer) are a mechanism to program the time or amount of water at which a pump is shut-off. As it helps to automatically or remotely turn the pumps off, it conserves the excess water that might otherwise be pumped, particularly at night [2]. Timers can be employed across irrigation systems for a variety of crops.

Cover crops (cover) are plants that are typically planted during the off season to cover the soil rather than for the purpose of being harvested. They can help sustain and improve soil health [20], microbial populations, and water infiltration, as well as provide benefits in terms of weed control [21]. This is the only conservation practice being analyzed that is not part of AAWEP.

1.2. Factors Affecting Conservation Practices

This article analyzes data from a regional Crop Irrigation Survey that collected 148 valid observations and include data on farmer practices, perceptions and attitudes, and socio-economic status. The factors selected are irrigated area in the operation, Groundwater (GW) use in irrigation, crop choice (rice), number of years farming, years of formal education, whether the farmer perceives a

GW problem at the farm or state level, average pumping cost in the county of residence, participation in a conservation program, and annual income levels. These factors obtained a sufficient number of valid responses in the sample and were mentioned in two recent comprehensive reviews of the literature [6,22] that identified factors associated with the adoption of agricultural conservation practices or in a recent study of conservation practice adoption in Arkansas [4].

The empirical analysis consists of testing how these factors correlate with adoption of the identified practices. The hypotheses with respect to this association are drawn from the existing literature. Specifically, we draw from a study using similar data in an adjacent area by Nian et al. [4], a comprehensive review of 102 papers in the agricultural conservation practice adoption literature by Prokopy et al. [22], and a 129 page Government Accountability Office report to Congress on Irrigated Agriculture by Pearsons and Morris [6]. In terms of the signs of the regression coefficients, the hypotheses are as follows:

(a)   Irrigated area: positive;

(b)   (GW) use: positive;

(c)   rice farming: negative (most conservation practices are geared towards row-cropping, so their adoption appears less likely for rice farmers.);

(d)   years farming: negative;

(e)   education: positive;

(f)   GW problem: positive;

(g)   pumping cost: positive;

(h)   conservation program: positive;

(i)   cracking soils: negative for surge irrigation, undetermined for other practices; and

(j)   income: positive.

2. Materials and Methods

2.1. Data

The data were from a survey of irrigators in Mississippi conducted by the Survey Research Laboratory at the Social Science Research Center at Mississippi State University [23]. A telephone-based survey secured a total of 148 completed interviews in Mississippi from a total 2216 telephone numbers acquired (861 were disconnected or inaccessible after 10 attempts) with an overall cooperation rate of 27.6 percent. The sample is representative of the Delta area of Mississippi with 131 respondents residing in the area and 3 more in neighboring counties. The survey instrument contained questions on growers' characteristics, cultural practices, irrigation management practices, and perceptions and attitudes regarding groundwater availability.

Except for irrigated area, years of education, and pumping cost, the variables included in the analysis were coded as categorical or indicator (dummy) variables. The pumping cost is calculated as [24]:

$$p = \theta_e p_e d, \tag{1}$$

where p_e is the price of the energy source for the power unit, d is depth to water used as a proxy for pumping lift, and θ_e is the amount of energy from source e needed to lift a cubic meter of water a distance of one meter. The distance to water was obtained from the U.S. Geological Survey [25] based on the respondent's claimed county of residence. Average energy prices were obtained from the U.S. Energy Information Administration (EIA).

Cracking soils is the percentage of soils with clay content dominated by smectite. Such soils crack on the surface when a moist soil shrinks due to drying. The data on soil composition is accessible through the USDA-NRCS Websoil survey [26]. This is an important control variable. For example, surge irrigation programs are more difficult to manage in this type of soil because the programming of the alternating cycles is more complicated. The typical program relies on visual cues to switch from one side of the field to the next based on the time water takes to reach the tail of the section. In cracking soils, water infiltrates through the cracks, and actual wetting occurs by 3 or more meters ahead of the wetting on the surface. The program is still applicable and carries the same water savings potential, but becomes harder for the farmer to realize.

Number of years of education was calculated based on a question that was originally categorical. The assigned values were as follows: 10 for less than completed high school, 12 for completed H.S., 13 for some college or vocational program, 14 for completed Associate's degree, 16 for completed Bachelor's, 18 for completed Master's, and 20 for more than Master's. This transformation is helpful in the estimation and interpretation of regression results with a sample that is small relative to the number of variables considered.

The variable GW problem is a dummy variable based on the combination of categorical responses to two different questions in the survey: "In your opinion, do you have a groundwater shortage problem on your farm?" and "In your opinion, do you have a groundwater shortage problem in your state?" Lastly, conservation program is a dummy variable based on the combination of responses to four different questions that would have otherwise yield 19 response categories (see Appendix A).

2.1.1. Choice Model

There are several ways to motivate the empirical strategy. Due to the nature of the survey and structure of the data, a scenario that allows for irrigators to adopt a single practice or a number of them simultaneously is needed. Hence, a model of individual practice adoption is adequate. An irrigator i adopts a water conservation practice w if the grower expects to receive a greater utility from using the practice (U_{iw}) than they would not using it (U_{iNg}). The probability of adopting practice w is the probability that $y_{iw}^* = U_{iw} - U_{iNg} > 0$ and depends on a vector of n identified factors X_i. Following Maddala [27], y_i^* is a latent, unobservable variable defined by the regression relationship:

$$y_{iw}^* = \beta' X_i + u_i \tag{2}$$

where u_i is the error term. The variable that is actually observed is whether a practice is adopted ($y = 1$) or not ($y = 0$).

From these relationships, the probability that any given practice w is adopted can be estimated using probit with the assumption that u_i follows a normal distribution:

$$Pr(y_w = 1) = Pr\left(\sum_j \beta_{jw} X_j > 0\right) = \Phi\left(\sum_j \beta_{jw} X_j\right), \tag{3}$$

where $\Phi(\cdot)$ is the cumulative normal distribution function.

To predict the effect a change in the value of a variable would have on the probability of adopting a given practice, the marginal effects are calculated as:

$$\frac{\partial}{\partial x_{ik}} \Phi\left(X_i'\beta\right) = \phi\left(X_i'\beta\right) \beta_k, \tag{4}$$

where $\phi(\cdot)$ is the normal probability density function. This marginal effect is denoted as dy/dx in the results.

3. Results and Discussion

3.1. The Sample

The survey instrument contained questions on growers' characteristics, cultural practices, irrigation management practices, and perceptions and attitudes regarding groundwater availability. Table 1 summarizes the information gathered on growers' land tenure, education, and income.

Responses were considered to be representative of irrigators operating in the Delta because 88.5 percent of respondents resided in that area and additional respondents lived in neighboring counties. Cracking soils were present in 83.1 percent of the counties where irrigators claimed residence with an average of 27.6 percent of soils classified as cracking.

Table 1. Summary statistics of farmer characteristics from an irrigation survey conducted in the Mississippi Delta in 2016.

Farmer Characteristics	N	%
Delta	131	88.5
Cracking soil	123	83.1
Avg. percentage		27.6
Operator	31	20.9
Landowner and operator	117	79.1
Education:		
Less than high school	5	3.4
Completed high school or GED	23	15.5
Some college	22	14.9
Completed Associate's	18	12.2
Completed Bachelor's	66	44.6
Completed Master's	11	7.4
Beyond Master's	2	1.4
Agriculture-related	63	42.6
Income per year:		
Less than USD50,000	13	8.7
USD50,000 to USD100,000	41	27.7
USD100,000 to USD150,000	17	11.5
USD150,000 to USD200,000	9	6.1
USD200,000 to USD250,000	6	4.1
USD250,000 to USD300,000	5	3.4
More than USD300,000	10	6.8
Unsure or no response	47	31.7

Almost 80 percent of respondents were land-owner operators, and the remaining growers were operators only. Nearly two-thirds of the farmers completed a post-secondary degree (65.6 percent), and 42.6 percent of respondents indicated that part of their formal education was related to agriculture.

Growers also identified the range of income they had achieved the previous year. A total of 101 valid responses were recorded with 31.7 percent of respondents refusing to provide an answer or being unsure with respect to which income bracket they belonged. Amongst the valid responses, fifty-three-point-five percent of farmers claimed an annual income of less than USD100,000. The median income in the sample was between USD75,000 and USD100,000 per year.

Income level is expected to be positively correlated with the adoption of conservation practices [4,6]. Also, the level of farmer education positively influences the adoption of irrigation-related precision agriculture practices [6].

3.1.1. Farming Practices

Data on growers' agricultural experience and practices are summarized in Table 2. In terms of farming experience, respondents represented a wide range of experience, from as little as three years to as much as 80 years of farming experience. Every measure indicated that these were seasoned farmers. On average, growers had 28 years of farming experience with a median and mode of 29 and 30 years of experience, respectively. More than 84 percent of farmers had 10 years or more of farming experience. Approximately two-thirds of the sample were farmers with more than 20 years of experience. The number of years farming was expected to be negatively associated with the adoption of agricultural innovations.

Table 2. Summary statistics of cultural practices.

	N	%	Min	Max	Mean	Std. Deviation
Years farming	148	100	3	80	28.03	14.761
			(Min, Max, and Mean in ha)			
Irrigated area (ha, all)	148	100.0	0	6070	896	1007
Irrigated crops						
Corn	106	71.6	2.02	1821	305	337.7
Cotton	49	33.1	18.1	2833	490	546.7
Rice	41	27.7	32.4	1558	373	407.2
Soybeans	131	88.5	27.1	3804	671	700.2
Cover crop	45	30.4	4.05	1335	-	-
Land leveling	124	83.8				
Zero grade	26	17.6	2.02	971	177	239.6
Precision grade	116	78.4	3.24	4654	733	848.5
Warped or OptiSurface	33	22.3	10.12	1619	266	385.9
Not leveled	76	51.4	4.05	1714	202	240.3
Conservation programs	111	75				
CRP	58	39.2	-	-	-	-
EQIP	87	58.8	-	-	-	-
RCPP	14	9.5	-	-	-	-
CSP	31	20.9	-	-	-	-
NRCS	8	5.4	-	-	-	-
Other	8	5.4	-	-	-	-

Note: CRP is NRCS Conservation Reserve Program; EQIP is NRCS Environmental Quality Incentives Program; RCPP is NRCS Regional Conservation Partnership Program; CSP is NRCS Conservation Stewardship Program; NRCS is USDA Natural Resources Conservation Service unspecified program.

The size of the farming operation is an important factor in deciding the adoption of agricultural practices in general. The average operation involved 896 ha of irrigated farmland with a median of 567 ha and as much as 6070 irrigated ha. More than three-fourths of the responding growers operated 1133 ha or less. Hence, the number of irrigated hectares was expected to be positively correlated with the adoption of conservation practices.

Crop choice is oftentimes associated with the choice of irrigation technology [28,29]. The largest number of growers reported producing irrigated soybeans (n = 131), which occupied the largest cultivated area among the irrigated crops reported: 671 ha on average and as much as 3804 ha. Irrigated corn was the second most popular crop choice with 106 farmers reporting an average of 305 ha and as much as 1821 ha of irrigated farmland dedicated to that crop. Cotton is a traditional crop in the Delta region of Mississippi. About a third of the respondents grew cotton with irrigation dedicating an average of 490 ha and as much as 2833 ha to its production. These are typically row-crops that employ the same or similar irrigation setups when the fields are prepared for furrow irrigation.

Kebede et al. [2] reported that irrigated rice consumes more water than any other crop in the region. Growers in this sample reported an average of 373 ha of irrigated rice farmland with as much as 1558 ha of rice under irrigation. Rice production was expected to be negatively correlated with the conservation practices considered in this article, which were better suited for row-crop irrigation.

Cover crops are typically not harvested. Consequently, these crops were not considered as part of crop choices, but rather as a conservation practice in this article. Nearly a third (30.1 percent) of growers in the sample claimed to plant cover crops. Responses to the survey varied widely in terms of crop and area matching. Wheat and radishes received the highest reported area of 1335 ha, while the least was reported for Asian mustard greens (4 ha). The adoption of this practice was tested against the identified explanatory factors.

Land leveling is a relatively common practice in the area with 84 percent of growers reporting having at least a part of their fields land-leveled in some way. It is also no surprise that 51.4 percent of the participating growers reported that some fields were not leveled because the Delta in Mississippi is unusually flat. Precision grade was the most common land leveling method with 78 percent of growers employing it on an average of 733 ha and in up to 4654 ha of their operation.

Awareness and participation in conservation programs were expected to positively influence the adoption of irrigation water conservation practices [4,6]. Three-fourths of the growers claimed participation in a conservation program. The program most commonly cited was the NRCS Environmental Quality Incentives Program (EQIP) with 59 percent participation.

3.1.2. Irrigation and Water Conservation Practices

Grower irrigation practices are summarized in Table 3. Groundwater was the principal source of water for irrigation with 93 percent of respondents identifying it as a source. On average, eight-hundred eighty-nine hectares were irrigated with groundwater with a maximum of 4856 ha relying on that source for irrigation. Surface water was also employed for irrigation including streams and bayous, which are the source for 178 ha on average. The surface sources can also be complemented with OFWS, 11.5 percent of responses, and TWS in 14.2 percent of responses. Some growers built OFWS and TWS capable of fully supplying the irrigation water needs for some of their fields. Producers relying on groundwater from a depletable aquifer were expected to be more inclined to adopt water conservation practices.

The predominant irrigation practice was furrow irrigation for row-crops, which was employed by 86 percent of respondents on an average of 823 ha. Practices that improve the performance of furrow irrigation are deep tilling, employed by 71 percent of the respondents, computerized hole selection (CHS), adopted by 59 percent of growers, and surge irrigation, adopted by 24 percent of farmers in the sample. The last two are considered water-conserving practices for which we sought to find determining adoption factors. Irrigation systems with higher application efficiency are also considered water conservation practices. Micro irrigation was very rare with only 3.4 percent of respondents employing it on an average of 65 ha; while center pivot sprinkler use was more widespread with 60 percent of respondents having used it on an average of 370 ha.

Irrigation scheduling is a crucial component of irrigation water management. The use of Soil Moisture Sensors (SMS) for scheduling has the potential to save as much as 50 percent of total water applied [30]. Agronomic studies in the area showed that SMS could help improve water use efficiency in furrow irrigated soybeans [7] and corn [10] by reducing water use without reductions in yields when compared to conventional farmer-managed scheduling.

Flow-meters at the irrigation wells are another important management tool. Nearly 70 percent of participant growers owned them. A voluntary metering program in Mississippi encourages their use, and participation in NRCS incentive programs makes participation in that program mandatory for their cooperators. Another pump accessory is the pump timer, which allows the irrigation events to be started or stopped automatically. Almost 44 percent of respondents employed pump timers on an average of 11 pumps.

Finally, the energy source for the pump power units varied with most growers having more than one type of energy source. Electricity was the most common energy source with 85 percent of farmers claiming it. Diesel was the second most common source with almost 80 percent of respondents using it. The energy source mix is important in calculating irrigation pumping costs. The average cost of pumping was estimated at USD 0.0538 per megaliter.

Table 3. Summary statistics of irrigation practices.

	N	%	Min (ha)	Max (ha)	Mean (ha)	Std. Dev.
Irrigation by source of water						
Groundwater	137	92.6	2.83	4856	889	909
Stream or bayou	39	26.4	7.69	599	178	142
Stream or bayou and OFWS	17	11.5	2.02	977	182	253
Stream or bayou and TWS	21	14.2	16.2	707	178	185
No outside source w/OFWS/TWS	16	10.8	4.45	304	81	84
Irrigation by practice						
Flood (row-crops)	69	46.6	6.07	4047	737	939
Furrow (row-crops)	127	85.8	1.62	4452	823	942
Deep till (furrow)	104	70.7	1.62	2428	512	522
Computerized hole selection	87	58.8	1.62	3462	734	706
Surge	35	23.6	12.1	607	140	146
Border (row-crops)	26	17.6	10.1	769	158	274
Micro (row-crops)	5	3.4	2.02	223	65	106
Pivot (row-crops)	88	59.5	2.43	2428	370	391
Irrigation scheduling						
Soil moisture sensors	72	48.6	0.4	6070	554	1027
Visual crop stress	103	69.6	-	-	-	-
Computerized scheduling	5	3.4	80.9	243	146	102
Routine	29	19.6	-	-	-	-
Probe/feel	27	18.2	-	-	-	-
ET or Atmometer	4	2.7	0	202	99	84
Watch other farmer	6	4.1	-	-	-	-
			Min (units)	Max (units)	Mean (units)	Std. Dev.
Irrigation pumps	146	98.6	1	120	21	24.0
Pump timers	65	43.9	1	90	11	16.6
Flow-meters	103	69.6	1	46	8	7.9
Power unit						
Electric	126	85.1	1	80	11	14.8
Diesel	117	79.1	1	85	13	14.0
Propane	24	16.2	1	45	7	9.3
Natural gas	3	2.0	1	5	3	2.0
Pumping cost (USD/ML)	94	-	0.02	0.088	0.054	0.022

Note: OFWS is On-Farm Water Storage; TWS is Tailwater Recovery System.

3.1.3. Grower Perceptions and Attitudes

The data collected on farmer perceptions and attitudes towards groundwater availability issues are summarized in Table 4. Less than one-third of growers observed a change in their wells' depth to water while over two-thirds of respondents indicated they perceived there was not a problem with the groundwater supply at their farm or in the state. A test of independence in these responses indicated that they had a statistically significant dependence: those who perceived there had been a change in their wells depth to water were more likely to believe there was a groundwater problem in their farm or at the state level.

The U.S. Geological Survey (USGS) has recently published a map of the Potentiometric Surface of the Mississippi River Valley Alluvial Aquifer for the Spring 2016 [25] that shows the location and gradient of the aquifer's cone of depression. A cross-tabulation of farmer perceptions and their location in the center of the cone of depression is presented in Table 5. A Pearson test of independence of the responses showed evidence that farmers located in the cone of depression were more likely to observe a change in their well levels and think there was a groundwater problem at the farm or state level. Half of those located in the center of the cone of depression believed there was a groundwater problem as opposed to 29 percent amongst those located outside that area. Similarly, forty-six percent of those in the cone of depression area observed a change in the depth-to-water in their wells while only 26 percent of those outside the area noticed such a change.

Table 4. Summary statistics of farmer perceptions and attitudes.

Thinks There Is a GW Problem			
Frequency	**No**	**Yes**	**Total**
Well Depth to Water:			
No change	60	24	**84**
Increased	9	11	**20**
Decreased	12	13	**25**
Do not know	16	2	**18**
Refused	1	0	**1**
Total	**98**	**50**	**148**
Percentage	**No**	**Yes**	**Total**
Change in Depth to Water:			
No/cannot tell	51	17	**68.9**
Changed	14	16	**30.4**
Refused	1	0	**0.7**
Total	**66.2**	**33.8**	**100**
Pearson χ^2_4 = 13.4 with Pr = 0.009.			

Note: GW is Groundwater.

Table 5. Aquifer "cone of depression" and farmer perceptions and attitudes.

Cone of Depression			
Percentages	**No**	**Yes**	**Total**
Depth to Water:			
No change	56	14	69
Changed	20	11	31
Total	76	24	100
Pearson χ^2_1 = 4.3 with Pr = 0.038			
Groundwater Problem:			
No	54	12	**66**
Yes	22	12	**34**
Total	76	24	**100**
Pearson χ^2_1 = 5.6 with Pr = 0.018.			

3.2. Probit Regression Analysis

The estimated models of practice adoption fit the data relatively well with pseudo-R^2 (McFadden's) ranging between 0.156 to 0.545. Except for micro-irrigation (only five adopters), the conservation practices being analyzed had at least one factor with a statistically significant coefficient. The probit regression coefficients are detailed in Table 6 for all factors except cracking soils and income, which are detailed in Table 7.

Tailwater Recovery System (TWS) adoption was positively and significantly influenced by the farmers perception that a groundwater problem existed (GW prob.). The marginal effect (dy/dx) indicated that a producer who becomes aware of the groundwater problems in the Delta area would be associated with a 25 percent higher likelihood of adopting TWS. The data indicated that farmers who do not use groundwater for irrigation have not adopted TWS.

For OFWS, the number of irrigated hectares under operation (Irr.area) was positively and significantly associated with the adoption of OFWS. The calculated marginal effect indicated that for an additional 40 hectares of farmed land, the probability of a farmer adopting OFWS was 0.8

percent higher. The data indicated that farmers who did not use groundwater for irrigation have not adopted OFWS.

Table 6. Results from probit regressions (coefficients by income level in Table 7). Irr, Irrigation.

	Irr. Area	GW Use	Rice	Years Farm	Educ.	GW Prob.	P. Cost	Cons.pr.
TWS	0.0001	(a)	−0.398	0.009	0.177	0.857 **	0.464	0.691
s.e.	(0.0001)		(0.435)	(0.015)	(0.132)	(0.448)	(1.052)	(0.588)
dy/dx	0.00003		−0.115	0.003	0.051	0.25 **	0.134	0.199
McFadden's $R^2 = 0.228$								
OFWS	0.0003 **	(a)	−0.007	0.027	0.214	0.779	1.45	0.186
s.e.	(0.005)		(0.507)	(0.018)	(0.159)	(0.54)	(1.18)	(0.67)
dy/dx	0.00008 ***		−0.002	0.006	0.052	0.189	0.353	0.045
McFadden's $R^2 = 0.324$								
CHS	0.0002 *	0.59	−0.176	−0.01	0.153	1.13 **	−0.58	−0.24
s.e.	(0.0001)	(1.11)	(0.49)	(0.015)	(0.124)	(0.51)	(1.03)	(0.495)
dy/dx	0.00005 *	0.16	−0.048	−0.003	0.041	0.30 **	−0.157	−0.065
McFadden's $R^2 = 0.271$								
Surge	0.00007	(a)	−0.516	−0.019	0.258	1.172 *	−2.462 **	0.735
s.e.	(0.00011)		(0.508)	(0.017)	(0.151)	(0.601)	(1.23)	(0.76)
dy/dx	0.00002		−0.116	−0.004	0.058 **	0.26 **	−0.55 **	0.165
McFadden's $R^2 = 0.362$								
SMS	0.0009 **	(a)	−2.72 **	−0.016	0.03	-0.81	−1.1	2.45 **
s.e.	(0.0003)		(0.94)	(0.019)	(0.168)	(0.9)	(1.31)	(1.06)
dy/dx	0.0002 ***		−0.5 ***	−0.003	0.005	−0.15	−0.204	0.45 **
McFadden's $R^2 = 0.545$								
Micro	−0.00003	(a)	(b)	0.02	0.47	(b)	−4.42	(a)
s.e.	(0.0004)			(0.05)	(0.63)		(10.6)	
dy/dx				0.0003	0.009		−0.074	
McFadden's $R^2 = 0.464$								
Pivot	0.0001	(a)	−0.622	0.0005	0.24 *	−0.028	1.28	−0.29
s.e.	(0.0001)		(0.471)	(0.015)	(0.13)	(0.48)	(1.08)	(0.58)
dy/dx	0.00003		−0.18	0.001	0.068 **	−0.0087	0.363	−0.083
McFadden's $R^2 = 0.156$								
Timer	0.0003 **	(a)	−0.84	0.003	−0.14	1.67 **	−0.43	(b)
s.e.	(0.0001)		(0.52)	(0.016)	(0.13)	(0.54)	(1.23)	
dy/dx	0.00006 **		−0.2 *	0.001	−0.034	0.41 ***	−0.1	
McFadden's $R^2 = 0.355$								
Flow-meter	0.0005 **	0.181	−0.43	0.017	0.167	0.932	−0.56	0.33
s.e.	(0.0003)	(1.15)	(0.61)	(0.016)	(0.143)	(0.67)	(1.26)	(0.56)
dy/dx	0.0001 **	0.037	−0.09	0.004	0.034	0.19	−0.12	0.068
McFadden's $R^2 = 0.34$								
Cover	0.0001	−0.76	0.43	−0.04 **	0.068	−0.44	−0.015	0.78
s.e.	(0.0001)	(1.55)	(0.47)	(0.018)	(0.125)	(0.50)	(0.96)	(0.59)
dy/dx	0.00003	−0.21	0.12	−0.011 **	0.02	−0.12	−0.004	0.22
McFadden's $R^2 = 0.21$								

Note: Educ. is years of formal education; GW Prob. is an indicator variable for perception of a groundwater problem at the farm or state level; P. Cost is the cost of pumping; and Cons. pr. is participation in a conservation program. Standard errors are in parentheses. *,**,***: significant at $p < 0.1$, $p < 0.05$, $p < 0.01$, respectively. (a) negative cases predict failure; (b) positive cases predict failure.

Computerized Hole Selection (CHS) was positively and significantly associated with the number of irrigated hectares, the perception of the existence of a groundwater problem, and having an income between $100,000 and $150,000. The marginal effects indicated a 0.5 percent higher probability of CHS adoption for an additional 40 ha of land irrigated, and the probability of adoption increased by 30 percent when a farmer realized there was a groundwater problem at the farm or state level.

Table 7. Results from probit regressions (continued) on "cracking" soils and income levels.

	Cracking	50 k to 100 k	100 k to 150 k	150 k to 200 k	200 k to 250 k	250 k to 300 k	$\geq$300 k
TWS	0.18	−0.49	−0.5	−1.195	0.302	(dropped)	0.604
s.e.	(0.02)	(0.6)	(0.637)	(1.02)	(1.063)		(1.185)
dy/dx	0.005	-0.14	−0.144	−0.345	0.087		0.174
OFWS	−0.008	−1.021	−0.7	(dropped)	0.615	(dropped)	0.186
s.e.	(0.03)	(0.684)	(0.676)		(1.33)		(0.671)
dy/dx	−0.002	−0.248	−0.17		0.15		0.045
CHS	0.035	0.611	1.325 *	0.59	0.273	0.613	1.33
s.e.	(0.02)	(0.625)	(0.694)	(0.878)	(1.05)	(1.25)	(1.24)
dy/dx	0.009	0.165	0.358 **	0.159	0.074	0.166	0.359
Surge	0.028	−1.26 *	−0.241	0.66	(dropped)	(dropped)	1.44
s.e.	(0.03)	(0.766)	(0.669)	(0.963)			(1.165)
dy/dx	0.006	−0.28 *	−0.05	0.148			0.324
SMS	0.014	0.99	2.48 **	−0.496	0.258	−0.019	−1.38
s.e.	(0.03)	(0.95)	(1.06)	(0.982)	(1.091)	(1.26)	(2.0)
dy/dx	0.003	0.183	0.458 **	−0.092	0.048	−0.004	−0.255
Micro	−0.052	(dropped)	(dropped)	(dropped)	(dropped)	(dropped)	(dropped)
s.e.	(0.06)						
dy/dx	-0.001						
Pivot	0.021	−0.45	−0.49	−0.365	(b)	0.548	(a)
s.e.	(0.02)	(0.64)	(0.64)	(0.927)		(1.183)	
dy/dx	0.06	−0.13	−0.14	−0.1		0.156	
Timer	0.046	−0.486	−0.129	0.421	−0.028	(c)	(c)
s.e.	(0.029)	(0.711)	(0.78)	(0.9)	(1.03)		
dy/dx	0.01	−0.12	−0.03	0.1	−0.01		
Flow meter	0.03	−0.88	0.09	−1.15	(c)	(c)	(c)
s.e.	(0.03)	(0.78)	(0.87)	(1.07)			
dy/dx	0.005	−0.18	0.02	−0.24			
Cover	−0.003	0.15	0.55	0.36	1.97 *	0.57	1.23
s.e.	(0.019)	(0.68)	(0.68)	(0.94)	(1.06)	(1.15)	(1.16)
dy/dx	−0.001	0.041	0.155	0.1	0.554 **	0.16	0.34

Note: base income is \$50,000 or less; k represents thousands of dollars. Standard errors are in parentheses. *,**: significant at $p < 0.1$, $p < 0.05$, respectively. (a) negative cases predict failure; (b) positive cases predict failure; (c) positive cases predict success.

The adoption of surge irrigation (surge) was positively and significantly associated with the perception of the existence of a groundwater problem and negatively by the pumping cost. The negative influence of the pumping cost variable was a departure from the hypothesized relations in the GAO report. Because this variable is a combination of various data with a fundamental rooting in the county of residence claimed by the farmer, there may be confounding of factors, the identification of which is beyond the scope of this study. However, the result was driven in part by the fact that nobody claiming to reside in the cone of depression (highest pumping cost) used surge irrigation. Surge irrigation is harder to manage in the cracking clays that are a common soil type in that area , but this effect did not appear statistically significant in this regression. The marginal effect calculations suggested that the probability of adoption of surge increased by 0.2 percent for an additional 40 hectares of irrigated land added to the operation, but an increase of one percent in the cost of pumping would decrease the adoption probability by 0.55 percent. The data indicated that farmers who do not use groundwater for irrigation have not adopted surge.

The use of SMS was significantly associated with irrigated area (positive at five percent), rice production (negative at five percent), participation in a conservation program (positive at five percent), and increasing income (from the baseline to the \$100 k to \$150 k income bracket, positive at five percent). The estimates suggested that an increase of 40 hectares in irrigated land would result in a two percent higher probability of adopting SMS; the choice of growing rice would reduce that probability by 50 percent, and the participation in a conservation program would add 45 percent to the SMS adoption probability. The data indicated that farmers who do not use groundwater for irrigation have not adopted SMS.

With respect to micro-irrigation, the probit regressions did not find statistically significant effects. This may be due in part to the relatively few respondents who claimed to practice it. For center pivot irrigation, however, there was enough variability to show a statistically significant and positive effect of the number of years of farmer formal education. For every additional year of formal education completed, the farmer was 0.68 percent more likely to adopt center pivot irrigation. The data indicated that farmers who do not use groundwater for irrigation have not adopted center pivot.

The adoption of a pump timer was positively and significantly associated with the number of irrigated hectares and the perception that a groundwater problem existed at the farm or state level. An additional 40 hectares or irrigated land was associated with a 0.6 percent higher probability of adoption, and the realization that a problem with groundwater stock existed in the state implied a 41 percent higher probability of employing a pump timer. All farmers in the sample with incomes above $250,000 used pump timers.

Flow meter adoption was also positively and significantly associated with the number of irrigated hectares. An additional 40 hectares or irrigated land were associated with a one percent higher probability of adoption. All farmers in the sample with incomes above $200,000 used flow meters.

Growing cover crops was negatively and significantly associated with farmer experience as represented by the number of years farming and positively associated with the $200,000 to $250,000 income bracket. An additional year of farming experience was associated with a 1.1 percent lower probability of growing cover crops. Since the agronomic benefits of cover crops payoff over a longer time-horizon, a farmer getting closer to retirement may be less eager to invest in a practice for which she/he will not see most of the benefits.

In terms of identifying factors that are associated with the adoption of conservation practices, the results indicated the following: irrigated area (positive), GW use (positive), rice (negative), years farming (negative), education years (positive), perception of GW problem (positive), pumping costs (negative), conservation program participation (positive), and income (positive). These results confirmed the stated hypotheses with respect to factor association except for the effect of pumping cost, which was expected to have a positive association with the adoption of conservation practices. It is not clear from the data what drives this result, which is limited to the adoption of surge irrigation only and does not appear statistically significant for other practices.

4. Discussion

Promoting the adoption of water conservation practices in irrigated agriculture has been a principal initiative to slow down the decline of the Mississippi River Valley Alluvial Aquifer (MRVAA). Ongoing agronomic research from scientists at Mississippi State University's Delta Research and Extension Center (DREC) and the USDA Agricultural Research Service in Stoneville, MS, continue to show the potential for these practices to reduce water use while maintaining farm yield and revenue levels. The Mississippi Department of Environmental Quality and the Yazoo Mississippi Delta Joint Water Management District require the use of a minimum number of these practices to issue groundwater well drilling and use permits in the area. Grower associations such as the Delta Farmers Advocating Resource Management (F.A.R.M.) sponsor and promote the use of these practices among their members. Financial and technical assistance from USDA Natural Resource Conservation Service (NRCS) is geared towards minimizing farmer risk exposure associated with the implementation of these conservation practices. However, little is known about the farmer factors that drive their decision to adopt any given conservation practice. This article provides insights that help identify and understand the determinants of conservation practice adoption in the Delta region in Mississippi.

The regression analyses indicated that no single factor consistently predicted the adoption of every conservation practice, but many factors influence a farmer's decision to adopt a given practice. The size of the farming operation is an important factor in deciding the adoption of agricultural practices in general. Indeed, it is the factor with the most statistically significant coefficients (positive) in the regression analyses. From an economic perspective, this may be attributed to the fact that practices that are marginally beneficial on a per hectare basis may not be worth the managerial cost to small operations, but could add-up to significant economies of scale for larger operations in which fixed and overhead costs associated with a practice can be spread over more hectares.

Results from the 2016 Mississippi Irrigation Survey indicated that groundwater is a source for irrigation for almost 93 percent of growers in the sample. Yet, only a third of the growers think there is a groundwater problem at the farm or state level. The analysis presented suggests that this lack of awareness is significantly related to whether the growers observe a change in the depth to water in their wells or not.

Perception of the existence of a GW problem at the farm or state level was the second most important factor identified. This is an encouraging finding because the results show a strong influence of this variable on the probability of the adoption of several adoption practices and because this is, essentially, an awareness issue. Perception of water quantity issues in regions of high rainfall can be difficult to overcome. This suggests that additional incentives are necessary to bring those who do not perceive a problem to the realization that it actually exists.

Increasing producer awareness is a task that fits, for example, the mission of university extension services, which can aid the communications efforts of federal and state research and regulatory entities in that regard. Regulatory agencies and universities can work together to have a consistent message regarding the projection of groundwater in the state and to increase grower awareness of the issue with a focus on county-based expected changes in wells' depth-to-water distances. However, such efforts have been carried out by these organizations, which suggests there is a need for an additional signal to help convince farmers of the seriousness of the situation.

At nearly 70 percent of respondents, this sample far surpasses the national average of 30 percent flow-meter use [31]. However, only 10 to 15 percent of permittees report individual water-use every year. There is promising evidence from areas with mandatory water use reporting that producers become better informed and more concerned with the health of the aquifer.

For example, growers near Sheridan county in Kansas widely supported a self-imposed Local Enhancement Management Area to create a five year allocation of groundwater that resulted in a 26 percent reduction in water use [32]. A similar case where the threat of regulation from the state level induced irrigators to form the Groundwater Subdistrict No. 1 in the San Luis Valley of Colorado and "formulate a homegrown governance response" that reduced water use by 33 percent in the district [33].

The survey results in our study indicated that farmers may require stronger signals that the aquifer problem is real and important. Furthermore, the experience documented in the aforementioned studies suggested that aquifer problem awareness, resulting largely from individual water use monitoring, and the threat of top-down regulation can induce more active farmer participation in water conservation.

Although participation in conservation programs was a statistically significant factor for only one of the practices (SMS), Figure 1 shows how NRCS expenditures [34] and practice adoption in the region track similar time trends. In particular, the rate of adoption of CHS, SMS, and surge starting in 2009 are noteworthy. This suggests there is an effect of NRCS technical and financial assistance in terms of adoption. However, perhaps due to insufficient data, this association cannot be validated empirically in this study.

These results can inform policy makers, regulatory entities, university extension services, and producers about the salient aspects of conservation practice adoption in Mississippi. Conservation agencies can use the insights in this study to better target their incentive programs, for instance focusing on incentivizing relatively young farmers to adopt practices with long-term benefits such as cover crops. Similarly, further research, extension, and incentives are necessary to devise incentives and training to facilitate practice adoption by smaller farms. Lastly, periodic surveys (2–5 year intervals) may be necessary to track trends and assess the effectiveness of conservation practice adoption programs.

Our results are largely consistent with the most recent literature [4,6,22]; see Table 8 for a comparison. The GAO report in particular describes policy options at the federal level including their benefits and challenges. With respect to irrigation technology, it recommends that policy makers promote: (1) the use of more efficient irrigation technology and practices, in conjunction with appropriate agreements to use the technology and practices to conserve water; and (2) the use of precision agriculture technologies, in conjunction with appropriate agreements to use the precision agriculture technologies to conserve water. These recommendations are consistent with the ongoing efforts mentioned above, and this study lends empirical validity to the recommendations for the case of Mississippi.

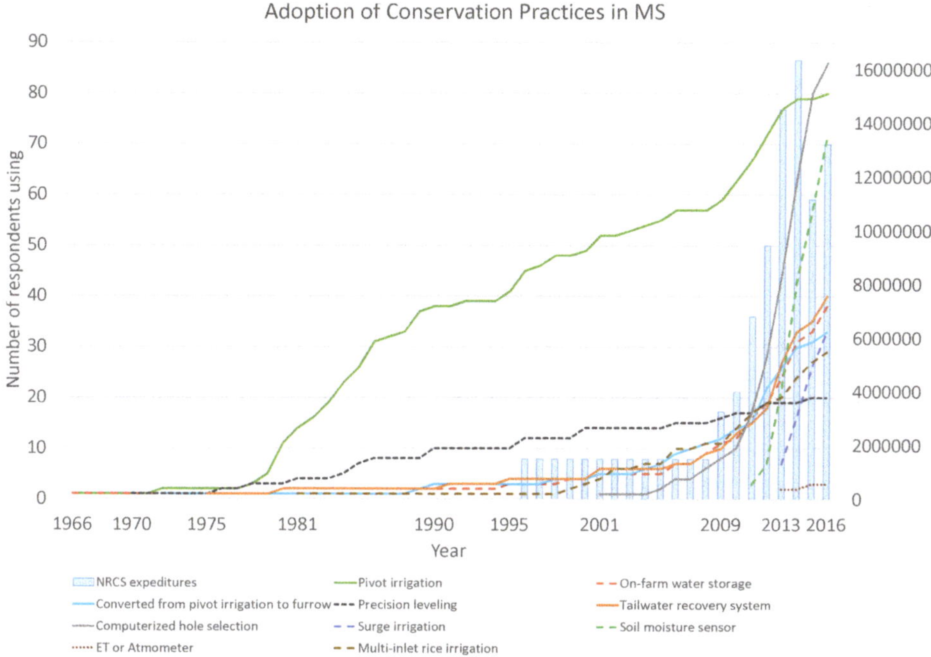

Figure 1. Timeline of the adoption of conservation practices and USDA-NRCS EQIP expenditures in the MS Delta area.

Table 8. Impact of conservation practice adoption factors compared to existing literature.

Factor	Results	Nian et al. Morris (2020)	Prokopy et al. (2019)	Pearsons and (2019)
Irrigated area	Pos	Pos	Pos	Pos
GW use	Pos	Pos	Pos	
Rice farming	Neg		Mix	
Years farming	Neg	Neg	Neg	Neg
Education	Pos		Mix	Pos
Perceives GW problem	Pos	Pos	Pos	Pos
Pumping cost	Neg		Pos	Neg
Participation in conservation programs	Pos	Pos	Pos	Pos
Cracking soils	Mix			Neg
Farmer income	Mix	Pos	Pos	Pos

Note: Pos denotes a Positive influence of the factor on adoption; Neg denotes a Negative influence of the factor on adoption; Mix denotes Mixed or undetermined influence of the factor on adoption.

The main limitation of this study is the small sample relative to the number of practices and factors being considered. This limitation determined the choice of probit regression models rather than multivariate and sequential modeling. There is evidence that practices, especially add-ons to furrow irrigation, are adopted in bundles. This feature is not incorporated in this study and constitutes a promising avenue for future research.

Author Contributions: Conceptualization, N.Q.-A. and D.M.G.; methodology, N.Q.-A.; software, N.Q.-A.; validation, D.M.G. and L.J.K.; formal analysis, N.Q.-A.; investigation, L.J.K. and C.G.H.; resources, L.J.K., C.G.H., and T.C.; data curation, N.Q.-A.; writing, original draft preparation, N.Q.-A.; writing, review and editing, N.Q.-A., D.M.G., and L.J.K.; funding acquisition, L.J.K., C.G.H., and T.C. All authors read and agreed to the published version of the manuscript.

Funding: This publication is a contribution of the National Center for Alluvial Aquifer Research and the Mississippi Agricultural and Forestry Experiment Station. This material is based on work that was funded jointly by the Agricultural Research Service, United States Department of Agriculture, under Cooperative Agreement Number 58-6001-7-001. Financial support for the 2016 Mississippi Survey of Irrigators came from the Mississippi Soybean Board, Mid-South Soybean Board, and United Soybean Board.

Acknowledgments: The authors acknowledge and thank Paul Rodrigue, USDA-NRCS Supervisory Engineer Area 4, for helpful advice and regional NRCS expenditure data. Additionally, the authors are grateful for financial support for the survey from the Mississippi Soybean Board, Mid-South Soybean Board, and United Soybean Board.

Conflicts of Interest: The authors declare no conflict of interest.

Abbreviations

The following abbreviations are used in this manuscript:

AAWEP	Acceptable Agricultural Water Efficiency Practices
AWEP	Agricultural Water Enhancement Program
ARS	USDA Agricultural Research Service
CHS	Computerized hole selection
CRP	Conservation Reserve Program
DREC	Mississippi State University Delta Research and Extension Center
EIA	U.S. Energy Information Administration
EQIP	NRCS Environmental Quality Incentives Program
F.A.R.M.	Delta Farmers Advocating Resource Management
FSA	Farm Service Agency
GAO	Government Accountability Office
GW	Groundwater
MRVAA	Mississippi River Valley Alluvial Aquifer
NRCS	USDA Natural Resources Conservation Service
OFWS	On-Farm Water Storage
RCPP	Regional Conservation Partnership Program
SMS	Soil Moisture Sensors
TWS	Tailwater Recovery System
USACE	U.S. Army Corps of Engineers
USDA	U.S. Department of Agriculture
WRP	Wetlands Reserve Program

Appendix A. Conservation Programs in Mississippi

The Delta region in Mississippi has a diversity of habitats in which urban, agricultural, and wildlife habitat landscapes coexist. Growers and landowners in the area actively seek guidance in attaining land and water resource stewardship. Hunting and fishing clubs are among the highest priced memberships in the areas. Agricultural land and water projects often include wildlife habitat enhancing features that further increase the value of such clubs. The National Wildlife Federation work in the area focuses on protecting and restoring healthy rivers and estuaries, conserving wetlands, springs, and aquifers, and protecting wildlife habitats. These goals overlap with producer interests and are prominent features in existing conservation programs in the area. Table A1 contains a list and brief description of the conservation programs in which survey respondents claimed to participate.

Table A1. List and brief description of the conservation programs in which survey respondents claimed to participate.

Program	Sponsor	Description
AWEP	NRCS	Agricultural Water Enhancement Program is a conservation initiative that provides financial and technical assistance to agricultural producers to implement agricultural water enhancement activities on agricultural land for the purposes of conserving surface and groundwater and improving water quality.
CRP	FSA	Conservation Reserve Program is a land conservation program to remove environmentally sensitive land from agricultural production and plant species that will improve environmental health and quality.
CSP	NRCS	Conservation Stewardship Program participants earn performance-based CSP payments: higher payment to higher performance.
Delta F.A.R.M.	Public-private partnership	Farmers Advocating Resource Management is an association of growers and landowners that strive to implement recognized agricultural practices, which will conserve, restore, and enhance the environment.
EQIP	NRCS	Environmental Quality Incentives Program provides incentive payments and cost-sharing for conservation practice adoption.
RCPP	NRCS	Regional Conservation Partnership Program promotes coordination of NRCS conservation activities with partners that offer value-added contributions to expand their collective ability to address on-farm, watershed, and regional natural resource concerns.
Rice stewardship	Public-private partnerships	USA Rice-Ducks Unlimited Rice Stewardship Partnership provides financial assistance for conserving water and wildlife in ricelands.
Soil erosion	Unspecified	Unspecified
Unspecified	USACE	The U.S. Army Corps of Engineers may enroll farmers adjacent to their projects as part of environmental or habitat enhancement features.
WRP	NRCS	Wetlands Reserve Program offers landowners the opportunity to protect, restore, and enhance wetlands on their property.

References

1. Christy, D. (Yazoo Mississippi Delta Joint Water Management District, Mississippi State University, Starkville, MS, USA). Personal communication, 2014.
2. Kebede, H.; Fisher, D.K.; Sui, R.; Reddy, K.N. Irrigation methods and scheduling in the Delta region of Mississippi: Current status and strategies to improve irrigation efficiency. *Am. J. Plant Sci.* **2014**, *5*, 2917. [CrossRef]
3. Barlow, J.R.; Clark, B.R. *Simulation of Water-Use Conservation Scenarios for the Mississippi Delta Using an Existing Regional Groundwater Flow Model*; U.S. Geological Survey Scientific Investigations Report 2011–5019; US Department of the Interior, US Geological Survey: Washington, DC, USA, 2011; 14p.
4. Nian, Y.; Huang, Q.; Kovacs, K.F.; Henry, C.; Krutz, J. Water Management Practices: Use Patterns, Related Factors and Correlations with Irrigated Acres. *Water Resour. Res.* **2020**, *56*, e2019WR025360. [CrossRef]
5. Balogh, P.; Bujdos, Á.; Czibere, I.; Fodor, L.; Gabnai, Z.; Kovách, I.; Nagy, J.; Bai, A. Main Motivational Factors of Farmers Adopting Precision Farming in Hungary. *Agronomy* **2020**, *10*, 610. [CrossRef]
6. Persons, T.M.; Morris, S.D. Irrigated Agriculture: Technologies, Practices, and Implications for Water Scarcity. In *Report to Congressional Requesters, Technology Assessment*; GAO-20-128SP; U.S. Government Accountability Office: Washington, DC, USA, 2019.
7. Bryant, C.; Krutz, L.; Falconer, L.; Irby, J.; Henry, C.; Pringle, H.; Henry, M.; Roach, D.; Pickelmann, D.; Atwill, R.; et al. Irrigation water management practices that reduce water requirements for Mid-South furrow-irrigated soybean. *Crop Forage Turfgrass Manag.* **2017**, *3*, 1–7. [CrossRef]

8. Henry, W.B.; Krutz, L.J. Water in agriculture: Improving corn production practices to minimize climate risk and optimize profitability. *Curr. Clim. Chang. Rep.* **2016**, *2*, 49–54. [CrossRef]

9. Wood, C.; Krutz, L.; Falconer, L.; Pringle, H.; Henry, B.; Irby, T.; Orlowski, J.; Bryant, C.; Boykin, D.; Atwill, R.; et al. Surge irrigation reduces irrigation requirements for soybean on smectitic clay-textured soils. *Crop Forage Turfgrass Manag.* **2017**, *3*, 1–6. [CrossRef]

10. Spencer, G.; Krutz, L.; Falconer, L.; Henry, W.; Henry, C.; Larson, E.; Pringle, H.; Bryant, C.; Atwill, R. Irrigation Water Management Technologies for Furrow-Irrigated Corn that Decrease Water Use and Improve Yield and On-Farm Profitability. *Crop Forage Turfgrass Manag.* **2019**, *5*, 1–8. [CrossRef]

11. Carey, J.M.; Zilberman, D. A model of investment under uncertainty: modern irrigation technology and emerging markets in water. *Am. J. Agric. Econ.* **2002**, *84*, 171–183. [CrossRef]

12. Yazoo Mississippi Delta Joint Water Management District, Permit Application. Available online: https://www.ymd.org/permitting (accessed on 15 June 2020).

13. Krutz, L. *Utilizing Moisture Sensors to Increase Irrigation Efficiency*; Mississippi State University Extension Service: Starkville, MS, USA, 2016.

14. Krutz, L. *Pipe Planner: The Foundation Water Management Practice for Furrow Irrigated Systems*; Mississippi State University Extension Service: Starkville, MS, USA, 2016.

15. Roach, D. *Flow Meters Available at County Extension Offices*; Mississippi State University Extension Service: Starkville, MS, USA, 2018.

16. Krutz, L. *Surge Valves Increase Application Efficiency*; Mississippi State University Extension Service: Starkville, MS, USA, 2016.

17. Tagert, M.L.; Paz, J.; Reginelli, D. *On-Farm Water Storage Systems and Surface Water for Irrigation*; Mississippi State University Extension Service: Starkville, MS, USA, 2018.

18. Zotarelli, L.; Fraisse, C.; Dourte, D. *Agricultural Management Options for Climate Variability and Change: Microirrigation*; EDIS UF/IFAS (HS1203); University of Florida: Gainesville, FL, USA, 2015.

19. Coblentz, B.A. *Pivot Irrigation, Not Furrows, Is Most Economical for Delta*; Mississippi State University Extension Service: Starkville, MS, USA, 2014.

20. Burdine, B. *Cover Crops: Benefits and Limitations*; Mississippi State University Extension Service: Starkville, MS, USA, 2019.

21. Coblentz, B.A. *Cover Crops Present Challenges, Advantages*; Mississippi State University Extension Service: Starkville, MS, USA, 2018.

22. Prokopy, L.S.; Floress, K.; Arbuckle, J.G.; Church, S.P.; Eanes, F.; Gao, Y.; Gramig, B.M.; Ranjan, P.; Singh, A.S. Adoption of agricultural conservation practices in the United States: Evidence from 35 years of quantitative literature. *J. Soil Water Conserv.* **2019**, *74*, 520–534. [CrossRef]

23. Edwards, J.F. *Crop Irrigation Survey*; Final Report; Mississippi State University, Social Science Research Center, Survey Research Laboratory: Starkville, MS, USA, 2016; unpublished.

24. Rogers, D.H.; Alam, M. *Comparing Irrigation Energy Costs*; Agricultural Experiment Station and Cooperative Extension Service, Kansas State University: Manhattan, KS, USA, 2006; MF-2360.

25. McGuire, V.L.; Seanor, R.C.; Asquith, W.H.; Kress, W.H.; Strauch, K.R. *Potentiometric surface of the Mississippi River Valley Alluvial Aquifer, Spring 2016: U.S. Geological Survey Scientific Investigations Map 3439*; Technical Report; US Geological Survey: Washington, DC, USA, 2019; 14p, 5 sheets. [CrossRef]

26. USDA-NRCS. Soil Survey Staff, Natural Resources Conservation Service, United States Department of Agriculture. In *Soil Survey Geographic (SSURGO) Database for Sunflower County, Mississippi*; USDA-NRCS: Washington, DC, USA, 2010.

27. Maddala, G.S. *Limited-Dependent and Qualitative Variables in Econometrics*; Cambridge University Press: Cambridge, UK, 1986; Volume 3.

28. Pfeiffer, L.; Lin, C.Y.C. Does efficient irrigation technology lead to reduced groundwater extraction? Empirical evidence. *J. Environ. Econ. Manag.* **2014**, *67*, 189–208.

29. Fenichel, E.P.; Abbott, J.K.; Bayham, J.; Boone, W.; Haacker, E.M.; Pfeiffer, L. Measuring the value of groundwater and other forms of natural capital. *Proc. Nat. Acad. Sci. USA* **2016**, *113*, 2382–2387. [CrossRef] [PubMed]

30. Hassanli, A.M.; Ebrahimizadeh, M.A.; Beecham, S. The effects of irrigation methods with effluent and irrigation scheduling on water use efficiency and corn yields in an arid region. *Agric. Water Manag.* **2009**, *96*, 93–99. [CrossRef]

31. USDA. *Irrigation: Results from the 2013 Farm and Ranch Irrigation Survey;* Census Agricultyre Highlights ACH12-16; USDA: Washington, DC, USA, 2014.
32. Drysdale, K.M.; Hendricks, N.P. Adaptation to an irrigation water restriction imposed through local governance. *J. Environ. Econ. Manag.* **2018**, *91*, 150–165.
33. Smith, S.M.; Andersson, K.; Cody, K.C.; Cox, M.; Ficklin, D. Responding to a groundwater crisis: The effects of self-imposed economic incentives. *J. Assoc. Environ. Resour. Econ.* **2017**, *4*, 985–1023. [CrossRef]
34. USDA-NRCS. *The PROTRACTS Database, Mississippi*; USDA-NRCS: Washington, DC, USA, 2019.

Article

Assessment of Landsat-Based Evapotranspiration Using Weighing Lysimeters in the Texas High Plains

Ahmed A. Hashem [1,2,3], Bernard A. Engel [3,*], Vincent F. Bralts [3], Gary W. Marek [4], Jerry E. Moorhead [5], Sherif A. Radwan [2] and Prasanna H. Gowda [6]

1   College of Agriculture, Arkansas State University, 422 University Loop W, Jonesboro, AR 72401, USA; ahashem@astate.edu
2   Agricultural Engineering Department, Suez Canal University, Kilo 4.5 Ring Road, Ismailia 41522, Egypt; Sherifabdelhak@hotmail.com
3   Agricultural & Biological Engineering Department, Purdue University, 225 South University Street, West Lafayette, IN 47907, USA; bralts@purdue.edu
4   USDA-ARS Conservation and Production Research Laboratory, 300 Simmons Road, Unit 10, Bushland, TX 79012, USA; Gary.Marek@usda.gov
5   Lindsay Corporation, 8948 Centerport Blvd, Amarillo, TX 79108, USA; Jed.moorhead@lindsay.com
6   USDA, ARS Southeast Area, 141 Experimental Station Road, Stoneville, MS 38776, USA; prasanna.gowda@usda.gov
*   Correspondence: engelb@purdue.edu; Tel.: +1-765-494-8362

Received: 8 October 2020; Accepted: 27 October 2020; Published: 30 October 2020

Abstract: Evapotranspiration (ET) is one of the largest data gaps in water management due to the limited availability of measured evapotranspiration data, and because ET spatial variability is difficult to characterize at various scales. Satellite-based ET estimation has been shown to have great potential for water resource planning and for estimating agricultural water use at field, watershed, and regional scales. Satellites with low spatial resolution, such as NASA's MODIS (Moderate Resolution Imaging Spectroradiometer), and those with higher spatial resolution, such as Landsat (Land Satellite), can potentially be used for irrigation water management purposes and other agricultural applications. The objective of this study is to assess satellite based-ET estimation accuracy using measured ET from large weighing lysimeters. Daily, seven-day running average, monthly, and seasonal satellite-based ET data were compared with corresponding lysimeter ET data. This study was performed at the USDA-ARS Conservation and Production Research Laboratory (CPRL) in Bushland, Texas, USA. The daily time series Landsat ET estimates were characterized as poor for irrigated fields, with a Nash Sutcliff efficiency (NSE) of 0.37, and good for monthly ET, with an NSE of 0.57. For the dryland managed fields, the daily and monthly ET estimates were unacceptable with an NSE of −1.38 and −0.19, respectively. There are various reasons for these results, including uncertainties with remotely sensed data due to errors in aerodynamic resistance surface roughness length estimation, surface temperature deviations between irrigated and dryland conditions, poor leaf area estimation in the METRIC model under dryland conditions, extended gap periods between satellite data, and using the linear interpolation method to extrapolate daily ET values between two consecutive scenes (images).

Keywords: BEARS; bushland; climate; evapotranspiration; groundwater management; irrigation water management; Ogallala aquifer region; remote sensing; lysimeter ET assessment

---

1. Introduction

Evapotranspiration (ET) is an important component of the water cycle and the surface energy balance, where water changes phase from liquid to vapor, and a change in energy takes place [1]. The development of reliable long-term estimates of ET is needed to improve agricultural water use efficiency (WUE). WUE is defined as the assimilated carbon amount as biomass/grain produced per consumed unit of water by a crop. Paulson [2] reported that climate elements and wind conditions affect regional and seasonal ET estimates. For the irrigation scheduling decision-making process, knowledge of ET variability is important for proper water resources management. This is especially true in arid regions, where crop water requirements exceed rainfall, and irrigation is essential for crop production.

To sustain agricultural production in regions that are dependent on irrigation, blue water (surface water) is extracted from streams and aquifers (groundwater) to help sustain the crop and to meet water demand. Over time, farmers are required to dig deeper wells and extract water from deeper aquifers to ensure adequate water supply for their irrigation needs [3]. Unfortunately, the recharge rate of some aquifers is not sufficient to meet this demand, and historical declines in underground water table have occurred due to extensive water withdrawal [4]. In some cases, this has resulted in land subsidence and localized infrastructure damage [3]. ET is the main driver in irrigation water management and planning. The more accurate the ET estimates, the more likely associated management strategies are to achieving potential crop yields. ET is also an important component in estimating soil water to facilitate improved water use efficiency.

Remote sensing-based ET models are effective for crop water requirement estimations at the field and regional scales [5]. Several ET algorithms have been developed to utilize airborne and satellite data for irrigation scheduling and management purposes. ET can be measured over a surface using the Bowen ratio (BR), the eddy covariance (EC), and lysimeter systems at the field-scale. In all of these cases, spatial variability does not apply, since each method represents a very small scale measurement (typically less than 150 m in an agricultural setting). In addition, these methods only provide a single, averaged value that may not adequately capture the variability across a region. Satellite-based ET models produce regional-scale crop water use [6]. Many remote sensing-based ET algorithms have been developed, assessed, and widely used for estimating regional ET [6–9].

Park et al. [10], Jackson [11], and Choudhury et al. [12] reported that regional and watershed-scale spatially distributed ET can be better represented using remote sensing compared to traditional ET estimation methods. There are typically two approaches that are used to provide remote sensing ET estimates: (1) the land surface energy balance (EB) approach, and (2) the reflectance-based crop coefficient (Kc) and reference ET approach. The first approach is based on ET being a change of the state of water, using available energy in the environment for vaporization [13].

Multiple satellite platforms are available to use with energy balance ET models, such as Land Satellite (Landsat), The Advanced Spaceborne Thermal Emission and Reflection Radiometer (ASTER), Geostationary Operational Environmental Satellite (GEOS), moderate resolution imaging spectroradiometer (MODIS), and others, based on surface energy balance [14]. However, the spatiotemporal resolution is complex and cannot be combined with only one satellite. For daily estimates of ET with a high spatial resolution (field-scale), results depend on data assimilation, model applicability, and accuracy [15]. ET calibration based on remote sensing ET estimates has been conducted for a single source method such as the Surface Energy Balance Algorithm for Land (SEBAL) and the Mapping Evapotranspiration with Internalized Calibration model (METRIC) [16,17], and two source methods such as the Two-Source Energy Balance model (TSEB) [18,19]. Energy balance models based on thermal remote sensing include TSEB [18,20], SEBAL [8], METRIC [6,17], and several others [7,21].

A detailed assessment of existing EB models [7,22] stated that daily ET estimates varied 3 to 35% in comparison to Bowen ratio and EC ET measurements. Error sources included (a) modeling uncertainties, and (b) measurement errors and discrepancies in model-measurement scales. Other studies that assessed multiple airborne high-resolution sensor platforms indicated good agreement with measured ET [23].

The BEAREX08 experiment [15] was a robust remote sensing experiment that involved mass balance measurements of ET using four weighing lysimeters [24,25]. A wide array of instrumentation was installed for supportive data, including neutron probe (NP) access tubes for soil water measurements, multi-level canopy temperature measurements, and above and below canopy irradiance measurements. The purpose of these measurements was to provide as many measured data as possible to reduce estimations and provide more points of comparison. These data were used to address the lack of studies where EB model estimates of ET are compared with mass balance measurements, reducing uncertainty sources in EB models. Gowda et al. [7] reported that few remotely sensed ET estimates are used in irrigation scheduling and in-field management due to the absence of daily data with field-scale resolution. Many methods have been proposed to overcome this issue, such as use of infrequent, high spatial resolution data, including Landsat with 60 m spatial resolution, 16-day temporal resolution, in combination with lower spatial resolution and more frequent data, such as the Moderate Resolution Imaging Spectroradiometer (MODIS) with 1000 m spatial resolution and 1-day temporal resolution. The Disaggregated Atmosphere-Land Exchange Inverse (DisALEXI) model demonstrated this concept [14,26,27], but no evaluation of this approach for management has been successful thus far.

Spatial and temporal resolution problems could be resolved using aircraft. However, cost, data processing, and lack of experienced users have prevented their widespread use in providing such platforms as potential imagery sources for water management. The EB method typically provides an instantaneous value of ET, and interpolated to daily ET using the evaporative fraction method, [28], or reference ET approach [6,29]. Each ET estimation approach has its uncertainties; however, measurement data quality assessment and quality control reduce such modeling uncertainties [30,31]. Multi-time scale remotely-sensed ET estimation accuracy is essential for water management and irrigation planning. Extensive studies have been conducted to assess the METRIC model performance on irrigated fields [32–35]; however, none of these studies evaluated the model performance under dryland conditions for extended periods with various crops and temporal resolutions [36]. This study assessed daily interpolated ET estimation accuracy using the linear interpolation method on dryland and irrigated lysimeters for a ten year period with various crops at multiple time scales. Such assessment is crucial for water management policies, researchers, hydrologists, and irrigators when using such technologies for real time irrigation decisions.

The objective of this research was to assess the daily, seven-day running average, monthly, and seasonal satellite-based ET estimation accuracy versus large weighing lysimeter ET measurements in the Texas High Plains under irrigated and dryland conditions during the growing and non-growing seasons.

2. Materials and Methods

2.1. Study Site

Data used in this study from 2001–2010 were obtained at the USDA-ARS Conservation and Production Research Laboratory (CPRL) in Bushland, Texas, USA (35.19° N, 102.10° W). Four square fields were selected for this study; each field was ~4.7 ha. Four large precise weighing lysimeters were installed towards the center of each field [24,25]. Two lysimeters were managed as irrigated (NE and SE), and the other two lysimeters were managed as dryland (NW and SW). The irrigated lysimeter field was equipped with a linear-move irrigation system with Nelson sprinklers (Nelson Irrigation Corporation, Walla Walla, WA, USA) [37]. Irrigation scheduling was performed using neutron probe data, and in 2001 50% of crop water requirements (CWR) was replenished for cotton (deficit irrigation), and 100% of CWR was added for the remaining study period. Crop management data were collected and summarized by the CPRL [38,39].

Leaf area index (LAI) data were collected from the study site between 2001 and 2010. Plant samples were collected from locations within fields near the lysimeters biweekly during the growing season through destructive plant sampling. Leaves were separated from stems, and the average LAI values were obtained from at least three samples. Plant samples were not collected from the lysimeters, as the sampling was destructive and would impact lysimeter ET measurements. A digital scanning bed leaf area meter (model LI-3100) was used to measure the leaf area index of the leaf samples. The LAI values were calculated as the ratio of the upper side leaf area (m^2) to the ground area (m^2) [22,40].

For analysis simplicity, one lysimeter was selected for each management condition (irrigated and dryland). Statistical assessment was achieved for the estimated ET values for the dryland lysimeter (NW) and the irrigated lysimeter (NE). The soil characteristics for the study field are deep, well-drained Pullman silty clay loam (fine, mixed, superactive, thermic torrertic paleustoll) [40]. The local climate is classified as semi-arid, with large daily air temperature variations. Cotton, soybean, grain, and silage sorghum, sunflower, and cotton were the predominant crops for the research fields during the ten year study period [38,41].

2.2. Bushland Evapotranspiration and Agricultural Remote Sensing (BEARS)

Bushland Evapotranspiration and Agricultural Remote Sensing (BEARS) is an image processing and geographic information system (GIS) software developed by researchers at the USDA ARS CPRL in Bushland, TX, used for deriving hourly, daily, and seasonal ET maps, and other energy exchanges between land and atmosphere using Landsat 5,7, and 8 [42]. It is an open-source Java software, Version 1.0.1 available for download at: https://data.nal.usda.gov/dataset/bushland-evapotranspiration-and-agricultural-remote-sensing-system-bears-software. The software allows for custom models and equations but provides the option to select one of five default energy balance-based ET methods: Mapping Evapotranspiration at High Resolution with Internalized Calibration (METRIC), Surface Energy Balance Algorithm for Land (SEBAL), Surface Energy Balance System (SEBS), Two-Source Model (TSM), and Simplified Surface Energy Balance (SSEB) [42].

2.3. Image Analysis

In this study, the METRIC model was used to analyze Landsat satellite imagery, producing ET time-series datasets for the study period. The METRIC model description and required inputs based on Landsat satellite datasets are summarized in the literature [6,41]. Several hourly and daily outputs were obtained at the completion of each image analysis. Four daily outputs were obtained including daily ET (mm day^{-1}), evaporation fraction (EF) (unitless), leaf area index (LAI) (m^2 m^{-2}), and normalized difference vegetation index (NDVI) (unitless). Hashem et al. [41] assessed hourly estimation accuracy for ET (mm hr^{-1}), net radiation (Rn) (W m^{-2}), soil heat flux (Go) (W m^{-2}), and surface air temperature (Ts) (°C). A detailed description of the hourly assessment procedure was summarized [38,41].

2.4. Landsat Satellite Dataset and Processing

Landsat 5 TM cloud-free images were selected for analysis, which were obtained through Earth Explorer (https://earthexplorer.usgs.gov/), with Paths 30 and 31 and Row 36 from 2001 to 2010. A total of 129 images were analyzed with a spatial resolution of 30 m and a temporal resolution of 16 days. The number of clear images vary from year to year based upon geographic location. For this study site, 2001 had the lowest number of annual clear images, with six images, and 2008 had the highest number of clear images with 16 images, as summarized in Table 1. Processed image date and day of the year (DOY) are summarized in Table 1.

Table 1. Landsat image dates and corresponding day of year (DOY).

#	2001 (Cotton-Limited Irrigation)	2002 (Cotton-Limited Irrigation)	2003 (Soybean)	2004 (Soybean)	2005 (Sorghum)	2006 (Forage Corn)	2007 (Forage Sorghum)	2008 (Cotton)	2009 (Sunflower)	2010 (Cotton)
	Date (DOY)	Date (DOY)	Date (DOY)	Date (DOY)	Date (DOY)	Date (DOY)	Date (DOY)	Date (DOY)	Date (DOY)	Date (DOY)
1	March 12 (71)	January 26 (26)	January 13 (13)	February 17 (48)	January 25 (25)	January 18 (28)	January 8 (8)	January 18 (18)	January 13 (13)	April 29 (119)
2	May,22 (142)	February 11 (42)	April 10 (100)	March 20 (80)	February 3 (34)	February 13 (44)	February 25 (56)	February 19 (50)	January 20 (20)	June 25 (176)
3	June 16 (167)	March 3 (90)	May 5 (125)	March 27 (87)	March 7 (66)	April 18 (108)	March 4 (63)	March 22 (82)	January 29 (29)	July 18 (199)
4	June 23 (174)	May 9 (129)	May 28 (148)	April 21 (112)	June 18 (169)	May 20 (140)	March 29 (88)	April 7 (95)	February 5 (36)	August 3 (215)
5	July 9 (190)	May 18 (138)	June 24 (205)	May 14 (135)	June 27 (178)	June 5 (156)	June 8 (159)	May 2 (123)	February 21 (52)	August 12 (224)
6	July 25 (206)	June 10 (161)	July 17 (198)	May 30 (151)	July 20 (201)	July 23 (204)	July 26 (207)	May 18 (139)	March 18 (77)	August 19 (231)
7	August 19 (231)	June 19 (170)	September 17 (260)	December 1 (336)	August 30 (242)	August 8 (220)	August 11 (223)	June 3 (155)	April 3 (93)	September 4 (247)
8	September 27 (270)	July 21 (202)	September 26 (269)	December 12 (352)	September 22 (265)	August 24 (236)		June 10 (162)	June 22 (173)	September 29 (272)
9	October 13 (286)	September 23 (266)	October 19 (292)		October 1 (274)	September 18 (261)		July 21 (203)	July 8 (189)	October 15 (288)
10	November 7 (311)		November 29 (333)		October 24 (297)	September 25 (268)		August 6 (219)	August 16 (228)	November 16 (320)
11	December 9 (343)				November 18 (315)	October 11 (284)		August 22 (235)	November 4 (308)	November 23 (327)
12	December 12 (359)					October 27 (300)		September 30 (274)	November 20 (324)	December 12 (355)
13						November 28 (332)		October 25 (299)		December 25 (359)
14								November 1 (306)		
15								November 17 (322)		
16								December 28 (363)		

2.5. Landsat ET Gap Filling

Linear interpolation methods were used to estimate the satellite-based ET during the gap period using the satellite-based daily evaporation fraction between consecutive images. The gap between consecutive images was split equally into two periods of eight days; the first eight days used the first image evaporation fraction value, and the second eight-day period used the next image evaporation fraction value.

To estimate the actual ET time series, the evaporation fraction (k_c) for each day was multiplied by the reference ET [43]. The daily reference ET was estimated using the ASCE−2005 [44] standardized reference evapotranspiration equation. Hourly ET was calculated for the study period, and the daily ET was obtained by accumulating the hourly ET in a given day. The average frequency and the maximum gap periods (days) were calculated using Equations (1) and (2), respectively. The year of 2006 had the longest average frequency with 52 days, and 2008 had the shortest average frequency with 23 days. The longest and shortest maximum gap period was 184 days and 40 days for 2004 and 2008, respectively, as shown in Table 2. Areas of interest (AOIs) of 3 by 3 grids (30 m spatial resolution), with a total of 9 pixels and a surface area of 8100 m^2 each, were created, and each lysimeter was located at the center of each AOI. ArcGIS 10.2.2 model builder tool software was utilized to extract the average ET for each AOI.

$$Average\ frequency\ =\ \frac{Number\ of\ days\ in\ the\ growing\ season\ (April\ to\ October)}{Number\ of\ clear\ images\ used\ in\ the\ analysis} \tag{1}$$

$$Average\ frequency\ =\ \frac{Number\ of\ days\ in\ the\ entire\ year\ (January\ to\ December)}{Number\ of\ clear\ images\ used\ in\ the\ analysis} \tag{2}$$

Table 2. Average frequency and maximum gap each year for the study site.

Year	Average Frequency (Days)	Maximum Gap (Days)	Dates of Maximum Gap	DOY of Maximum Gap
2001	30	70	March 12–May 22	DOY 71 To DOY 142
2002	41	99	September 23–December 31	DOY 266 To DOY 365
2003	37	86	January 13–April 10	DOY 13 To DOY 100
2004	46	184	May 30–December 1	DOY 151 To DOY 336
2005	33	102	March 7–June 18	DOY 66 To DOY 169
2006	28	63	February 13–April 18	DOY 44 To DOY 108
2007	52	141	August 19–December 31	DOY 231 To DOY 365
2008	23	40	June 10–July 21	DOY 162 To DOY 203
2009	30	56	August 16–November 11	DOY 228 To DOY 315
2010	28	118	January 1–April 29	DOY 1 To DOY 119

The years 2004 and 2007 were omitted from the analysis due to limited clear image availability. Day of the year (DOY).

2.6. Dry and Wet Pixel Determination

Obtaining the most accurate wet (cold) and dry (hot) agricultural pixels is a significant step that is essential to produce accurate ET maps [34,45]. Wet and dry pixels represent the extreme conditions per scene. The manual selection method was used, along with the surface temperature and NDVI histogram distribution, to obtain these pixels for each image. NDVI and surface temperature threshold values are summarized in Table 3 to obtain the hot and dry pixel characteristics. The hot pixel is defined as bare agricultural soil with high temperature and low ET value, and the cold pixel is defined as cultivated agricultural soil with low temperature and high ET value [6,46]. The histogram distribution for surface temperature and NDVI help to determine the range of temperature and NDVI values per image.

Table 3. Hot and cold pixel conditions.

Pixel	Constraint		Condition
	T_s	**NDVI**	
Hot (dry)	High	≤ 0.2	Bare agricultural soil
Cold (Wet)	Low	≥ 0.7	Cultivated agricultural soil

2.7. Statistical Analysis

Statistical comparisons were performed including multiple statistical coefficients, including the coefficient of determination (R^2), regression line slope and intercept, root mean square error (*RMSE*), percent of root mean square error (% *RMSE* error), mean bias error (*MBE*), and Nash Sutcliff efficiency (*NSE*), between the measured and the calculated satellite-based ET. A detailed description of each coefficient and its range and interpretation of values were provided [38,41,47]. The mean bias error (*MBE*) and % *MBE* provided a means to determine the deviation between the measured and satellite-based estimates, with *MBE* = 0 indicating no bias in the estimation within the satellite data. The *MBE* and % *MBE* were calculated using Equations (3) and (4), respectively.

$$MBE = \frac{\sum_{i=1}^{n} (y_t - \hat{y}_t)}{n} \tag{3}$$

where y_t = the *i*th measured value, $\hat{y}_t$ = the *i*th simulated value, and n = the total number of observations.

$$\%MBE = \frac{MBE}{\bar{x}_{obs}} \times 100 \tag{4}$$

where $\bar{x}_{obs}$ is mean of measured values.

3. Results

3.1. Dryland Lysimeter ET Estimation

The northwest dryland lysimeter daily measured ET, Landsat ET, and seven-day running average values were plotted from 2001 to 2005, as shown in Figure 1a, and from 2006 to 2010 as shown in Figure 1b. The seven-day running average was used for visual comparison. The 1:1 graph between the daily measured and calculated Landsat ET can be seen in Figure 2, and the daily and monthly summary statistics are summarized in Table 4. The monthly average ET values were plotted, as shown in Figure 3. Detailed growing and non-growing season summary statistics are reported in Table 5.

Table 4. Daily and monthly summary statistics for the NW dryland lysimeter with Landsat ET.

		Daily	**Monthly**
	RMSE (mm)	1.8	1.2
	% *RMSE* error	144.3	105.7
	NSE	−1.38	−0.19
	Measured average ET (mm d^{-1})	1.3	1.1
	Landsat average ET (mm d^{-1})	1.7	1.4
Regression	R^2	0.01	-
	Slope	0.09	-

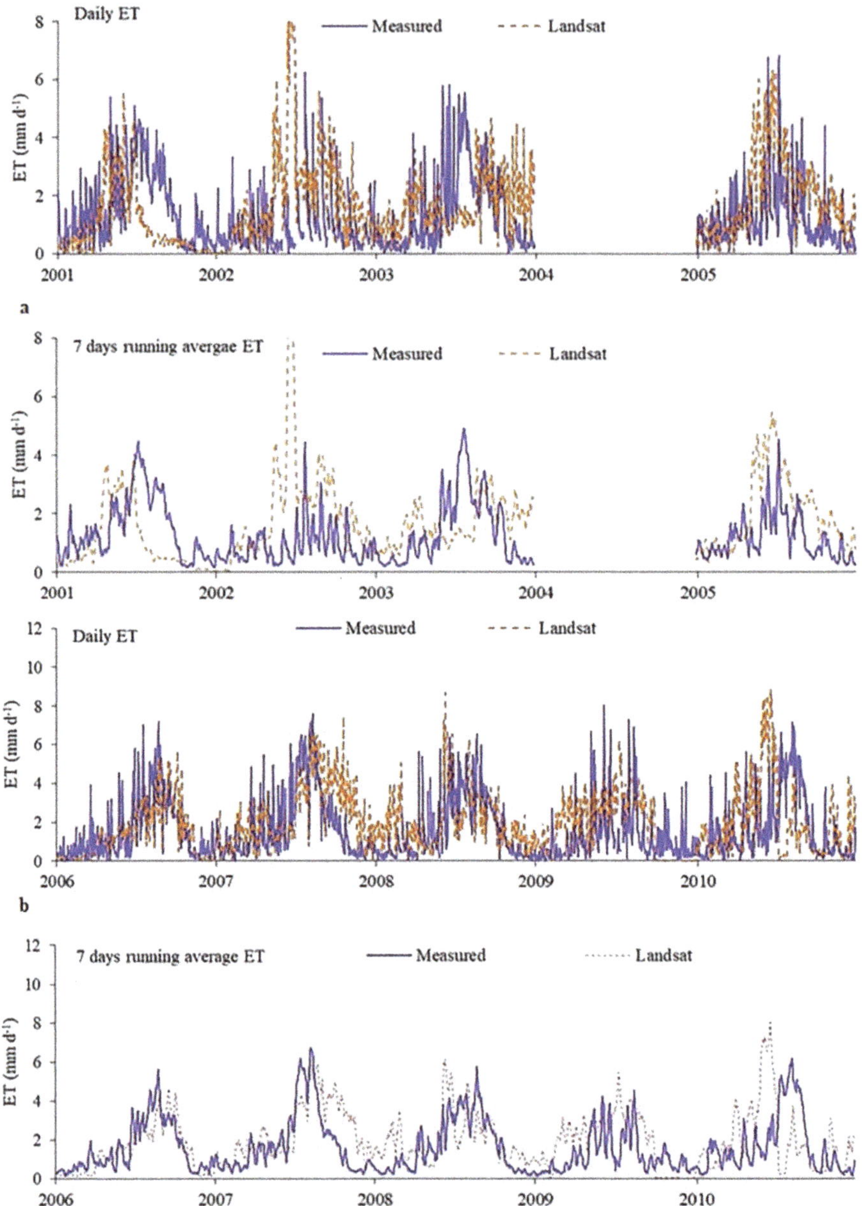

Figure 1. Daily and seven-day running average measured and Landsat evapotranspiration (ET) from (**a**) 2001–2005, and (**b**) 2006–2010 for the northwest (NW) dryland lysimeter. The years 2004 and 2007 had limited clear remote sensing observations during the growing season.

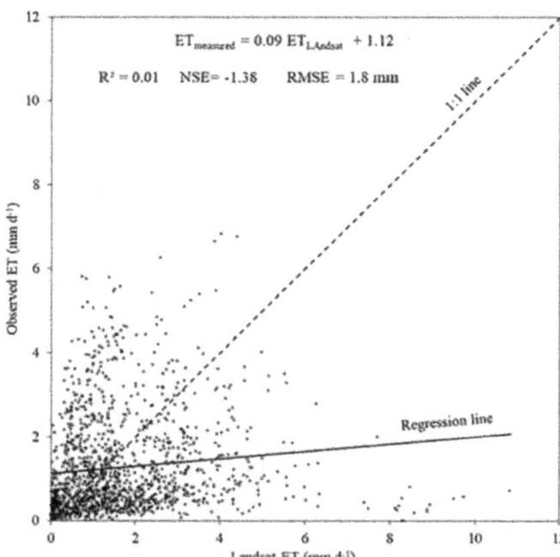

Figure 2. Daily 1:1 graph of measured and Landsat ET for NW dryland lysimeter. The years 2004 and 2007 were omitted from the analysis due to limited clear image availability.

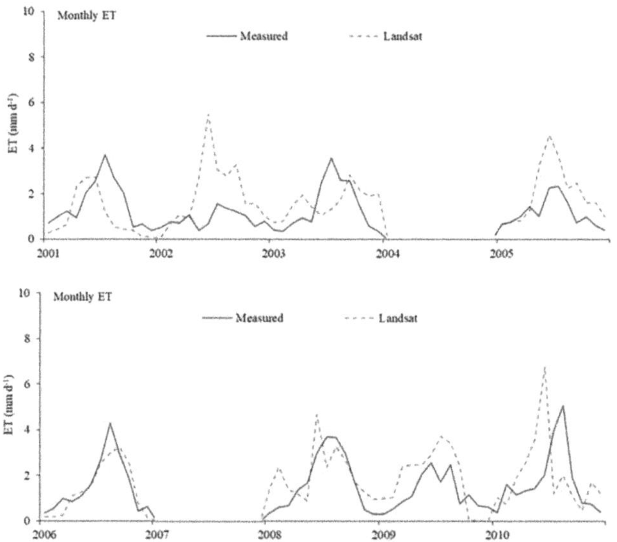

Figure 3. Monthly average measured and Landsat ET for the NW dryland lysimeter. The years 2004 and 2007 had limited clear remote sensing observations during the growing season.

Table 5. Seasonal summary statistics for the NW dryland lysimeter with Landsat ET.

Crop		RMSE (mm)	% RMSE Error	Measured Average ET (mm d^{-1})	Landsat Average ET (mm d^{-1})
Cotton	2001 GS	1.7	83.4	2.0	1.1
	2001 NG	1.2	123.1	0.9	0.9
Fallow	2002 GS	-	-	-	-
	2002 NG	2.5	288.4	0.9	2.0

Table 5. *Cont.*

Crop		RMSE (mm)	% RMSE Error	Measured Average ET (mm d^{-1})	Landsat Average ET (mm d^{-1})
Sorghum	2003 GS	1.9	71.4	2.7	1.7
	2003 NG	1.4	171.0	0.8	1.5
Fallow	2005 GS	-	-	-	-
	2005 NG	1.7	143.6	1.2	2.0
Sorghum	2006 GS	1.3	48.9	2.7	2.6
	2006 NG	0.8	104.5	0.8	0.7
Cotton	2008 GS	1.8	70.5	2.5	2.6
	2008 NG	1.5	177.6	0.8	1.4
Fallow	2009 GS	-	-	-	-
	2009 NG	1.8	144.3	1.2	1.
Soybean	2010 GS	3.3	99.1	3.3	2.0
	2010 NG	2.2	215.9	1.0	2.0

GS: growing season, NG: non-growing season. The years 2004 and 2007 were omitted from the analysis due to limited clear image availability.

The measured mean ET was 1.3 mm d^{-1}, and the mean Landsat ET was 1.7 mm d^{-1}. Due to the daily ET value deviations, the summary statistics provided a poor match during the comparison period with an R^2 value of 0.01, NSE value of −1.38, RMSE of 1.8, and an RMSE error of ~144.3%, which is considered a high error value. These statistical parameters indicated that there was no correlation between measured and calculated values [48].

The growing season dryland measured LAI was plotted versus calculated Landsat estimates for the days where Landsat images were available, as shown in Figure 4. It can be clearly seen that the Landsat LAI underestimated the LAI during the study period for all cultivated crops. Dryland conditions in Bushland, TX typically have much less vegetation than irrigated areas. The lower amounts of plant biomass allow the soil background to be more prominent at the Landsat spatial resolution. The increased soil background can skew the reflectance and LAI, causing lower values. Another potential source of errors is due to uncertainties with the remotely sensed datasets under dryland conditions, due to surface roughness length and ET extrapolation methods that have been incorporated in the METRIC model [32,38,49]. Chavez et al. [7] reported that the METRIC model estimation error was (0.7 ± 0.9 mm d^{-1}), explaining that the variations were due to errors associated with the surface roughness length and aerodynamic resistance. Another potential source of uncertainties is that the METRIC model uses a SURFACE NDVI vs. LAI relationship, where it is generated by fitting a generalized equation to six LAIs compared to NDVI functions [35], defined in the MODIS LAI backup [45], indicating a higher source of LAI estimation under dryland conditions.

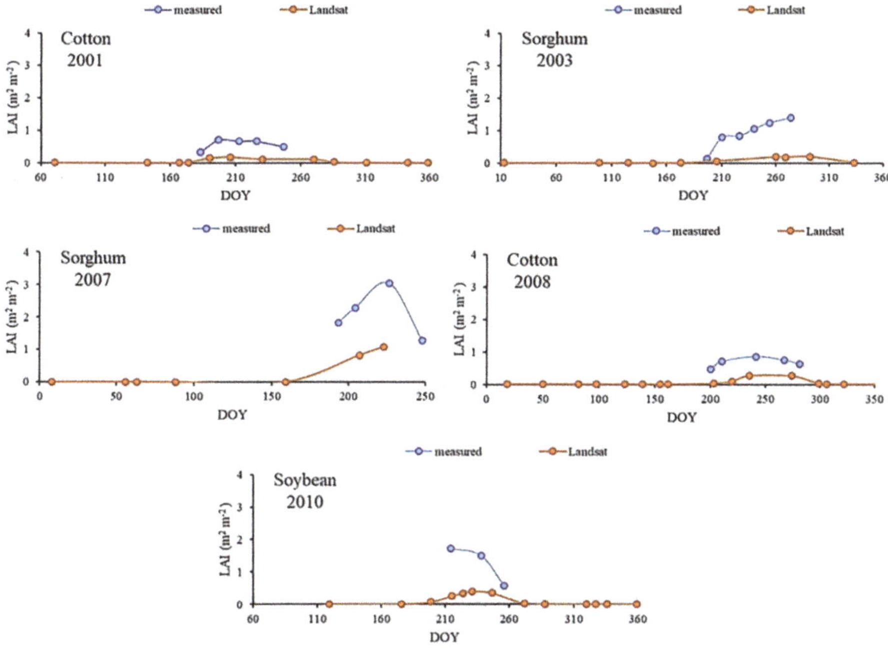

Figure 4. Measured and calculated Landsat leaf area index (LAI) for the NW dryland lysimeter. The year 2004 graph omitted because Landsat LAI values were not available.

3.2. Irrigated Lysimeter ET Estimation

Figure 5a,b presents the northeast irrigated lysimeter daily measured ET, Landsat ET, and seven-day running average from 2001 to 2010. The 1:1 graph between the daily measured and calculated Landsat ET is shown in Figure 6, and the daily and monthly summary statistics are summarized in Table 6. The growing and non-growing season summary statistics are reported in Table 7. The daily 1:1 graph is shown in Figure 6, and the monthly ET is shown in Figure 7.

Table 6. Daily and monthly summary statistics for the NE irrigated lysimeter with Landsat ET.

		Daily	Monthly
	RMSE (mm)	2.1	1.5
	% RMSE error	86.4	56.7
	NSE	0.37	0.57
	Measured average ET (mm d^{-1})	2.4	1.9
	Landsat average ET (mm d^{-1})	2.4	1.9
Regression	R^2	0.38	-
	Slope	0.86	-

The years 2004 and 2007 were omitted from the statistical analysis due to limited clear image availability.

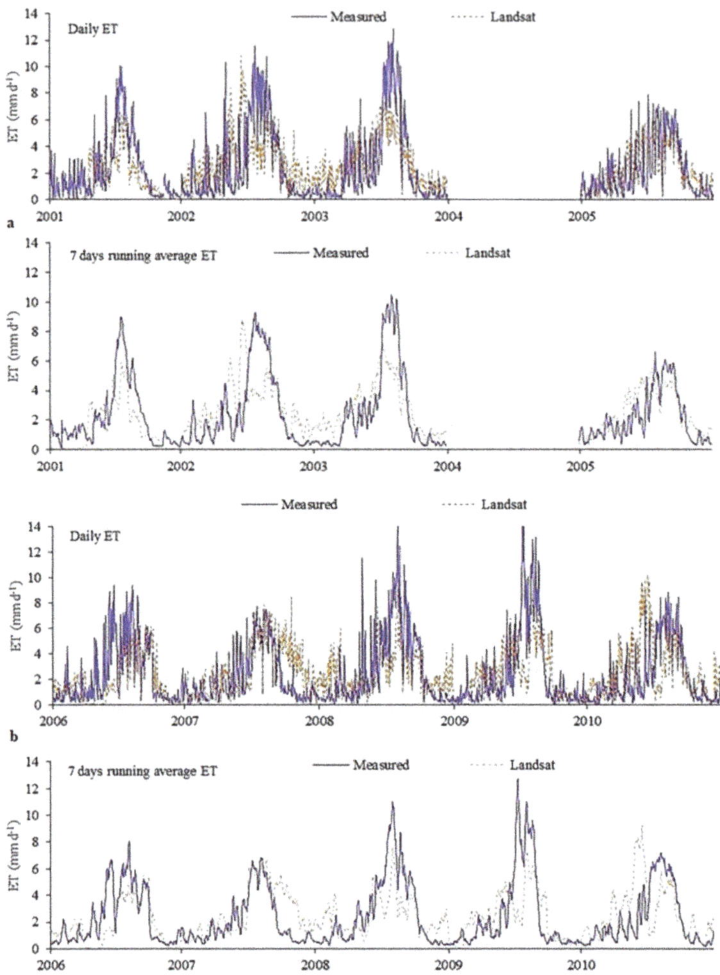

Figure 5. Daily and seven-day running average measured and calculated Landsat ET from (**a**) 2001–2005, and (**b**) 2006–2010 for the northeast (NE) irrigated lysimeter. The years 2004 and 2007 had limited clear remote sensing observations during the growing season.

Table 7. Seasonal summary statistics for the NE irrigated lysimeter with Landsat ET.

Crop		RMSE (mm)	%RMSE Error	Measured Average ET (mm d^{-1})	Landsat Average ET (mm d^{-1})
Cotton	2001 GS	2.0	66.2	3.1	2.1
	2001 NG	1.4	132.1	1.1	0.5
Cotton	2002 GS	3.5	81.9	4.2	4.0
	2002 NG	1.8	156.3	1.1	1.9
Soybean	2003 GS	2.7	55.2	4.9	4.0
	2003 NG	1.4	156.0	0.9	1.8

Table 7. *Cont.*

Crop		RMSE (mm)	%RMSE Error	Measured Average ET (mm d^{-1})	Landsat Average ET (mm d^{-1})
Sorghum	2005 GS	1.8	53.4	3.4	3.6
	2005 NG	1.1	114.2	1.0	1.3
Forage corn	2006 GS	2.7	62.3	4.3	3.1
	2006 NG	1.3	138.4	0.9	0.9
Cotton	2008 GS	2.7	56.1	4.9	3.3
	2008 NG	1.8	186.9	0.9	1.8
Sunflower	2009 GS	3.9	75.7	5.1	3.8
	2009 NG	1.4	139.9	0.9	1.4
Cotton	2010 GS	3.5	87.8	3.9	3.9
	2010 NG	1.8	185.5	0.9	1.8

GS: growing season, NG: non-growing season. The years 2004 and 2007 were omitted from the analysis due to limited clear image availability.

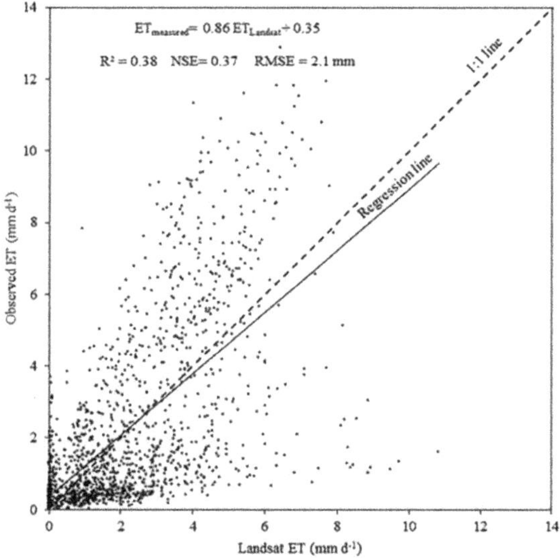

Figure 6. Daily 1:1 graph of measured and Landsat ET for the NE irrigated lysimeter. The years 2004 and 2007 were omitted from the analysis due to limited clear image availability.

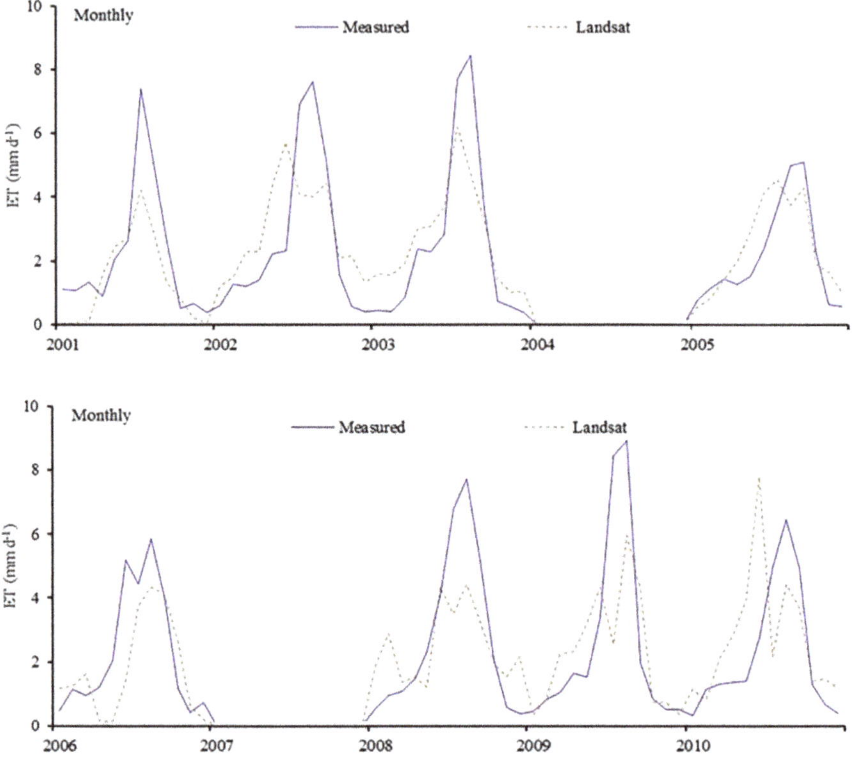

Figure 7. Monthly average measured and Landsat estimated ET for the NE irrigated lysimeter. The years 2004 and 2007 had limited clear remote sensing observations during the growing season.

The daily mean ET was 2.4 mm d^{-1} and 2.4 mm d^{-1} for the Landsat and measured, respectively. The summary statistics improved with the irrigated field and provided a weak correlation with an R^2 value of 0.38, NSE of 0.37, RMSE of 2.1, and RMSE ~86.4%.

The growing season irrigated measured LAI was plotted versus Landsat estimates for the days where Landsat images were available during the year, as shown in Figure 8. Landsat better estimated LAI under irrigated conditions compared to the dryland conditions. Consequently, higher NDVI values were obtained [38], producing higher LAI values for irrigated fields, and resulted in better ET estimates.

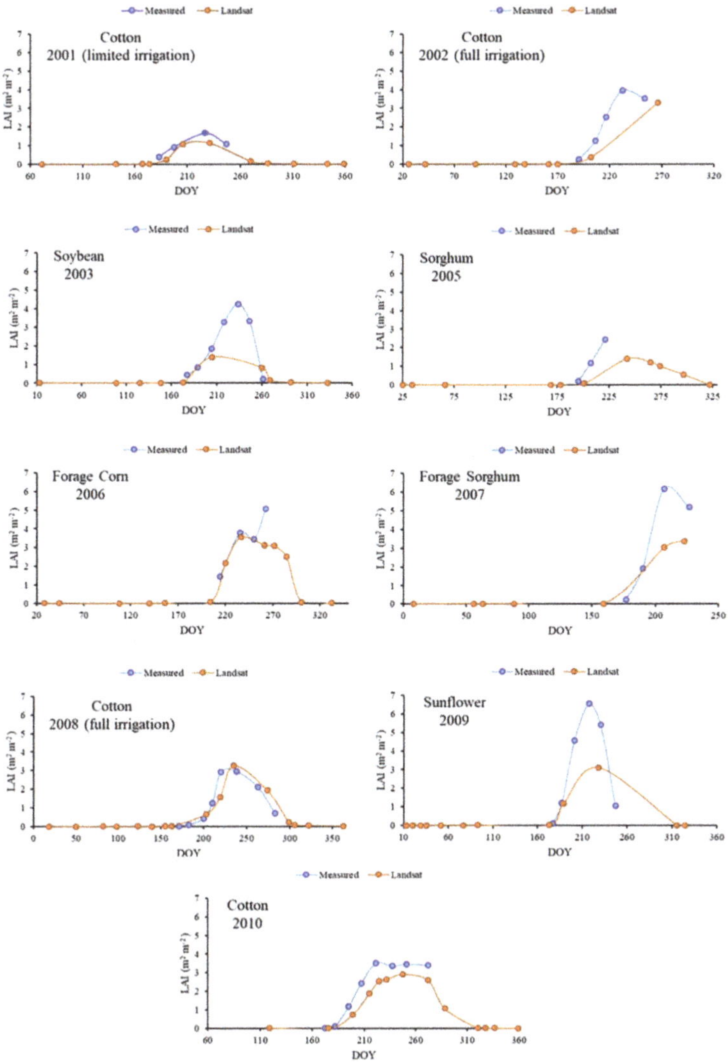

Figure 8. Measured and Landsat leaf area index (LAI) for the NE irrigated lysimeter. The year 2004 graph omitted because Landsat LAI values were not available.

4. Discussion

4.1. Dryland Daily ET Comparison

The relationship between measured and Landsat ET for the dryland lysimeter showed significant deviation with periods of both over and underestimation of ET throughout the year for the entire study period. The satellite-based LAI was assessed versus the measured LAI (Figure 4), and the LAI assessment summary is summarized in Hashem [38]. The daily time series ET deviations were related to errors in LAI estimation [38,41,47], where Landsat LAI estimates were significantly lower than measured LAI during the growing season for the dryland lysimeter. The higher the NDVI values, the more the LAI values increase, resulting in greater ET values.

In 2002, 2005, and 2009, the lysimeter field was fallow, and Landsat overestimated ET in each of the three years (Figure 1), and these results agree with Allen et al. [46]. Cotton was cultivated in 2001 and 2008, and Landsat estimates of ET closely matched the measured ET at the beginning of each year. However, towards the end of 2001, Landsat significantly underestimated ET due to low NDVI values and, consequently, underpredicting LAI [38]. ET data in 2004 and 2007 were omitted from the analysis, as the Landsat data overestimated the ET compared to the measured ET due to the large gap period in Landsat data and linear interpolation method used to fill the gap. In 2003, when sorghum was cultivated, the satellite-based-ET overestimated measure ET in both the beginning and towards the end of the year, and underestimated towards the middle of the growing season.

A detailed statistical analysis was performed for the growing and non-growing seasons (Table 5), where the growing season was defined as the days between planting and harvest. Monthly statistics showed better statistical performance, with monthly RMSE and NSE of −0.19 and 1.2 compared to values of −1.38 and 1.8 mm for the daily assessment [38,46,50]. The RMSE during the growing season was greater compared to the non-growing season (Table 5), with values almost double for the growing season compared to the non-growing season due to low measured ET values during the non-growing season. Hence, there was less variation between the measured and satellite-based ET values. However, the %RMSE error was higher during the non-growing season than the growing season, and these results agree with Allen et al. [46].

The satellite was able to distinguish between bare soil and vegetation in the field, providing useful information on when the field was fallow versus when a crop was growing. However, the overall LAI estimation from Landsat was lower than the measured LAI for all cultivated crops during this study under dryland conditions. Potential reasons for the LAI undercalculations are the water stress during the growing season producing low NDVI values under dryland conditions, uncertainties with aerodynamic resistance surface roughness length [36], long gap periods, and using the linear interpolation method to generate daily ET time series [38].

4.2. Irrigated Daily ET Comparison

The relationship between measured and Landsat ET for the irrigated lysimeter provided overall better agreement compared to the dryland field [38,41,51,52]. The Landsat ET estimates were closely matched most of the year, except the middle of the growing seasons, during the peak crop water requirements. The satellite-based approach underestimated ET toward the middle of the growing season for cotton and soybeans, and overestimated the ET early and late during the growing season.

A detailed statistical analysis was performed for the daily and monthly ET (Table 6). The irrigated daily ET estimates were considered poor with an NSE of 0.37, RMSE of 2.1 mm d^{-1}, and %RMSE of 86.4%. However, there was a statistical improvement with the monthly ET values with an NSE of 0.57, RMSE of 1.5 mm d^{-1}, and % RMSE of 56.7%. Similar to the dryland lysimeter, the RMSE during the growing season was greater compared to the non-growing season (Table 7), with values almost double for the growing season compared to the non-growing season due to low ET measured values during the non-growing season. Hence, there was less variation between the measured and satellite-based ET values. However, the %RMSE error was higher during the non-growing season than the growing season, and these results agree with Allen et al. [46]. Allen et al. [6] illustrated that the use of reference ET considers the advective effects on a METRIC model performance, which can make the METRIC ET overestimate the ET from irrigated fields, exceeding daily net radiation in arid and semi-arid conditions. Allen et al. [46] reported that daily ET had the largest differences due to ET fluctuating the most during the growing season, and the monthly and season ET lumped most of the daily variations [38,46], and this is in agreement with the current study results.

Similarly, for the dryland lysimeter, the deviation between Landsat and measured ET was related to higher LAI estimation [32,36,38,47], advective condition effects under irrigated conditions [6, 36], and extended gap periods [38]. In addition, most of the studies conducted evaluated the ET on the current scene (image) days with minimal EB closure errors [36,49,53], and no studies evaluated the extrapolated daily ET assessment for dryland conditions with clumped crops [36]. However, the irrigated field difference magnitude was far less than for the dryland field.

Landsat LAI estimates were better for the irrigated lysimeter, as the METRIC model performance was affected with the wet and cold pixel determination, and the METRIC model performed better with full canopy (full irrigated), compared to dryland (partial canopy) [32,36,38,49]. As irrigated fields produced more vegetation vigor, higher NDVI values were obtained [36,38], and consequently higher estimates of LAI were obtained and resulted in better estimates of Landsat ET for areas managed under irrigated conditions (Figure 8) [32,36,38,41].

The overall Landsat LAI estimation somewhat matched the measured LAI for most cultivated crops from 2001 to 2010 (2004 and 2007 omitted due to large gaps in clear Landsat data during the growing season). Three indicators that the satellite imagery was able to differentiate between irrigated (full canopy) and dryland fields (partial canopy) as well as identify the growing season were as follows:

1. The estimated LAI values for the irrigated field were much larger than that of the dryland field.
2. All LAI values were zero in the beginning and end of each year, and this reflects that the field was bare soil. However, there were LAI values recorded during the growing season for the same field, providing useful information on when the field was fallow versus when a crop was growing.
3. The magnitudes of NDVI values for irrigated fields were higher than those for the dryland fields [36,38].

The reason behind this is likely that LAI is better estimated for the irrigated field than the dryland field [36,38] due to more vegetation coverage, resulting in higher NDVI values and consequently ET values. In 2007 and 2009, when forage sorghum and sunflowers were cultivated, respectively, the LAI estimated using Landsat was slightly lower than the measured LAI for the irrigated lysimeter.

5. Conclusions

Remote sensing-based ET estimation is considered a promising tool for irrigation water management. However, uncertainties associated with satellite-based ET estimation still exist, especially with various remotely sensed platforms due to variations in spatial and temporal resolution. In this study, satellite-based ET was evaluated using Landsat under semi-arid conditions in Texas under irrigated and dryland conditions.

Ten years of lysimeter measured ET data were used in this study. The Landsat-based ET overestimated the measured ET early and late in the growing season and underestimated ET during the peak of the growing season. The daily and monthly ET for the dryland lysimeter was unacceptable with negative NSE (-1.38 and -0.19), indicating there was no correlation between the estimates and measured ET; however, the daily and monthly ET for the irrigated lysimeter values showed better statistics with an NSE of 0.37 and 0.57, respectively. Seasonal ET showed more variations during the growing season compared to the non-growing season, because higher ET values were estimated during the growing season.

Under dryland conditions, there was significant LAI underestimation compared to the measured LAI values due to water stress during the growing season. LAI plays a significant role in evapotranspiration; where greater values of NDVI were obtained, consequently greater LAI was obtained under irrigated conditions, resulting in more ET for irrigated conditions. There are several reasons behind uncertainties of LAI and ET estimation, including the following: (1) METRIC model uncertainties with partial canopy estimates, (2) dryland plants' rapid modification of LAI based on available soil water (partial canopy), and (3) uncertainties with aerodynamic resistance surface roughness length as well as surface temperature deviations between irrigated and dryland conditions.

Extended gap periods are another significant challenge, and the selection of the filling method can account for ET estimation errors. In this study, gap periods reached up to 184 days in 2004, and the minimum was in 2008 with 40 days. The linear interpolation method was utilized to extrapolate the daily ET estimates between every two consecutive images in this study.

More satellite-based ET assessment under arid and semi-arid conditions is required, where the magnitude and frequency of precipitation are erratic, and irrigation is the only source under arid conditions to replenish crop water needs. With advances in remote sensing, more frequent satellite imagery will be available, with high spatial resolutions. Other extrapolation methods should be considered to generate daily time-series ET datasets. This would likely improve overall ET estimation accuracy by improving the overall spatial and temporal resolution.

Future research opportunities that include the assessment of ET relationship with crop physiology, yield, and yield components (number of flowers, grain quality, etc.) would provide potential information on crop response under dryland and irrigated conditions. Economic analysis of commodity market prices would be another research project due to groundwater decline in the Ogallala aquifer.

Author Contributions: Writing—original draft, A.A.H.; methodology, A.A.H. and P.H.G.; software, J.E.M., G.W.M., and P.H.G.; resources, B.A.E. and P.H.G.; review and editing, B.A.E., G.W.M., V.F.B., S.A.R., and J.E.M.; supervision, B.A.E. All authors have read and agreed to the published version of the manuscript.

Funding: This research was funded by the Egyptian Government General Mission Scholarship Program administrated by the Egyptian Cultural and Education Bureau, Washington, DC; the Purdue Research Foundation, and the Agricultural and Biological Engineering Department, Purdue University.

Acknowledgments: The authors express their sincere thanks to (1) the Egyptian government general mission scholarship administrated by the Egyptian Cultural and Education Bureau, Washington, DC, for partially supporting this research; (2) the Purdue Research Foundation and the Agricultural and Biological Engineering Department for funding support during this research; and (3) the USDA-ARS at Bushland, Texas, USA for sharing the lysimeter data and data analysis.

Conflicts of Interest: The authors declare no conflict of interest.

References

1. Sellers, P.J.; Randall, D.A.; Collatz, G.J.; Berry, J.A.; Field, C.B.; Dazlich, D.A.; Zhang, C.; Collelo, G.D.; Bounoua, L. A revised land surface parameterization (SiB2) for atmospheric GCMs. Part I: Model formulation. *J. Clim.* **1996**, *9*, 676–705. [CrossRef]

2. Paulson, R.W. *Evapotranspiration and Droughts. National Water Summary 1988–1989: Hydrologic Events and Floods and Droughts*; US Government Printing Office: Washington, DC, USA, 1991; Volume 2375, pp. 1–147.

3. Famiglietti, J.S. The global groundwater crisis. *Nat. Clim. Chang.* **2014**, *4*, 945–948. [CrossRef]

4. Yaeger, M.A.; Massey, J.H.; Reba, M.L.; Adviento-Borbe, M.A.A. Trends in the construction of on-farm irrigation reservoirs in response to aquifer decline in eastern Arkansas: Implications for conjunctive water resource management. *Agric. Water Manag.* **2018**, *208*, 373–383. [CrossRef]

5. Gowda, P.H.; Chávez, J.L.; Colaizzi, P.D.; Evett, S.R.; Howell, T.A.; Tolk, J.A. Remote Sensing Based Energy Balance Algorithms for Mapping ET: Current Status and Future Challenges. *Trans. ASABE* **2007**, *50*, 1639–1644. [CrossRef]

6. Allen, R.G.; Tasumi, M.; Morse, A.; Trezza, R.; Wright, J.L.; Bastiaanssen, W.; Kramber, W.; Lorite, I.; Robison, C.W. Satellite-Based Energy Balance for Mapping Evapotranspiration with Internalized Calibration (METRIC)—Applications. *J. Irrig. Drain. Eng.* **2007**, *133*, 395–406. [CrossRef]

7. Gowda, P.H.; Chavez, J.L.; Colaizzi, P.D.; Evett, S.R.; Howell, T.A.; Tolk, J.A. ET mapping for agricultural water management: Present status and challenges. *Irrig. Sci.* **2007**, *26*, 223–237. [CrossRef]

8. Bastiaanssen, W.; Pelgrum, H.; Wang, J.; Ma, Y.; Moreno, J.; Roerink, G.; van Der Wal, T. A remote sensing surface energy balance algorithm for land (SEBAL). *J. Hydrol.* **1998**, 213–229. [CrossRef]

9. French, A.N.; Hunsaker, D.J.; Thorp, K.R. Remote sensing of evapotranspiration over cotton using the TSEB and METRIC energy balance models. *Remote. Sens. Environ.* **2015**, *158*, 281–294. [CrossRef]

10. Park, A.B.; Colwell, R.N.A.; Meyers, V.F. Resource Survey by Satellite; sci-ence fiction coming true. In *Yearbook of Agriculture*; US Government Printing Office: Washington, DC, USA, 1968; pp. 13–19.

11. Jackson, R.D. Remote Sensing of Vegetation Characteristics for Farm Management. *1984 Tech. Symp. East* **1984**, *475*, 81–97. [CrossRef]

12. Choudhury, B.; Idso, S.; Reginato, R. Analysis of an empirical model for soil heat flux under a growing wheat crop for estimating evaporation by an infrared-temperature based energy balance equation. *Agric. For. Meteorol.* **1987**, *39*, 283–297. [CrossRef]

13. Su, H.; McCabe, M.; Wood, E.F.; Su, Z.; Prueger, J.H. Modeling Evapotranspiration during SMACEX: Comparing Two Approaches for Local- and Regional-Scale Prediction. *J. Hydrometeorol.* **2005**, *6*, 910–922. [CrossRef]

14. Anderson, M.C.; Kustas, W.P.; Norman, J.M.; Hain, C.R.; Mecikalski, J.R.; Schultz, L.; González-Dugo, M.P.; Cammalleri, C.; D'Urso, G.; Pimstein, A.; et al. Mapping daily evapotranspiration at field to continental scales using geostationary and polar orbiting satellite imagery. *Hydrol. Earth Syst. Sci.* **2011**, *15*, 223–239. [CrossRef]

15. Evett, S.R.; Kustas, W.P.; Gowda, P.H.; Anderson, M.C.; Prueger, J.H.; Howell, T.A. Overview of the Bushland Evapotranspiration and Agricultural Remote sensing EXperiment 2008 (BEAREX08): A field experiment evaluating methods for quantifying ET at multiple scales. *Adv. Water Resour.* **2012**, *50*, 4–19. [CrossRef]

16. Bastiaanssen, W.G.M.; Noordman, E.J.M.; Pelgrum, H.; Davids, G.; Thoreson, B.P.; Allen, R. SEBAL Model with Remotely Sensed Data to Improve Water-Resources Management under Actual Field Conditions. *J. Irrig. Drain. Eng.* **2005**, *131*, 85–93. [CrossRef]

17. Tasumi, M.; Allen, R.G. Satellite-based ET mapping to assess variation in ET with timing of crop development. *Agric. Water Manag.* **2007**, *88*, 54–62. [CrossRef]

18. Norman, J.; Kustas, W.; Humes, K. Source approach for estimating soil and vegetation energy fluxes in observations of directional radiometric surface temperature. *Agric. For. Meteorol.* **1995**, *77*, 263–293. [CrossRef]

19. Colaizzi, P.D.; Evett, S.R.; Howell, T.A.; Gowda, P.H.; Shaughnessy, S.A.; Tolk, J.A.; Kustas, W.P.; Anderson, M.C. Two-Source Energy Balance Model: Refinements and Lysimeter Tests in the Southern High Plains. *Trans. ASABE* **2012**, *55*, 551–562. [CrossRef]

20. Kustas, W.P.; Norman, J.M. Evaluation of soil and vegetation heat flux predictions using a simple two-source model with radiometric temperatures for partial canopy cover. *Agric. For. Meteorol.* **1999**, *94*, 13–29. [CrossRef]

21. Kalma, J.D.; McVicar, T.R.; Matthew, F. UNSW Faculty of Engineering Matthew Francis McCabe Estimating Land Surface Evaporation: A Review of Methods Using Remotely Sensed Surface Temperature Data. *Surv. Geophys.* **2008**, *29*, 421–469. [CrossRef]

22. Moorhead, J.E.; Marek, G.W.; Gowda, P.H.; Lin, X.; Colaizzi, P.D.; Evett, S.R.; Kutikoff, S. Evaluation of Evapotranspiration from Eddy Covariance Using Large Weighing Lysimeters. *Agronomy* **2019**, *9*, 99. [CrossRef]

23. Anderson, M.; Norman, J.; Kustas, W.; Houborg, R.; Starks, P.; Agam, N. A thermal-based remote sensing technique for routine mapping of land-surface carbon, water and energy fluxes from field to regional scales. *Remote Sens. Environ.* **2008**, *112*, 4227–4241. [CrossRef]

24. Marek, T.H.; Schneider, A.D.; Howell, T.A.; Ebeling, L.L. Design and Construction of Large Weighing Monolithic Lysimeters. *Trans. ASAE* **1988**, *31*, 477–484. [CrossRef]

25. Howell, T.A.; Schneider, A.D.; Dusek, D.A.; Marek, T.H.; Steiner, J.L. Calibration and Scale Performance of Bushland Weighing Lysimeters. *Trans. ASAE* **1995**, *38*, 1019–1024. [CrossRef]

26. Norman, J.M.; Anderson, M.C.; Kustas, W.P.; French, A.N.; Mecikalski, J.; Torn, R.; Diak, G.R.; Schmugge, T.J.; Tanner, B.C.W. Remote sensing of surface energy fluxes at 101-m pixel resolutions. *Water Resour. Res.* **2003**, *39*, 39. [CrossRef]

27. Anderson, M.C.; Norman, J.M.; Mecikalski, J.R.; Otkin, J.A.; Kustas, W.P. A climatological study of evapotranspiration and moisture stress across the continental United States based on thermal remote sensing: 2. Surface moisture climatology. *J. Geophys. Res. Space Phys.* **2007**, *112*, 112. [CrossRef]

28. Anderson, M.; Kustas, W.P.; Norman, J.M. Upscaling Flux Observations from Local to Continental Scales Using Thermal Remote Sensing. *Agron. J.* **2007**, *99*, 240–254. [CrossRef]

29. Colaizzi, P.D.; Evett, S.R.; Howell, T.A.; Tolk, J.A. Comparison of Five Models to Scale Daily Evapotranspiration from One-Time-of-Day Measurements. *Trans. ASABE* **2006**, *49*, 1409–1417. [CrossRef]

30. Allen, R.G.; Pereira, L.S.; Howell, T.A.; Jensen, M.E. Evapotranspiration information reporting: I. Factors governing measurement accuracy. *Agric. Water Manag.* **2011**, *98*, 899–920. [CrossRef]

31. Allen, R.G.; Pereira, L.S.; Howell, T.A.; Jensen, M.E. Evapotranspiration information reporting: II. Recommended documentation. *Agric. Water Manag.* **2011**, *98*, 921–929. [CrossRef]

32. Chávez, J.L.; Gowda, P.H.; Howell, T.A.; Garcia, L.A.; Copeland, K.S.; Neale, C.M.U. ET Mapping with High-Resolution Airborne Remote Sensing Data in an Advective Semiarid Environment. *J. Irrig. Drain. Eng.* **2012**, *138*, 416–423. [CrossRef]
33. Mkhwanazi, M.; Chávez, J.L.; Rambikur, E.H. Comparison of Large Aperture Scintillometer and Satellite-based Energy Balance Models in Sensible Heat Flux and Crop Evapotranspiration Determination. *Int. J. Remote Sens. Appl.* **2012**, *2*, 24.
34. Morton, C.G.; Huntington, J.L.; Pohll, G.M.; Allen, R.G.; McGwire, K.C.; Bassett, S.D. Assessing Calibration Uncertainty and Automation for Estimating Evapotranspiration from Agricultural Areas Using METRIC. *JAWRA J. Am. Water Resour. Assoc.* **2013**, *49*, 549–562. [CrossRef]
35. Trezza, R.; Allen, R.; Tasumi, M. Estimation of Actual Evapotranspiration along the Middle Rio Grande of New Mexico Using MODIS and Landsat Imagery with the METRIC Model. *Remote Sens.* **2013**, *5*, 5397–5423. [CrossRef]
36. Chávez, J.L.; Gowda, P.H.; Howell, T.A.; Copeland, K.S. Evaluating Three Evapotranspiration Mapping Algorithms with Lysimetric Data in the Semi-arid Texas High Plains. In Proceedings of the 28th annual international irrigation show, San Diego, CA, USA, 9–11 December 2007.
37. Dusek, D.A.; Howell, T.A.; Schneider, A.D.; Copeland, K.S. Bushland weighing lysimeter data acquisition systems for evapotranspiration research. *Am. Soc. Agric. Eng.* **1987**, *1061*, 81–2506.
38. Hashem, A.A. Irrigation Water Management Using Remote Sensing and Hydrologic Modeling. Ph.D. Thesis, Purdue University, West Lafayette, IN, USA, 2018.
39. Marek, G.W.; Gowda, P.; Evett, S.R.; Baumhardt, R.L.; Brauer, D.; Howell, T.A.; Marek, T.H.; Srinivasan, R. Calibration and Validation of the SWAT Model for Predicting Daily ET over Irrigated Crops in the Texas High Plains Using Lysimetric Data. *Trans. ASABE* **2016**, *59*, 611–622. [CrossRef]
40. Marek, G.W.; Gowda, P.; Evett, S.R.; Baumhardt, R.L.; Brauer, D.K.; Howell, T.A.; Marek, T.H.; Srinivasan, R. Estimating Evapotranspiration for Dryland Cropping Systems in the Semiarid Texas High Plains Using SWAT. *JAWRA J. Am. Water Resour. Assoc.* **2016**, *52*, 298–314. [CrossRef]
41. Hashem, A.A.; Engel, B.; Bralts, V.F.; Marek, G.W.; Moorhead, J.E.; Rashad, M.; Radwan, S.; Gowda, P.H. Landsat Hourly Evapotranspiration Flux Assessment using Lysimeters for the Texas High Plains. *Water* **2020**, *12*, 1192. [CrossRef]
42. Moorhead, J.E.; Gowda, P.H.; Ponder, B.A.; Brauer, D.K. Bushland Evapotranspiration and Agricultural Remote Sensing System (BEARS) software. In Proceedings of the Managing Global Resources for a Secure Future, Tampa, FL, USA, 22–25 October 2017.
43. Allen, R.G.; Pereira, L.S.; Raes, D.; Smith, M. *Crop Evapotranspiration—Guidelines for Computing Crop Water Requirements*; FAO Irrigation and Drainage Paper 56: Rome, Italy, 1998; ISBN 92-5-104219-5.
44. Blonquist, J.; Allen, R.; Bugbee, B. An evaluation of the net radiation sub-model in the ASCE standardized reference evapotranspiration equation: Implications for evapotranspiration prediction. *Agric. Water Manag.* **2010**, *97*, 1026–1038. [CrossRef]
45. Allen, R.G.; Burnett, B.; Kramber, W.; Huntington, J.L.; Kjaersgaard, J.; Kilic, A.; Kelly, C.; Trezza, R. Automated Calibration of the METRIC-Landsat Evapotranspiration Process. *JAWRA J. Am. Water Resour. Assoc.* **2013**, *49*, 563–576. [CrossRef]
46. Allen, R.G.; Irmak, A.; Trezza, R.; Hendrickx, J.M.H.; Bastiaanssen, W.G.M.; Kjaersgaard, J. Satellite-based ET estimation in agriculture using SEBAL and METRIC. *Hydrol. Process.* **2011**, *25*, 4011–4027. [CrossRef]
47. Hashem, A.A.; Engel, B.A.; Marek, G.W.; Moorhead, J.E.; Rashad, M.; Flanagan, D.C.; Radwan, S.; Bralts, V.F.; Gowda, P.H. Evaluation of SWAT Soil Water Estimation Accuracy Using Data from Indiana, Colorado, and Texas. *Trans. ASABE* **2020**, *5*, 1539–1559.
48. Moriasi, D.N.; Arnold, J.G.; van Liew, M.W.; Bingner, R.L.; Harmel, R.D.; Veith, T.L. Model Evaluation Guidelines for Systematic Quantification of Accuracy in Watershed Simulations. *Trans. ASABE* **2007**, *50*, 885–900. [CrossRef]
49. Madugundu, R.; Al-Gaadi, K.A.; Tola, E.; Hassaballa, A.A.; Patil, V.C. Performance of the METRIC model in estimating evapotranspiration fluxes over an irrigated field in Saudi Arabia using Landsat-8 images. *Hydrol. Earth Syst. Sci.* **2017**, *21*, 6135–6151. [CrossRef]
50. Senay, G.B.; Friedrichs, M.; Singh, R.K.; Velpuri, N.M. Evaluating Landsat 8 evapotranspiration for water use mapping in the Colorado River Basin. *Remote Sens. Environ.* **2016**, *185*, 171–185. [CrossRef]

51. Moorhead, J.E.; Marek, G.W.; Colaizzi, P.D.; Gowda, P.; Evett, S.R.; Brauer, D.; Marek, T.H.; Porter, D.O. Evaluation of Sensible Heat Flux and Evapotranspiration Estimates Using a Surface Layer Scintillometer and a Large Weighing Lysimeter. *Sensors* **2017**, *17*, 2350. [CrossRef] [PubMed]
52. Gowda, P.H.; Howell, T.A.; Paul, G.; Colaizzi, P.D.; Marek, T.H.; Su, Z.; Copeland, K.S. Deriving Hourly Evapotranspiration Rates with SEBS: A Lysimetric Evaluation. *Vadose Zone J.* **2013**, *12*, 1–11. [CrossRef]
53. Numata, I.; Khand, K.; Kjaersgaard, J.; Cochrane, M.A.; Silva, S.S. Evaluation of Landsat-Based METRIC Modeling to Provide High-Spatial Resolution Evapotranspiration Estimates for Amazonian Forests. *Remote Sens.* **2017**, *9*, 46. [CrossRef]

 *agronomy*

Article

Modeling the Effects of Irrigation Water Salinity on Growth, Yield and Water Productivity of Barley in Three Contrasted Environments

Zied Hammami [1,*], Asad S. Qureshi [1], Ali Sahli [2], Arnaud Gauffreteau [3], Zoubeir Chamekh [2,4], Fatma Ezzahra Ben Azaiez [2], Sawsen Ayadi [2] and Youssef Trifa [2]

[1] International Center for Biosaline Agriculture (ICBA), Dubai P.O. Box 14660, UAE; a.qureshi@biosaline.org.ae
[2] Laboratory of Genetics and Cereal Breeding, National Agronomic Institute of Tunisia, Carthage University, 43 Avenue Charles Nicole, 1082 Tunis, Tunisia; sahli_inat_tn@yahoo.fr (A.S.); zoubeirchamek@gmail.com (Z.C.); zahra.azaiez@gmail.com (F.E.B.A.); sawsen.ayadi@gmail.com (S.A.); youssef.trifa@gmail.com (Y.T.)
[3] INRA–INA-PG–AgroParisTech, UMR 0211, Avenue Lucien Brétignières, F-78850 Thiverval Grignon, France; arnaud.gauffreteau@inrae.fr
[4] Carthage University, National Agronomic Research Institute of Tunisia, LR16INRAT02, Hédi Karray, 1082 Tunis, Tunisia
*   Correspondence: z.hammami@biosaline.org.ae

Received: 25 August 2020; Accepted: 17 September 2020; Published: 24 September 2020

Abstract: Freshwater scarcity and other abiotic factors, such as climate and soil salinity in the Near East and North Africa (NENA) region, are affecting crop production. Therefore, farmers are looking for salt-tolerant crops that can successfully be grown in these harsh environments using poor-quality groundwater. Barley is the main staple food crop for most of the countries of this region, including Tunisia. In this study, the AquaCrop model with a salinity module was used to evaluate the performance of two barley varieties contrasted for their resistance to salinity in three contrasted agro-climatic areas in Tunisia. These zones represent sub-humid, semi-arid, and arid climates. The model was calibrated and evaluated using field data collected from two cropping seasons (2012–14), then the calibrated model was used to develop different scenarios under irrigation with saline water from 5, 10 to 15 dS m^{-1}. The scenario results indicate that biomass and yield were reduced by 40% and 27% in the semi-arid region (KAI) by increasing the irrigation water salinity from 5 to 15 dS m^{-1}, respectively. For the salt-sensitive variety, the reductions in biomass and grain yield were about 70%, respectively, although overall biomass and yield in the arid region (MED) were lower than in the KAI area, mainly with increasing salinity levels. Under the same environmental conditions, biomass and yield reductions for the salt-tolerant barley variety were only 16% and 8%. For the salt-sensitive variety, the biomass and grain yield reductions in the MED area were about 12% and 43%, respectively, with a similar increase in the salinity levels. Similar trends were visible in water productivities. Interestingly, biomass, grain yield, and water productivity values for both barley varieties were comparable in the sub-humid region (BEJ) that does not suffer from salt stress. However, the results confirm the interest of cultivating a variety tolerant to salinity in environments subjected to salt stress. Therefore, farmers can grow both varieties in the rainfed of BEJ; however, in KAI and MED areas where irrigation is necessary for crop growth, the salt-tolerant barley variety should be preferred. Indeed, the water cost will be reduced by 49% through growing a tolerant variety irrigated with saline water of 15 dS m^{-1}.

Keywords: salinity; environments; AquaCrop model; water productivity; scenarios; tolerant

1. Introduction

The world food supply is affected by environmental abiotic stresses, which damages up to 70% of food crop yields [1–3]. In the Near East and North Africa (NENA) region, physical water scarcity is already affecting food production [4]. The NENA region is characterized by an arid climate with a total annual rainfall much lower than the evapotranspiration of the field crops. In the Arab World, more than 85% of the available water resources are used for agriculture [5]. Despite this high-water allocation for the agriculture sector, about 50% of food requirements are imported [4]. Crop irrigation uses poor quality groundwater, which is saline in nature. The uninterrupted application of groundwater for irrigation is replete, which leads to a severe increase in soil salinity and reduction in crop yields. Climate changes, namely the increase in global temperatures and the decline in rainfall, exacerbate soil salinization, resulting in loss of production in arable lands [6]. According to recent estimates, one-fifth of the irrigated lands in the world are affected by salinity. Every day, on average, 2000 ha of irrigated land in arid and semi-arid areas is adversely affected by salinity problems [7]. The annual economic loss due to these increases in soil salinity is about USD 27.3 billion [8].

Cereals are the main crops in the Mediterranean and NENA regions, contributing to food security and social stability. Barley is one of these staple crops in the area. However, its production is constrained by abiotic factors, such as the arid climate, low and erratic rainfall, and soil and water salinity. The anticipated climate changes will further increase the negative impacts of these factors in the future [9]. Barley (*Hordeum vulgare* L.) is a drought- and salt-tolerant crop with considerable economic importance in Mediterranean and NENA regions since it is a source of stable farm income [10]. Indeed, barley is a staple food for over 106 countries in the world [11]. Barley is characterized by its high adaptability from humid to arid and even Saharan environments. Barley is grown in many areas of the world and is used for feed, food, and malt production [2,12].

To improve barley production in these regions, plant scientists have adopted a strategy to identify tolerant genotypes for maintaining reasonable yield on salt-affected soils [13]. Crops physiologists and breeders are working to assess how efficient a genotype is in converting water into biomass or yield. To do so, they use production parameters, with which measurement in field experiments is difficult and time-consuming. However, these complex parameters can be determined with the help of crop growth simulation models [6,13]. Dynamic simulation models describe the growth and development of crops based on the interaction with soil, water, and climate parameters. Models can be used to simulate soil and water salinity and crop management practices on the growth and yield of crops under different agro-climatic conditions [6].

Models were used to test the impact of salinity on crops under different environmental conditions and different fertilization practices [14,15].

AquaCrop is a water-driven dynamic model (Vanuytrecht et al., 2014). AquaCrop is a simulation model to study crops' water productivity. As crop-water-productivity is affected by climatic conditions, it is crucial to understand water productivity's response to changing rainfall and temperature patterns [9].

Among the available models, AquaCrop is preferred due to its robustness, precision, and the limited number of variables to be introduced [16]. It uses a small number of explicit and intuitive parameters that require simple calculation [16]. AquaCrop is a software system developed by the Land and Water Division of FAO to estimate water use efficiency and improve agricultural systems' irrigation management practices [17,18].

Water productivity (WP) can be described as the ratio of crops' net benefits, including both rain and irrigation.

According to [19], irrigation management organizations are interested in the yield per unit of irrigation water applied, as they have to improve the yield through human-induced irrigation processes. However, the downside is that not all irrigation water is used to generate crop production. Therefore, FAO defines water productivity as a ratio between a unit of output and a unit of input. Here, water productivity is used exclusively to indicate the amount or value of the product over the volume or value of water that is depleted or diverted [20].

This model was developed by the Food and Agriculture Organization (FAO) [16,21]. AquaCrop simulates the response of crop yield to water and is particularly suited to regions where water is the main limiting factor for agricultural production. The model is based on the concepts of crops' yield response to water developed by Doorenbos and Kassam [22]. The AquaCrop model (v4.0) published in 2012 can estimate yield under salt stress conditions.

The AquaCrop model has been used to predict crop yields under salt stress conditions in different parts of the world [23,24]. Kumar et al. [23] successfully used the AquaCrop model to predict the water productivity of winter wheat under different salinity irrigation water regimes. Mondal et al. [24] used AquaCrop to evaluate the potential impacts of water, soil salinity, and climatic parameters on rice yield in the coastal region of Bangladesh. The AquaCrop model has also been widely used to simulate yields of various crops under diverse environments. For example, barley (*Hordeum vulgare* L.) [5,25,26], teff (*Eragrostis teff* L.) [5], cotton (*Gossypium hirsutum* L.) [27], maize (*Zea mays* L.) [28] wheat (*Triticum aestivum* L.) [3].

In this study, the AquaCrop model (v4.0) is used to assess the performance of two barley genotypes under three contrasted agro-ecosystems (soil, salinity, and climate). In these areas, groundwater is primarily used for irrigation. The salinity of irrigation water ranges from 3 to 15 dS m^{-1}. Farmers do not know which barley variety is most tolerant to producing a reasonable yield under these saline environments. Furthermore, model simulations were also performed to evaluate the impact of three irrigation water salinity levels (5, 10, and 15 dS m^{-1}) on the barley yield. A cost–benefit analysis was performed to determine the economic returns of each level of salinity water irrigation and genotype tolerance based on model simulation results. Those results should help recommend the farmers of saline areas to enhance barley yield and economic return.

2. Materials and Methods

Description of Field Trial Sites

Field experiments were conducted during the 2012–2014 period in three contrasting locations (Beja, Kairouan, Medenine) of Tunisia. The Beja site (36°44′01.13″ N; 9°08′14.30″ E) is sub-humid, Kairouan (35°34′34.97″ N; 10°02′50.88″ E) is located in the semi-arid area of central Tunisia, and Medenine (33°26′54″ N, 10°56′31″ E) is part of the South East arid region of Tunisia (Figure 1). Two barley varieties (Konouz from Tunisia and Batini 100/1 B from Oman) were used for field experiments. The Konouz variety is salt-sensitive [29,30], whereas Batini 100/1 B is salt-tolerant [29,31].

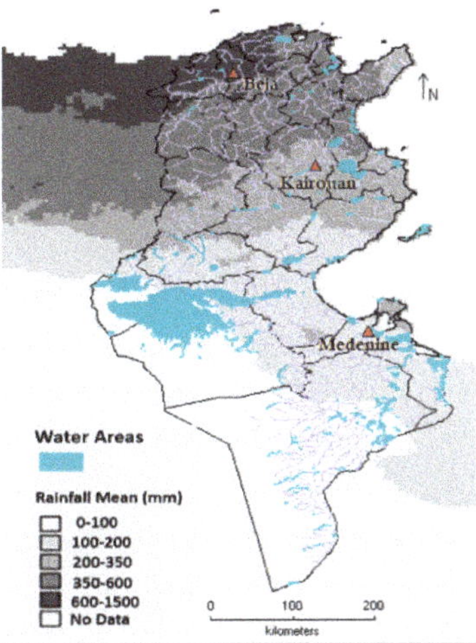

Figure 1. Location of field trial sites in different agro-climatic zones of Tunisia.

In Kairouan (KAI) and Medenine (MED) field trial sites were divided into two sub-plots. Each subplot was irrigated by one water salinity treatment (EC = 2 and 13 dS m^{-1}). Three blocks were defined perpendicularly to the sub-plots so that both treatments were observed in each block. As Beja is located in the rainfed cereal growing area of Tunisia, no irrigation was applied.

The weather data characterize the trials sites related to temperature, and rain was described by [29]. The irrigation water applied and reference evapotranspiration (ET$_0$) registered in the trials during the two growing seasons are presented in Table 1. The collected data from each site were used to estimate the reference evapotranspiration (ET) according to the Penman-Monteith Evapotranspiration FAO-56 Method, and then the total water supplied was determined for each site to obtain the water barley requirement. Irrigation was applied using a drip system. To ensure water supply homogeneity, line source emitters were installed at each planting row and 33-cm spacing between emitters on the same row.

Table 1. Rainfall, irrigation water applied and evapotranspiration (ET$_0$) in three trial sites.

Growing Season	Rainfall (mm)			Irrigation Water Applied (mm)			ET$_0$ (mm)		
	Sites			Sites			Sites		
	Beja	KAI	MED	Beja	KAI	MED	Beja	KAI	MED
2012–2013	472.2	151.9	81.1	0	360	455	393.8	364.7	327.6
2013–2014	413.5	180.0	156.1	0	360	405	390.1	363.7	328.4

Soil samples were taken from the trial sites, and physico-chemical analyses were performed. The site's soil characteristics are diverse, from soil rich in clay and organic matter in BEJ to sandy soil with impoverished organic matter continent in MED (Table 2).

Table 2. Soil properties in three field trial sites.

Site	Sand (%)	Clay (%)	Silt (%)	OM (%)	Na$^+$ Content (ppm)	K$^+$ Content (ppm)	Ca^{2+} Content (ppm)	PWP (% vol)	FC (% vol)
Beja	15.0	57.5	27.5	4.7	10–20	250–300	100–110	32.0	50.0
KAI	14.8	45.1	40.1	4.0	230–270	390–550	90–140	23.0	39.0
MED	55.5	20.5	24.0	0.9	120–200	30–70	30–55	6.0	13.0

(OM: organic matter, PWP: permanent wilting point; FC: field capacity).

Crops were sown during the last week of November. Seeds were hand sown at the rate of 200 viable grains per m^2. Nitrogen, potassium and phosphorus were applied separately at 85, 50, and 50 kg/ha rates, respectively.

At the five different stages, plants for each genotype, from three small areas (25 × 25 cm) were taken from each experimental unit and used to determine the biomass. At a final harvest stage, plot (1 × 2 m) was used for biomass and grain yield assessment. Water productivity (WP) was calculated as the ration between the collected yield expressed in kg ha^{-1} and the daily transpiration simulated by the model.

3. Description of the AquaCrop Model

The model describes soil, water, crop, and atmosphere interactions through four sub-model components: (i) the soil with its water balance; (ii) the crop (development, growth, and yield); (iii) the atmosphere (temperature, evapotranspiration, and rainfall), and carbon dioxide (CO$_2$) concentration; and (iv) the management, such as irrigation and crop fertilization soil fertility.

The AquaCrop model is based on the relationship between the relative yield and the relative evapotranspiration [22] as follows

$$\frac{Y_x - Y_a}{Y_x} = K_y\left(\frac{ET_x - ET_a}{ET_x}\right) \tag{1}$$

where Y_x is the maximum yield, Y_a is the actual yield, ET_x is the maximum evapotranspiration, ET_a is the actual evapotranspiration, and K_y is the yield response factor between the decrease in the relative yield and the relative reduction in evapotranspiration.

The AquaCrop model does not take into account the non-productive use of water for separating evapotranspiration (ET) into crop transpiration (T) and soil evaporation (E)

$$ET = E + Tr \tag{2}$$

where ET = actual evapotranspiration, E = soil evaporation and Tr = the sweating of crop.

At a daily time step, the model successively simulates the following processes: (i) groundwater balance; (ii) development of green canopy (CC); (iii) crop transpiration; (iv) biomass (B); and (v) conversion of biomass (B) to crop yield (Y). Therefore, through the daily potential evapotranspiration (ET_o) and productivity of water (WP*), the daily transpiration (Tr) is converted into vegetal biomass as follows

$$B_i = WP^* \left(\frac{T_{ri}}{ETo_i} \right) \tag{3}$$

where WP* is the normalized water productivity [32,33] relative to Tr. After the normalization of water productivity for different climatic conditions, its value can be converted into a fixed parameter [34]. The estimation and prediction of performance are based on the final biomass (B) and harvest index (HI). This allows a clear distinction between impact of stress on B and HI, in response to the environmental conditions

$$Y = HI * (B) \tag{4}$$

where: Y = final yield; B = biomass; HI = harvest index.

During the calibration and testing of the model, we calculated water productivity (WP) as presented by Araya et al. [5]

$$WP = \left[\frac{Y}{\sum Tr} \right] \tag{5}$$

where Y is the yield expressed in kg ha^{-1} and Tr is the daily transpiration simulated by the model.

3.1. Crop Response to Soil Salinity Stress

The electrical conductivity of saturation soil-past extracts from the root zone (ECe) is commonly used as an indicator of the soil salinity stress to determine the total reduction in biomass production, determines the value for soil salinity stress coefficient ($K_{s,\,salt}$).

The coefficient of soil salinity stress (Ks_{salt}) varied between 0 (full effect of stress of soil salinity) and 1 (no effect). The following equation determined the reduction in biomass

$$B_{rel} = 100\,(1 - Ks_{salt}) \tag{6}$$

B_{rel} represents the expected biomass production under given salinity stress relative to the biomass produced in the absence of salt stress. The coefficient is adjusted daily to the average ECe in the root zone [35].

Then, the thresholds values are given for the sensitive and tolerant barley genotype and expressed in dS m^{-1}. This allows the estimation of the lower limit (EC_{en}) to which the soil salinity stress begins to affect the production of biomass and the upper threshold (EC_{ex}), in which soil salinity stress has reached its maximum effect.

3.2. Soil Salinity Calculation

AquaCrop adopts the calculation procedure presented in BUDGET [36] to simulate the movement and retention of salt in the soil profile. The salts enter the soil profile as solutes after irrigation with saline water or through capillary rise from a shallow groundwater table (vertical downward and upward salt movement). The average ECe in the compartments of the effective rooting depth determines the effects of soil salinity on biomass production.

To explain the movement and retention of soil water and salts in the soil profile, AquaCrop divides the soil profile into 2 to 11 soil compartments called "cells", depending on the type of soil in each horizon (clay, sandy horizon) and its saturated hydraulic conductivity (Ksat in mm/day). The salt diffusion between two adjacent cells (cell j and cell j+1) is determined by the differences in salt concentration and expressed by the electrical conductivity (EC) of soil water.

AquaCrop determines the vertical salt movement in response to soil evaporation, considering the amount of water extracted from the soil profile by evaporation and the wetness of the upper soil layer. The relative soil water content of the topsoil layer determines the fraction of the dissolved salts that moves with the evaporating water.

AquaCrop determines the vertical salt movement because of the capillary rise. Finally, the salt content of a cell is determined by

$$\text{Salt}_{\text{cell}} = 0.64 \, W_{\text{cell}} EC_{\text{cell}} \tag{7}$$

$\text{Salt}_{\text{cell}}$ is the salt content expressed in grams salts per m^2 soil surface, Wcell its volume expressed in liter per m^2 (1 mm = 1 L/m^2), and 0.64 a global conversion factor used in AquaCrop to convert dS/m to g/L. The electrical conductivity of the soil water (ECsw) and of the electrical conductivity of saturation soil-past extract (ECe) at a particular soil depth (soil compartment) is calculated as

$$EC_{sw} = \frac{\sum_{j=1}^{n} \text{Salt}_{\text{cell.j}}}{0.64 \, (1000 \, \theta \Delta_z) \left\{ 1 - \frac{\text{Vol\%}_{\text{gravel}}}{100} \right\}} \tag{8}$$

$$EC_{e} = \frac{\sum_{j=1}^{n} \text{Salt}_{\text{cell.j}}}{0.64 \, (1000 \, \theta_{\text{sat}} \Delta_z) \left\{ 1 - \frac{\text{Vol\%}_{\text{gravel}}}{100} \right\}} \tag{9}$$

where n is the number of cells in each soil compartment; θ is the soil water content (m^3/m^3); θ_{sat} is the soil water content (m^3/m^3) at saturation; Δz (m) is the thickness of the soil compartment and Vol% gravel is the volume percentage of the gravel in the soil horizon of each compartment.

4. Model Calibration

4.1. Input and Output Variables of the Model

The model was calibrated using data from the growing season of 2012–2013 and evaluated using data from 2013–2014. Determining parameters for crop development and production, as well as water and salinity stress, was fundamental for calibrating the AquaCrop model. The parameters of climate, soil, and crop management used for the model calibration are presented in Table 3.

Table 3. Climate, soil and crop parameters used for the simulation model AquaCrop.

Climate		- Daily rainfall, daily ET$_o$, daily temperatures - CO$_2$ concentration
Crop	Limited set	Crop development and production parameters which include phenology and life cycle
	Crop parameters	- Harvest index - Root zone threshold at the end of the canopy expansion - Threshold root zone depletion for early senescence - Time for the maximum canopy cover - Maximum vegetation - Flowering time - Initial vegetative cover - Depletion threshold root zone for stomata closure - Extraction of water

Table 3. Cont.

	Field	- Soil fertility, mulch - Field practices (surface runoff presence, ground bond)
Soil	Soil profile	Characteristics of soil horizon (no of soil horizon, thickness, Permanent Wilting Point (PWP), Field Capacity (FC), Soil saturation (SAT), Ksat); soil surface (runoff, evaporation); Restrictive soil layer capillary rise).
	Soil water and groundwater	Constant depth; variable depth; water quality.

4.2. Statistical Parameters Used for the Calibration and Evaluation of Model

Several statistical indices were used to evaluate the performance of the model on the field measured data. These include Percentage Error (PE), Root Mean Square Error (RMSE), Model Efficiency (ME) and Coefficient of Determination (R^2).

Percentage Error (PE) was determined using the following equation

$$EP = \frac{(S_i - O_i)}{O_i} \times 100 \tag{10}$$

where S_i and O_i are simulated and observed values, respectively.

The root means square error (RMSE) [37] is presented by the following equation

$$RMSE = \sqrt{\frac{1}{n} \sum_{i=1}^{n} (S_i - O_i)^2} \tag{11}$$

with the values of RMSE close to zero indicate the best model fit.

The model efficiency (ME) [38] was applied to assess the effectiveness of the model. The ME indicator compares the variability of prediction errors by the model to those of collected data from the field. If the prediction errors are greater than the data error, then the indicator becomes negative. The upper ME bound is at 1.

$$ME = \frac{\sum_{i=1}^{n}(O_i - MO)^2 - \sum_{i=1}^{n}(S_i - O_i)^2}{\sum_{i=1}^{n}(O_i - MO)^2} \tag{12}$$

The coefficient of determination (R^2), as a result of regression analysis, is the proportion of the variance in the dependent variable (predict value) that is predictable from the independent variable (observed value) and is computed according to [35]

$$R^2 = \left\{ \frac{\sum_{i=1}^{n}\left(O_i - \overline{O}\right)\left(S_i - \overline{S}\right)}{\left[\sum_{i=1}^{n}(O_i - \overline{O})^2\right]^{0.5}\left[\sum_{i=1}^{n}(S_i - \overline{S})^2\right]^{0.5}} \right\}^2 \tag{13}$$

R^2 is between 0 and 1.

4.3. Parameters Used for Model Calibration

In total, 26 input parameters were used for the model calibration (Table 4). Out of these, 14 parameters were considered as "conservative" because they do not change with salinity and are independent of limiting or non-limiting conditions. These parameters include normalized crop water productivity and crop transpiration coefficient. The remaining 12 are site-specific (climate, water, and soil salinity) and crop-specific (tolerant or sensitive). These input parameters were adjusted during the calibration process to obtain better adequacy between the measured and simulation values.

Table 4. Final values of different model input parameters obtained after calibration for two genotypes under different salinity levels (S1 = 2 dS m^{-1}; S2 = 13 dS m^{-1}).

		Batini-100/1 B (Salt-Tolerant)			Konouz (Salt-Sensitive)			Remarks
		BEJ	KAI	MED	BEJ	KAI	MED	
Base temperature (°C)	S1	0	0	0	0	0	0	Conservative
	S2	-	0	0	-	0	0	
Upper temperature (°C)	S1	30	30	30	30	30	30	Conservative
	S2	-	30	30	-	30	30	
Initial canopy cover, CC0 (%)	S1	1.5	1.5	1.5	1.5	1.5	1.5	Conservative
	S2	-	1.5	1.5	-	1.5	1.5	
Canopy cover per seeding (cm^2/plant)	S1	0.75	0.75	0.75	0.75	0.75	0.75	Conservative
	S2	-	0.75	0.75	-	0.75	0.75	
Maximum coefficient for transpiration, KcTr, x	S1	0.90	0.90	0.90	0.90	0.90	0.90	Conservative
	S2	-	0.90	0.90	-	0.90	0.90	
Maximum coefficient for soil evaporation, Kex	S1	0.4	0.4	0.4	0.4	0.4	0.4	Conservative
	S2	-	0.4	0.4	-	0.4	0.4	
Upper threshold for canopy expansion, Pexp, upper	S1	0.30	0.30	0.30	0.20	0.20	0.20	Varietal effect
	S2	-	0.30	0.30	-	0.20	0.20	
Lower threshold for canopy expansion, Pexp, lower	S1	0.65	0.65	0.65	0.55	0.55	0.55	Varietal effect
	S2	-	0.65	0.65	-	0.55	0.55	
Leaf expansion stress coefficient curve shape	S1	4.5	4.5	4.5	4.5	4.5	4.5	Conservative
	S2	4.5	4.5	4.5	4.5	4.5	4.5	
Upper threshold for stomatal closure, Psto, upper	S1	0.6	0.6	0.6	0.55	0.55	0.55	Varietal effect
	S2	-	0.6	0.6	-	0.55	0.55	
Leaf expansion stress coefficient curve shape	S1	4.5	4.5	4.5	4.5	4.5	4.5	Conservative
	S2	4.5	4.5	4.5	4.5	4.5	4.5	
Canopy senescence stress coefficient, Psen, upper	S1	0.65	0.65	0.65	0.55	0.45	0.45	Varietal effect and site effect for the sensitive
	S2	-	0.65	0.65	-	0.45	0.45	
Senescence stress coefficient curve shape	S1	4.5	4.5	4.5	4.5	4.5	4.5	Conservative
	S2	4.5	4.5	4.5	-	4.5	4.5	

Table 4. *Cont.*

		Batini-100/1 B (Salt-Tolerant)			Konouz (Salt-Sensitive)			Remarks
		BEJ	KAI	MED	BEJ	KAI	MED	
Reference harvest index, HI0 (%)	S1	40	40	41	41	42	45	Varietal and salt stress effect
	S2		40	45		41	41	
Normalized crop water productivity, WP* (g/m^2)	S1	14	14	14	14	14	14	Conservative
	S2	14	14	14	14	14	14	
Time from sowing to emergence (day)	S1	7	7	7	7	7	7	Conservative
	S2	7	7	7	7	7	7	
Time from sowing to maximum CC (jours)	S1	60	60	60	62	60	57	Varietal and salt stress effect
	S2	-	60	58	-	59	55	
Time from sowing to maximum CC (day)	S1	145	145	145	145	145	145	Conservative
	S2	-	145	145	145	145	145	
Time from sowing to maturity (day)	S1	178	157	157	178	157	157	Varietal and salt stress effect
	S2	-	157	157	-	157	157	
Maximum canopy cover, CCx (%)	S1	87	87	87	87	75	63	Varietal and salt stress effect
	S2	-	87	70	-	60	40	
Canopy growth coefficient, CGC (%/day)	S1	12.5	12.5	12.5	12	12	12	Varietal effect
	S2		12.5	12.5		12	12	
Canopy decline coefficient, CDC (%/day)	S1	6	6	6	6	6	6	Conservative
	S2	-	6	6	-	6	6	
Maximum effective rooting depth, Zx (m)	S1	0,9	0.75	0.75	0,9	0.75	0.75	Site effect
	S2	-	0.75	0.75	-	0.75	0.75	
Salinity stress, lower threshold, ECen (dS m^{-1})	S1	3	3	3	1	1	1	Varietal effect
	S2	3	3	3	1	1	1	
Salinity stress, upper threshold, ECex (dS m^{-1})	S1	22	22	22	18	18	18	Varietal effect
	S2	22	22	22	18	18	18	
Shape factor for salinity stress coefficient curve	S1	1	1	1	1	1	1	Conservative
	S2	1	1	1	1	1	1	

For Bej only rainfall; for Kai, two levels of water salinity (S1 = 1.2 dS m^{-1}; S2 = 13 dS m^{-1} (S2)); for Med, two levels of water salinity (S1 = 2 dS m^{-1}; S2 = 13 dS m^{-1} (S2)).

4.4. Development of Different Scenarios

After calibration and evaluation, the model was used to assess the performance of two barley varieties under three water salinity conditions scenarios i.e., 5, 10, and 15 dS m^{-1}, using the weather data for the growing season 2013–2014.

4.5. The Economic Gain from the Use of a Unit of Water Consumed in the Tow Barley Varieties under Different Climatic Conditions

The economic productivity of two barley varieties was estimated using the average unit cost of one water cubic meter in Tunisia and the water use predicted by AquaCrop. The crop water economic productivity of the tolerant and the sensitive barley varieties as the measure of the biophysical and then economic gain from the use of a unit of water consumed were estimated by AquaCrop model in grain yield production [20]. This is expressed in productive crop units of kg/m^3 and money unit/m^3.

5. Results

5.1. Biophysical Environments Variability of Experimental Sites

The experiments are conducted in adaptability trials set up in three contrasting biophysical environments (from the sub-humid to the arid interior). These sites, namely Beja, Kairouan and Medenine, were selected on a North–South transect (Figure 1). The soils of the trial sites are very diverse, from soil rich in clay and poor in organic matter in BEJ to sandy soil with poor organic matter continent in MED (Table 2). Beja's sub-humid site received annual rainfall of 472 and 413 mm respectively during the two cropping seasons. However, in the semi-arid and arid sites, low rainfall was registered. The arid site of MED received an annual rainfall of 81 mm during the first cropping season and 156 mm during the second. At Kairouan, the rainfall for the 2012/2013 and 2013/2014 seasons was 152 and 180 mm, respectively (Table 1). As Beja is located in the rainfed cereal-growing area of Tunisia, no irrigation was applied. KAI and MED field trial sites, two different salinities (EC = 2 and 13 dS m^{-1}) of water were used for irrigation.

Soil calcium and potassium content was higher in KAI as compared to MED. Soil sodium content changes during the different experimentation period following irrigation with saline water in KAI and MED, where sodium is the dominant element present in the saline irrigation water. The variation between sites might be explained by the variation in the cationic exchange capacity of the sandy soil and torrential character of the rainfall in this area (Table 2).

5.2. Biomass, Grain Yield, and Water Use Efficiency

The correlation between grain yield, biomass, and water productivity values for two barley genotypes showed that the observed and simulated values are closely co-related, as evidenced by the high R^2 values, i.e., 0.91, 0.93, and 0.89 for grain yield, biomass, and water productivity, respectively (Figure 2).

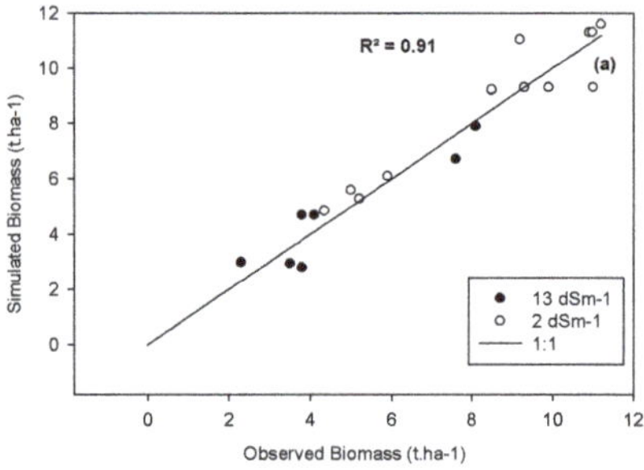

Figure 2. *Cont.*

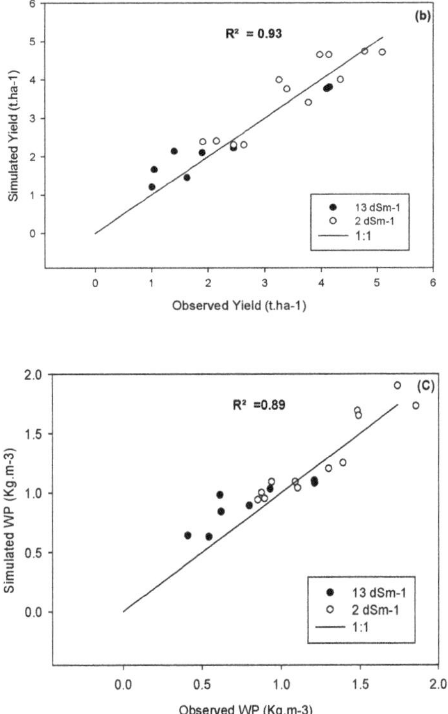

Figure 2. Correlation between observed and simulated (**a**) biomass yield; (**b**) grain yield; and (**c**) water productivity compared with 1:1 line.

The correlation between observed and simulated values of biomass yield for two barley genotypes at three locations showed proximity (Figure 3), which indicates the excellent ability of the AquaCrop model to predict biomass yield under different agro-climatic conditions. The results also show that the sensitive barley variety at MED produces the lowest biomass for both irrigation water qualities. Similar trends were observed for grain yield, where the tolerant barley variety performed better than the sensitive variety regardless of the location and the quality of irrigation water.

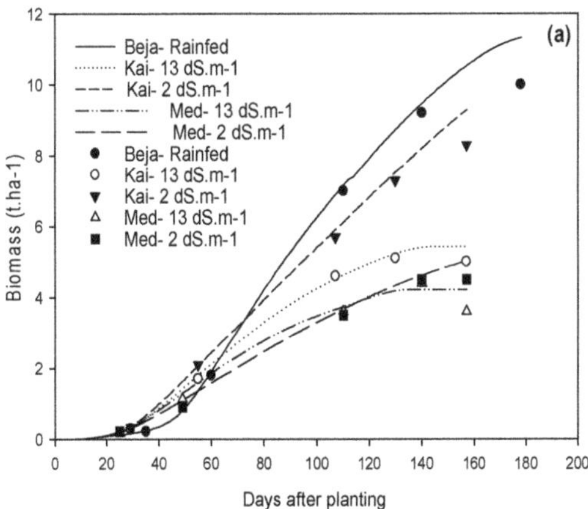

Figure 3. *Cont.*

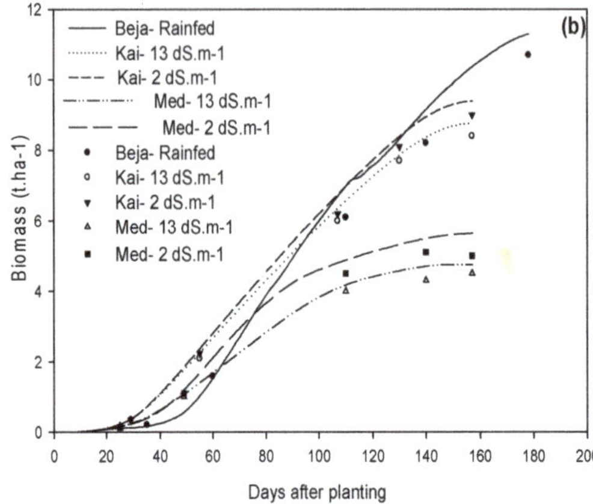

Figure 3. Simulated and observed biomass of (**a**) tolerant and (**b**) sensitive barley genotypes (dots represent observations; simulations are represented by lines).

5.3. Canopy Cover (CC)

The maximum and minimum CC were 85% and 30% in the sub-humid and arid areas, respectively. The salinity induces a 10% reduction in the CC in the sub-humid environment and 5–30% in the dry climate of MED. CC reduction under saline irrigation water is less noticeable in the tolerant variety than the sensitive variety for both salinity levels. However, in the rainfed area of Beja, the growth of both varieties was comparable.

Figure 4 shows a strong correlation between measured and simulated CC values for both varieties of barley (R^2 = 0.91 and R^2 = 0.93). In general, a good match between the observed and the simulated CC was observed in all three locations. However, the model somewhat over-estimated CC in the rainfed environment of Beja and slightly under-estimated it in the other two situations.

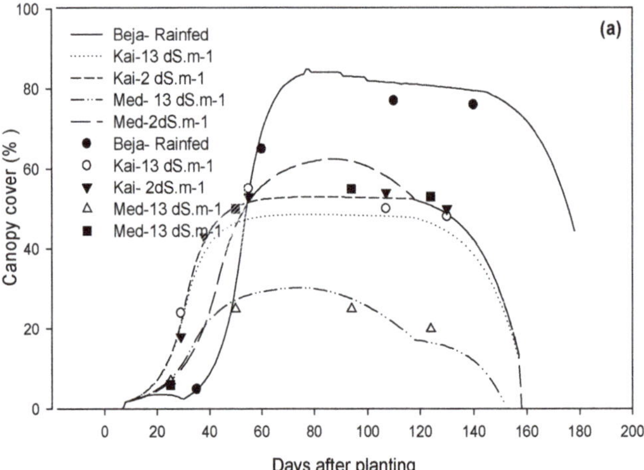

Figure 4. *Cont.*

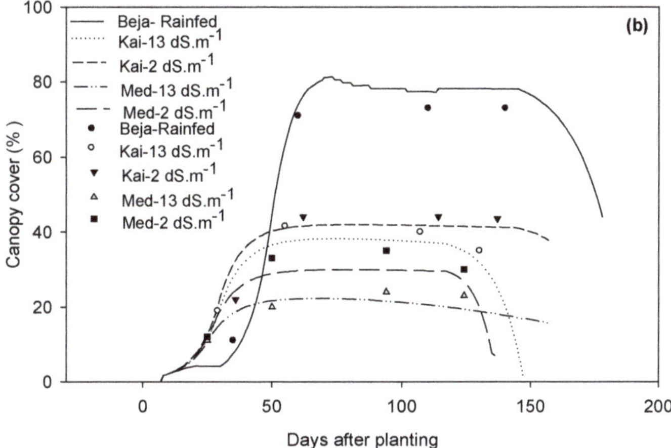

Figure 4. Simulated and observed canopy cover for (**a**) tolerant and (**b**) sensitive barley varieties.

5.4. Effects of Soil Salinity

The maximum soil salinity was in the arid and semi-arid areas irrigated with saline water, respectively. The soil salinisation dynamic depends on the salinity of irrigation water. However, in the rainfed area of Beja, we noted the absence of any salty issue.

Figure 5 shows that the simulated soil salinity trend in the root zone (up to a depth of 0.7 m) corresponds very well with the measured values under different saline water regimes across different environments throughout the growing season. The observed and modeled soil salinity correlated well, with an R^2 of 0.96. Figure 5 shows that the model reliably simulated average root zone salinity when the crop is irrigated with low-salinity water (2 dS m^{-1}). However, it slightly underestimated soil salinity under higher saline water conditions (13 dS m^{-1}), particularly for the late growing season.

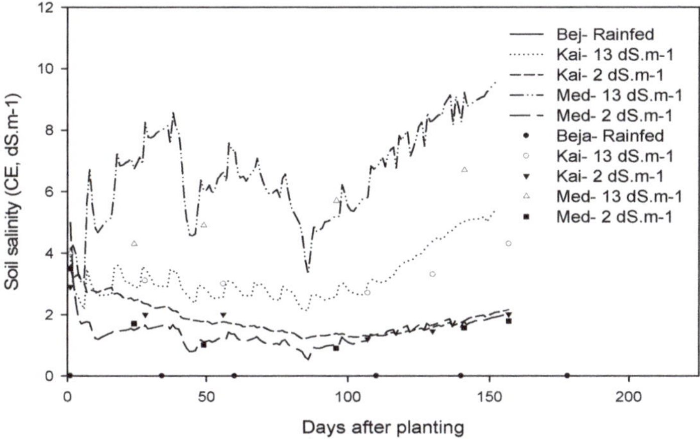

Figure 5. Simulated and observed soil salinity in the testing-cropping season under different saline water regimes and across different environments.

5.5. Statistical Indices for AquaCrop Model Evaluation

The statistical indices derived for evaluating the AquaCrop model's performance in predicting soil water content, yield, canopy cover percent, biomass, and water productivity (*WP*) of barley genotypes under different saline water regimes across different environments are given in Table 5. All statistical parameters depict a strong correlation between simulated and observed values for model calibration and evaluation periods. The correlation between all statistical parameters remained almost the same for the calibration and evaluation period, which indicates the robustness of the model prediction. Based on the model calibration and evaluation results, the model was found robust enough to calculate different scenarios.

Table 5. Statistical indices values for different parameters obtained for model calibration.

	Variable	RMSE	ME	R^2
	Grain yield (t ha^{-1})	0.40	0.89	0.91
	Biomass (t ha^{-1})	0.87	0.96	0.93
Calibration	water productivity (kg ha^{-1} mm^{-1})	0.15	0.84	0.89
	Soil salinity	0.34	0.91	0.95
	Canopy cover percent	1.5	0.89	0.91
	Grain yield (t ha^{-1})	0.45	0.87	0.89
	Biomass (t ha^{-1})	0.89	0.86	0.87
Evaluation	water productivity (kg ha^{-1} mm^{-1})	0.13	0.91	0.84
	Soil salinity	1.25	0.91	0.96
	Canopy cover percent	2.25	0.89	0.91

6. Development of Different Scenarios

Due to a shortage of surface water, farmers of KAI and MED regions have no option than to use groundwater for irrigation. The quality of groundwater ranges from 4 to15 dS m^{-1} in these two regions. Farmers are interested to know which barley varieties would be most suitable to grow under these groundwater quality conditions. The calibrated and evaluated model was used to assess the performance of two barley varieties under three water salinity conditions i.e., 5, 10, and 15 dS m^{-1}, and the results are presented in Table 6.

Table 6. Predicted values of biomass, yield, and water productivity of two barley varieties for different scenarios.

	BEJ	KAI			MED		
	Rainfed	5 dS m^{-1}	10 dS m^{-1}	15 dS m^{-1}	5 dS m^{-1}	10 dS m^{-1}	15 dS m^{-1}
			Tolerant genotype				
Biomass (t ha^{-1})	11.30	9.07	8.36	5.48	5.60	4.74	4.70
Yield (t ha^{-1})	4.70	3.65	3.44	2.20	2.29	2.13	2.10
WP (kg m^{-3})	1.73	1.29	1.19	0.85	1.25	1.18	1.00
			Sensitive genotype				
Biomass (t ha^{-1})	11.33	6.62	4.60	1.90	3.18	3.03	2.80
Yield (t ha^{-1})	4.64	2.70	1.90	0.80	1.40	1.30	0.80
WP (kg m^{-3})	1.65	1.12	0.85	0.45	0.74	0.72	0.51

The performance of both barley varieties in the KAI area is predicted to be much higher than MED area under all salinity levels due to prevailing climatic conditions. In the KAI area, biomass and grain yield reductions are much higher with the increasing water salinity for both varieties. For example, the biomass and yield reductions in the KAI area were about 40%with an increase in salinity from 5 to 10 and 15 dS m^{-1}. For the sensitive genotype, the biomass and yield reductions in the KAI area would be above 72% with a similar increase in the salinity levels. Although overall biomass and grain yields in the MED area were lower than in the KAI area, biomass and yield reductions for the salt-tolerant barley variety were only 16% and 8%, with an increase in salinity from 5 to 15 dSm^{-1}, respectively.

However, for the sensitive genotype, reductions in biomass and yield were 12% and 43%, respectively, with a similar increase in salinity levels. Similar trends are obtained for water productivities.

Without salt stress, both varieties have the same performance. However, the tolerant variety performs better than the sensitive variety under salt stress. This is because it has better potential. Therefore, farmers can grow both varieties in the rainfed areas of BEJ, while, in KAI and MED areas where irrigation is necessary for crop growth, the salt-tolerant barley variety should be preferred. The cultivation of the salt-sensitive barley variety in the MED area will be risky, as the yields will be low, and the development of soil salinity over time will remain a challenge. This situation will be very critical for long-term sustainable crop production in the area.

7. Economic Productivity of Barley Varieties under Different Climatic Conditions

The economic productivity of two barley varieties was estimated using the average unit cost of one water cubic meter in Tunisia and the water use predicted by AquaCrop. The results show that the production cost of 1 kg of barley is lowest in the BEJ area compared to those areas where it is irrigated with saline water.

In the KAI region, the cost will be reduced by 13.28% 28.72% and 47.19% by growing the tolerant variety irrigated with saline water of 5, 10, and 15 dS m^{-1}, respectively. In the arid region of MED, the benefit will be reduced by 40%, 38%, and 49% by growing the tolerant barley variety by irrigating with saline water of 5, 10 and 15 dS m^{-1}, respectively (Figure 6). However, in the sub-humid region of BEJ, there is no significant difference between susceptible and tolerant genotypes. The results show the economic interest for arid region farmers to grow the tolerant barley variety. This stresses the need for appropriate breeding programs for the saline environments for optimizing crop production instead of targeting potential yields.

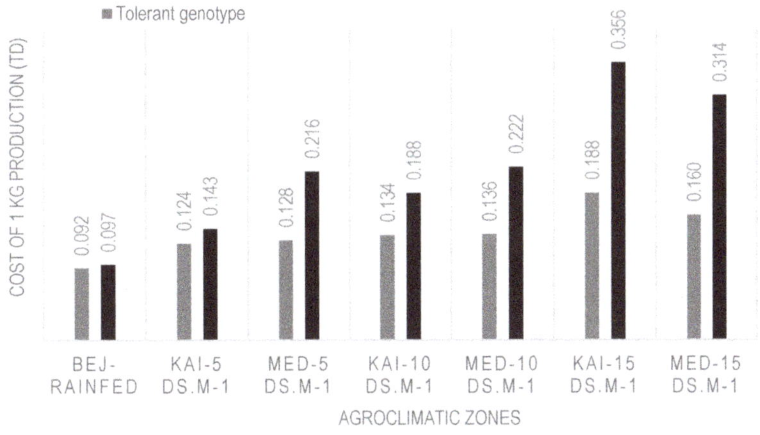

Figure 6. Economic productivity of two barley varieties under different climatic conditions.

8. Discussion

We evaluated the AquaCrop model for two barley varieties under contrasting environments and different water salinity levels. The simulated model values were close to the field measurements concerning biomass, yield and soil salinity. ME and R^2 parameters were close to 0.9, showing the model's ability to simulate the behavior of sensitive and resistant cultivars in contrasting environments and irrigation practices. Araya et al. [5] reported R^2 values of 0.80 when simulating barley biomass and grain yield using AquaCrop. El Mokh et al. [25] reported R^2 values of 0.88 when simulating barley yield under different irrigation regimes in a dry environment using AquaCrop. Mondal et al. [24] reported a 0.12 t ha^{-1} root mean square error after simulating the yield response of rice to salinity stress with the AquaCrop model. Our results also show a correct prediction with an RMSE of 0.45 t ha^{-1} (Table 5). This shows that the AquaCrop model simulates biomass production for all environments with an acceptable accuracy level.

AquaCrop model produces consistent simulation results for CC with an R^2 of 0.89 and RMSE of 2.25 (Table 5). The model also simulated soil salinity satisfactorily for all environments (R^2 = 0.96) for all situations. The R^2 values exceeding 0.8 are considered excellent for model performance [39]. The ability of AquaCrop to predict yield depends on the appropriate calibration of the canopy cover curve [1,40]. Indeed, after simulation of soil water balance at a daily time step, the model simulates CC and then simulates the transpiration of a crop, biomass above the soil, and converts biomass into yield. Therefore, it is essential to make accurate predictions of the canopy cover by the proper calibration of crop traits.

Therefore, through proper calibration, models can be used for additional solutions for the quantification of salinity build-up in the root zone [41].

We also noted the overestimation of the soil salinity at the end of the growing season when saline water is used for irrigation (Figure 5). This could be due to the excessive leaching of salts from the soil profile through irrigation, as reported by Mohammadi et al. [42]. Over- or underestimation at the end of the season could be the simplification of soil salt transport calculations in the model based on some empirical functions, including the parameters of Ks and the drainage coefficient for vertical downward salt movement. Furthermore, the occasional leaching of salts from the root zone using relatively better-quality water is also recommended. Changing cropping patterns is also a useful strategy for the rehabilitation and management of saline soils, especially when only saline water is available for irrigation.

The AquaCrop model was also capable of predicting water productivity under sub-humid, semi-arid, and arid environments and the effect of salinity. Plants subjected to salinity stress show a varying response in *WP*. The sensitive genotype was more exposed to varying responses in *WP*. Besides, heat stress induced by increased temperatures and the water deficit also decreases productivity, as demonstrated by Hatfield [43]. The observed and predicted water productivities were directly affected by climate aridity and the salinity of the irrigation water. However, the tolerant barley variety was less affected by these factors. These results are in agreement with the earlier studies [16,44].

Water scarcity is already hampering agricultural production in the MENA region. Therefore, the adoption of integrated management strategies will be useful for growing tolerant genotypes under saline water conditions and increasing the water use efficiency. For the sustainable management of crop growth in saline environments, soil-crop-water management interventions consistent with site-specific conditions need to be adopted [41]. These may include cyclic or conjunctive saline water use and freshwater through proper irrigation scheduling to avoid salinity development.

There are several traits available for screening genetic material for enhanced production and *WP* under different climate scenarios. This study shows that, under different water salinity conditions, sensitive barley genotype is more affected by the increasing water salinity than the tolerant barley genotype. The crop yields for both genotypes under all water salinity levels were higher in KAI area compared to the MED area. Therefore, this study recommends that farmers with higher salinity water for irrigation should grow tolerant barley genotypes, allowing them to reduce the cost, on average, by 30% (Figure 6). However, from a sustainability point of view, irrigation amounts should be kept to a minimum to optimize crop yields instead of targeting potential yields [45]. This exercise will help there be less accumulation of salts in the root zone. Besides, the occasional leaching of salts from the root zone using relatively better-quality water is also recommended. Changing cropping patterns is also regarded as a useful strategy for the rehabilitation and management of saline soils, especially when only saline water is available for irrigation [46,47].

9. Conclusions

The AquaCrop model with a salinity module was used to evaluate the agronomic performance of two barley varieties for the three different agro-climatic zones in Tunisia. These zones represent sub-humid, semi-arid, and arid climates. The model was calibrated and evaluated using field data from two years (2012 and 2014). The excellent correlation between the simulated and measured data of biomass, yield, and soil salinity confirms the ability of AquaCrop model to simulate crop growth under different climatic conditions. The scenario results using the calibrated model indicate that farmers with higher salinity water for irrigation should grow tolerant barley genotypes. However, from a sustainability point of view, irrigation amounts should be kept to a minimum to optimize crop yields instead of targeting potential yields.

Author Contributions: Conceptualization, Z.H.; Data curation, F.E.B.A. and S.A.; Formal analysis, Z.H.; Methodology, Z.H., A.S. and Z.C.; Supervision, Y.T.; Writing—Original draft, Z.H.; Writing—Review and editing, A.S.Q. and A.G. All authors have read and agreed to the published version of the manuscript.

Funding: This research was supported by the Laboratory of Genetics and Cereal Breeding, National Agronomic Institute of Tunisia, Carthage University, and International Center for Biosaline Agriculture (ICBA). And The APC was funded by ICBA.

Conflicts of Interest: The authors declare no conflict of interest.

References

1. Abi Saab, M.T.; Albrizio, R.; Nangia, V.; Karam, F.; Rouphael, Y. Developing scenarios to assess sunflower and soybean yield under different sowing dates and water regimes in the Bekaa valley (Lebanon): Simulations with Aquacrop. *Int. J. Plant Prod.* **2014**, *8*, 457–482.

2. Ahmed, M.; Goyal, M.; Asif, M. Silicon the non-essential beneficial plant nutrient to enhanced drought tolereance in wheat. In *Crop Plant*; Goyal, A., Ed.; Intech Publication House: London, UK, 2012; pp. 31–48.

3. Andarzian, B.; Bannayan, M.; Steduto, P.; Mazraeh, H.; Barati, M.E.; Barati, M.A.; Rahnama, A. Validation and testing of the AquaCrop model under fulland deficit irrigated wheat production in Iran. *Agric. Water Manag.* **2011**, *100*, 1–8. [CrossRef]

4. Araya, A.; Habtub, S.; Hadguc, K.; Kebedea, A.; Dejened, T. Test of AquaCrop model in simulating biomass and yield of water deficient and irrigated barley (Hordeum vulgare). *Agric. Water Manag.* **2010**, *97*, 1838–1846. [CrossRef]

5. Araya, A.; Keesstra, S.D.; Stroosnijder, L. Simulating yield response to water of Tef (Eragrostistef) with FAO's AquaCrop model. *Field Crop. Res.* **2010**, *116*, 1996–2204. [CrossRef]

6. Chauhdarya, J.N.; Bakhsh, A.; Ragab, R.; Khaliq, A.; Bernard, A.; Engeld, M.R.; Shahid, M.N.; Nawaz, Q. Modeling corn growth and root zone salinity dynamics to improve irrigation and fertigation management under semi-arid conditions. *Agric. Water Manag.* **2020**, *230*, 105952. [CrossRef]

7. FAO (Food and Agriculture Organization of the United Nations). *Advances in the Assessment and Mmonitoring of Salinization and Status of Biosaline Agriculture*; FAO: Rome, Italy, 2010.

8. Qadir, M.; Quillérou, E.; Nangia, V.; Murtaza, G.; Singh, M.; Thomas, R.J.; Drechsel, P.; Noble, A.D. Economics of salt-induced land degradation and restoration. *Nat. Resour. Forum* **2014**, *38*, 282–295. [CrossRef]

9. Hatfield, J.L.; Dold, C. Water-Use efficiency: Advances and challenges in a changing climate. *Front. Plant Sci.* **2019**. [CrossRef]

10. Zhou, G.; Johnson, P.; Ryan, P.R. Quantitative trait loci for salinity tolerance in barley (*Hordeum vulgare* L.). *Mol. Breed.* **2012**, *29*, 427–436. [CrossRef]

11. Newton, A.C.; Flavell, A.J.; George, T.S.; Leat, P.; Mullholland, B.; Ramsay, L.; RevoredoGiha, C.; Russell, J.; Steffenson, B.J.; Swanston, J.S.; et al. Crops that feed the world 4. Barley: A resilient crop? Strengths and weaknesses in the context of food security. *Food Secur.* **2011**, *3*, 141–178. [CrossRef]

12. Zhang, H.; Han, B.; Wang, T.; Chen, S.; Li, H.; Zhang, Y.; Dai, S. Mechanisms of Plant Salt Response: Insights from Proteomics. *J. Proteome Res.* **2012**, *11*, 49–67. [CrossRef]

13. Negrao, S.; Schmockel, S.M.; Tester, M. Evaluating physiological responses of plants to salinity stress. *Ann. Bot.* **2017**, *119*, 1–11. [CrossRef] [PubMed]

14. Heng, L.K.; Hsiao, T.; Evett, S.; Howell, T.; Steduto, P. Validating the FAO AquaCrop model for irrigated and water deficient field maize. *Agron. J.* **2009**, *101*, 488–498. [CrossRef]

15. Verma, A.K.; Gupta, S.K.; Isaac, R.K. Use of saline water for irrigation in monsoon climate and deep water table regions: Simulation modelling with SWAP. *Agric. Water Manag.* **2012**, *115*, 186–193. [CrossRef]

16. Soothar, R.K.; Wenying, Z.; Yanqing, Z.; Moussa, T.; Uris, M.; Wang, Y. Evaluating the performance of SALTMED model under alternate irrigation using saline and fresh water strategies to winter wheat in the North China Plain. *Environ. Sci. Pollut. Res.* **2019**, *26*, 34499–34509. [CrossRef] [PubMed]

17. Steduto, P.; Hsiao, T.C.; Raes, D.; Fereres, E. AquaCrop—The FAO crop model to simulate yield response to water. I. Concepts and underlying principles. *Agron. J.* **2009**, *101*, 426–437. [CrossRef]

18. Van Gaelen, H. *AquaCrop Training Handbooks–Book II Running AquaCrop*; Food and Agriculture Organization of the United Nations: Rome, Italy, 2016.

19. Doorenbos, J.; Kassam, A.H. *Yield Response to Water*; FAO Irrigation and Drainage Paper No. 33; FAO: Rome, Italy, 1979.

20. Kumar, A.; Sarangi, D.K.; Singh, R.; Parihar, S.S. Evaluation of aquacrop model in predicting wheat yield and water productivity under irrigated saline regimes. *Irrig. Drain.* **2014**, *63*, 474–487. [CrossRef]

21. Mondal, M.S.; Fazal, M.A.; Saleh, M.D.; Akanda, A.R.; Biswas, S.K.; Moslehuddin, Z.; Sinora, Z.; Attila, N. Simulating yield response of rice to salinity stress with the AquaCrop model. *Environ. Sci. Process. Impacts* **2015**, *17*, 1118–1126. [CrossRef]

22. El Mokh, F.; Nagaz, K.; Masmoudi, M.M.; Mechlia, N.B.; Fereres, E. *Calibration of AquaCrop Salinity Stress Parameters for Barley under Different Irrigation Regimes in a Dry Environment*; Springer: Cham, Germany, 2017. [CrossRef]

23. Hellal, F.; Mansour, H.; Mohamed, A.H.; El-Sayed, S.; Abdelly, C. Assessment water productivity of barley varieties under water stress by AquaCrop model. *AIMS Agric. Food* **2019**, *4*, 501–517. [CrossRef]

24. Tan, S.; Wang, Q.; Zhang, J.; Chen, Y.; Shan, Y.; Xu, D. Performance of AquaCrop model for cotton growth simulation under film-mulched drip irrigation in southern Xinjiang, China. *Agric. Water Manag.* **2018**, *196*, 99–113. [CrossRef]

25. Hammami, Z.; Gauffreteau, A.; BelhajFraj, M.; Sahlia, A.; Jeuffroy, M.H.; Rezgui, S.; Bergaoui, K.; McDonnell, R.; Trifa, Y. Predicting yield reduction in improved barley (*Hordeum vulgare* L.) varieties and landraces under salinity using selected tolerance traits. *Field Crop. Res.* **2017**, *211*, 10–18. [CrossRef]

26. Sbei, H.; Sato, K.; Shehzad, T.; Harrabi, M.; Okuno, K. Detection of QTLs for salt tolerance in Asian barley (*Hordeum vulgare* L.) by association analysis with SNP markers. *Breed. Sci.* **2014**, *64*, 378–388. [CrossRef] [PubMed]

27. Jaradat, A.A.; Shahid, M.; Al-Maskri, A.Y. Genetic diversity in the Batini barley landrace from Oman: Spike and seed quantitative and qualitative traits. *Crop. Sci.* **2014**, *44*, 304–315. [CrossRef]

28. Raes, D.; Steduto, P.; Hsiao, T.C.; Fereres, E. *Crop Water Productivity. Calculation Procedures and Calibration Guidance. AquaCrop Version 3.0. FAO*; Land and Water Development Division: Rome, Italy, 2009.

29. Trombetta, A.; Iacobellis, V.; Tarantino, E.; Gentile, F. Calibration of the AquaCrop model for winter wheat using MODIS LAI images. *Agric. Water Manag.* **2015**, *164*. [CrossRef]

30. Hanks, R.J. Yield and water-use relationships. In *Limitations to Efficient Water Use in Crop Production*; Taylor, H.M., Jordan, W.R., Sinclair, T.R., Eds.; ASA, CSSA, and SSSA: Madison, WI, USA, 1983; pp. 393–411.

31. Tanner, C.B.; Sinclair, T.R. Efficient water use in crop production: Research or re-search? In *Limitations to Efficient Water Use in Crop Production*; Taylor, H.M., Jordan, W.R., Sinclair, T.R., Eds.; ASA, CSSA, and SSSA: Madison, WI, USA, 1983; pp. 1–27.

32. Steduto, P.; Hsiao, T.C.; Fereres, E. On the conservative behavior of biomass water productivity. *Irrig. Sci.* **2007**, *25*, 189–207. [CrossRef]

33. Raes, D.; Steduto, P.; Hsiao, T.C.; Fereres, E. *AquaCrop Version 5.0 Reference Manual*; Food and Agriculture Organization of the United Nations: Rome, Italy, 2016.

34. Loague, K.; Green, R.E. Statistical and graphical methods for evaluating solute transport models: Overview and application. *J. Contam. Hydrol.* **1991**, *7*, 51–73. [CrossRef]

35. Minhas, P.S.; Tiago, B.; AlonBen-Gal, R.; Pereira, L.S. Coping with salinity in irrigated agriculture: Crop evapotranspiration and water management issues. *Agric. Water Manag.* **2020**, *227*, 105832. [CrossRef]

36. Pereira, L.S.; Paredes, P.; Rodrigues, G.C.; Neves, M. Modeling barley water use and evapotranspiration partitioning in two contrasting rainfall years. Assessing SIMDualKc and AquaCrop models. *Agric. Water Manag.* **2015**, *159*, 239–254. [CrossRef]

37. Iqbal, M.A.; Shen, Y.; Stricevic, R.; Pei, H.; Sun, H.; Amiri, E.; Rio, S. Evaluation of FAO Aquacrop model for winter wheat on the North China plain under deficit from field experiment to regional yield simulation. *Agric. Water Manag.* **2014**, *135*, 61–72. [CrossRef]

38. Mohammadi, M.; Ghahraman, B.; Davary, K.; Ansari, H.; Shahidi, A.; Bannayan, M. Nested validation of AquaCrop model for simulation of winter wheat grain yield soil moisture and salinity profiles under simultaneous salinity and water stress. *Irrig. Drain.* **2016**, *65*, 112–128. [CrossRef]

39. Hatfield, J.L. Increased temperatures have dramatic effects on growth and grain yield of three maize hybrids. *Agric. Environ. Lett.* **2016**, *1*, 1–5. [CrossRef]

40. Hsiao, T.C.; Heng, L.; Steduto, P.; Roja-Lara, B.; Raes, D.; Fereres, E. AquaCrop—The FAO model to simulate yield response to water: Parametrization and testing for maize. *Agron. J.* **2009**, *101*, 448–459. [CrossRef]

41. Wiegmanna, M.; William, T.B.; Thomasb, H.J.; Bullb, I.; Andrew, J.; Flavellc, J.; Annette, Z.; Edgar, P.; Klaus, P.; Andreas, M. Wild barley serves as a source for biofortification of barley grains. *Plant Sci.* **2019**, *283*, 83–94. [CrossRef] [PubMed]

42. Roberts, D.P.; Mattoo, A.K. Sustainable agriculture—Enhancing environmental benefits, food nutritional quality and building crop resilience to abiotic and biotic stresses. *Agriculture* **2018**, *8*, 8. [CrossRef]

43. Tavakoli, A.R.; Moghadam, M.M.; Sepaskhah, A.R. Evaluation of the AquaCrop model for barley production under deficit irrigation and rainfed condition in Iran. *Agric. Water Manag.* **2015**, *161*, 136–146. [CrossRef]

44. Eisenhauer, J.G. Regression through the origin. *Teach. Stat.* **2003**, *25*, 76–80. [CrossRef]

45. Teixeira, A.D.C.; Bassoi, L.H. HBassoi Crop Water Productivity in Semi-arid Regions: From Field to Large Scales. *Ann. Arid Zone* **2009**, *48*, 1–13.

46. FAO. The Irrigation Challenge. In *Increasing Irrigation Contribution To Food Security through Higher Water Productivity from Canal Irrigation Systems*; Issue paper; FAO: Rome, Italy, 2003.

47. Barnston, A. Correspondence among the Correlation [root mean square error] and Heidke Verification Measures; Refinement of the Heidke Score. *Notes Corresp. Clim. Anal. Cent.* **1992**, *7*, 699–709.

 *agronomy*

Article

Impact of Partial Root Drying and Soil Mulching on Squash Yield and Water Use Efficiency in Arid

Abdulhalim H. Farah [1], Hussein M. Al-Ghobari [1], Tarek K. Zin El-Abedin [2], Mohammed S. Alrasasimah [1] and Ahmed A. El-Shafei [1,2,*]

[1]   Agricultural Engineering Department, College of Food and Agricultural Sciences, King Saud University, Riyadh 11451, Saudi Arabia; enghersi61@gmail.com (A.H.F.); hghobari@ksu.edu.sa (H.M.A.-G.); m.alrsasmah1@gmail.com (M.S.A.)
[2]   Agricultural Engineering Department, Faculty of Agriculture, Alexandria University, Alexandria 21545, Egypt; drtkz60@gmail.com
*   Correspondence: aelshafei1bn.c@ksu.edu.sa; Tel.: +966-11-4678504

Abstract: Practical and sustainable water management systems are needed in arid regions due to water shortages and climate change. Therefore, an experiment was initiated in winter (WS) and spring (SS), to investigate integrating deficit irrigation, associated with partial root drying (PRD) and soil mulching, under subsurface drip irrigation on squash yield, fruit quality, and irrigation water use efficiency (IWUE). Two mulching treatments, transparent plastic mulch (WM) and black plastic mulch (BM), were tested, and a treatment without mulch (NM) was used as a control. Three levels of irrigation were examined in a split-plot design with three replications: 100% of crop evapotranspiration (ETc), representing full irrigation (FI), 70% of ETc (PRD70), and 50% of ETc (PRD50). There was a higher squash yield and lower IWUE in SS than WS. The highest squash yields were recorded for PDR70 (82.53 Mg ha^{-1}) and FI (80.62 Mg ha^{-1}). The highest IWUE was obtained under PRD50. Plastic mulch significantly increased the squash yield (34%) and IWUE (46%) and enhanced stomatal conductance, photosynthesis, transpiration, leaf chlorophyll fluorescence, and leaf chlorophyll contents under PRD plants. These results indicate that in arid and semi-arid regions, soil mulch with deficit PRD could be used as a water-saving strategy without reducing yields.

Keywords: squash; partial root drying; water use efficiency; soil mulch; growing seasons; gas exchange; fruit quality

Citation: Farah, A.H.; Al-Ghobari, H.M.; Zin El-Abedin, T.K.; Alrasasimah, M.S.; El-Shafei, A.A. Impact of Partial Root Drying and Soil Mulching on Squash Yield and Water Use Efficiency in Arid. *Agronomy* **2021**, *11*, 706. https://doi.org/10.3390/agronomy11040706

Academic Editor: Aliasghar Montazar

Received: 25 February 2021
Accepted: 4 April 2021
Published: 7 April 2021

Publisher's Note: MDPI stays neutral with regard to jurisdictional claims in published maps and institutional affiliations.

1. Introduction

Increasing the consumption of water in the agricultural sector, and a lack of preventative measures to permanently conserve water and avoid water shortages, make it vital to manage water resources rather than develop new ways to supply water. Therefore, the need to develop practical and sustainable management systems for water supply has become a subject of intense discussion. Drip irrigation is a promising irrigation strategy that reduces soil evaporation and deep drainage losses, while efficiently delivering water to plant roots [1]. Compared with conventional methods, drip irrigation has shown its utility for water-saving and the efficient use of fertilizers, especially fruit and vegetable crops [2].

Various methods are currently used to increase the efficiency of delivering water to plants. One of these is subsurface drip irrigation (SSDI), which is primarily utilized to decrease water loss during water delivery to plants. Compared with other drip irrigation methods, SSDI has gained more acceptance in the irrigation sector in its ability to increase crop yield and reduce plant diseases and soil erosion [3–5]. Other methods that are used to efficiently managing water irrigation include deficit irrigation (DI) and soil mulching. Ever since the focus of water irrigation shifted from increasing yield per planted area to increasing yield per unit volume of water applied [6], DI has become an important strategy in arid and semi-arid regions where water shortages are a major limitation to farming.

Agronomy **2021**, *11*, 706. https://doi.org/10.3390/agronomy11040706   https://www.mdpi.com/journal/agronomy

Therefore, the optimal goal of using DI is to save water, either by reducing the number of irrigation cycles or reducing the volume of water applied during each irrigation event [7]. Irrigation water use efficiency (IWUE) has been developed to indicate increasing crop yield while using less water, or maximizing yield in limited water sources [8]. IWUE was defined by the total yield to the total water applied [9,10]. In a physiological perspective, IWUE is used to describe the amount of carbon to the water lost through transpiration [11]. However, agronomists primarily focus on maximizing yield per water applied [12].

Partial root drying (PRD) is an improved form of DI strategy that involves applying water to the sides of a plant root zone, either by irrigating one side of the plant root (fixed PRD) or alternately watering both sides of the root (alternating PRD) [13]. Adequate water and nutrients are delivered to the plant on the wet side of the root, while the dry side is stimulated and releases chemical hormones. These chemical hormones cause stomata to partially close, which increases IWUE [13]. This strategy makes PRD more efficient than DI [13–15]. Barideh et al. [16] reported that alternating PRD saves more water than fixed PRD. Several studies have shown the advantages of DI and PRD over full irrigation (FI), in terms of IWUE without the reduction of yield [17–20]. A number of researchers working on different crops found that the PRD strategy increased IWUE by 38–53% compared with FI without a significant reduction in yield [18,21,22]. Ors et al. [23] reported that deficit irrigation (67%) had significantly reduced chlorophyll index value (7%), leaf water content (42%), stomatal conductance (69%), transpiration (62%), photosynthesis (62%) of squash. Al-Ghobari and Dewidar [24] indicated that deficit irrigation significant affected the fresh leaf, stem weight of tomato, compared with full irrigation. In terms of fruit quality, PRD preserved fruit quality, compared with deficit irrigation. Zhang et al. [25] reported that PRD was not affected by soluble solid contents of strawberry, while deficit irrigation considerably decreased soluble solid contents. Guang-Cheng et al. [26] indicated that both PRD50 and DI50 strategies considerably decreased dry weigh of shoot and root pepper compared with full irrigation. Furthermore, PRD50 reduced photosynthesis 19% while DI50 decreased 22%. In chlorophyll fluorescence (FV/Fm) PRD50 reduced by 9.5% while DI50 decreased 12.0%.

Another method that can increase IWUE is soil mulching, which can be used for many purposes in the agriculture sector. However, preserving soil moisture, improving soil physical properties, and controlling soil erosion are the most significant uses of soil mulching in arid and semi-arid regions [27,28]. Mulching materials positively affect the moisture of the soil by improving soil structure and soil retention, as well as decreasing soil evaporation [27,29–31]. Yaghi et al. [32] reported that combining drip irrigation with plastic mulch increased cucumber yield (45%) and IWUE (72%) compared with the treatment without mulch during two successive growing seasons in the arid region. Abhivyakti et al. [33] found that black plastic mulch increased the tomato yield by 30% compared with bare soil in open field conditions. Abd El-Mageed et al. [34] also reported that soil mulching increased both, the squash yield and IWUE by 26%, compared with the non-mulched treatments. Experimenting on broccoli, Verma et al. [35] observed that mulching increased the photosynthetic rate, stomatal conductance, intercellular CO_2 concentration, and transpiration rate. Additionally, Lira-Saldivar et al. [36] found that plastic mulch significantly increased photosynthetic activity in zucchini plants (17.9%) compared with non-mulched treatments.

In addition to water, growing season also influences both crop yield and IWUE. Numerous studies on zucchini squash have reported that growing season has a significant effect on crop yield and IWUE [9,34,37,38].

Despite numerous studies that have been conducted on SSDI, PRD irrigation and soil mulching, as water management strategies, combined with arid regions with different growing seasons, has not been fully investigated. Therefore, this study aimed to investigate the effect of DI levels, associated with PRD strategy and plastic mulch on squash yield and IWUE in winter and spring. This study also examined the combined effects of PRD, soil mulch, and growing season on gas exchanges, chlorophyll fluorescence, and the chlorophyll content index at different plant growth stages.

2. Materials and Methods

2.1. Experimental Design and Growth Conditions

Two experiments were conducted in two consecutive growing seasons: the winter season (WS) and spring season (SS) of 2018–2019 at the Educational farm, King Saud University, Riyadh, which is in an arid area. A meteorological station was set up to constantly measure weather parameters, namely, air temperature, relative humidity, solar radiation, evapotranspiration, and rainfall throughout the WS and SS (Figures 1 and 2). Field preparations were made, including plowing, grading, and leveling. Then, the irrigation layout was implemented according to the experimental design, as shown in Figure 3.

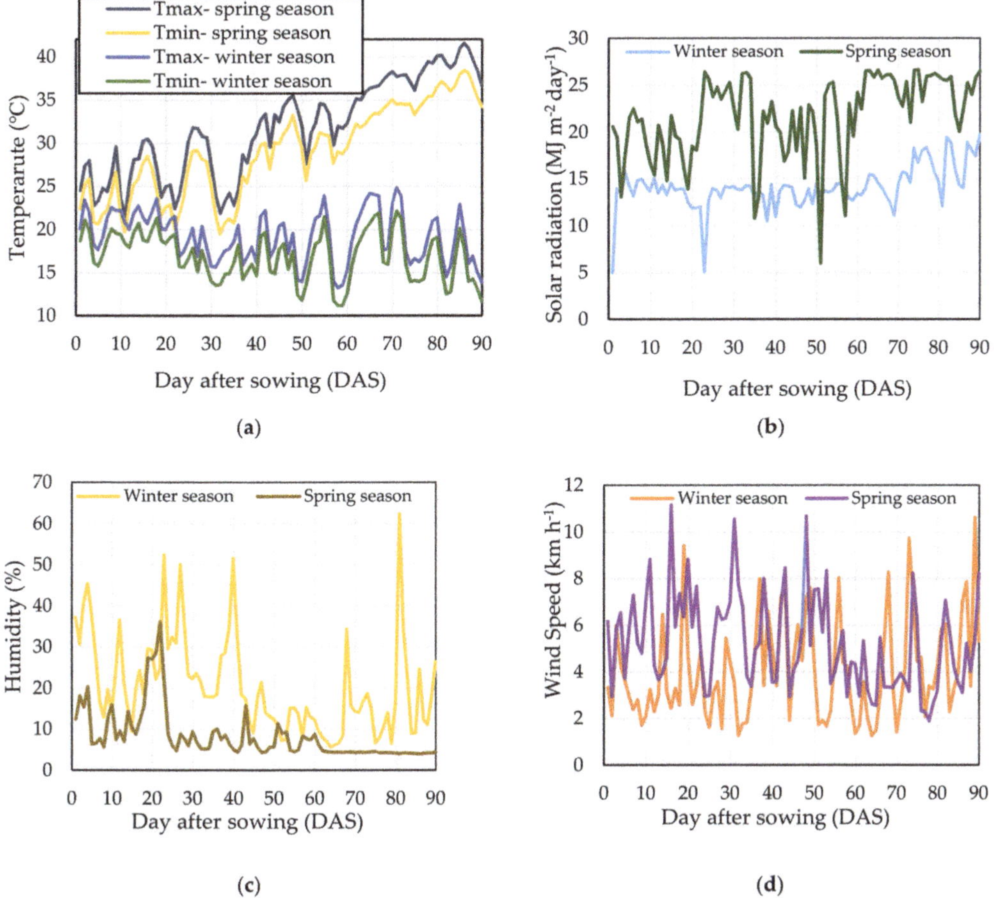

Figure 1. Daily climate parameters in the winter and spring of 2018–2019 during the squash growing seasons: (**a**) Daily maximum and minimum temperature, (**b**) solar radiation, (**c**) relative humidity, and (**d**) wind speed.

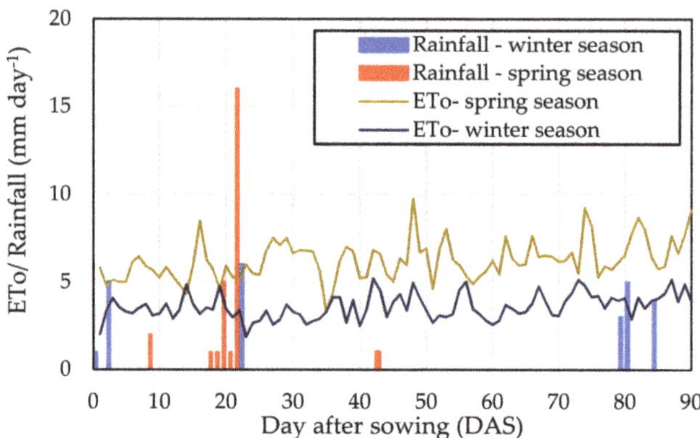

Figure 2. Seasonal reference evapotranspiration (ETo) at the experimental field throughout the winter and spring growing seasons.

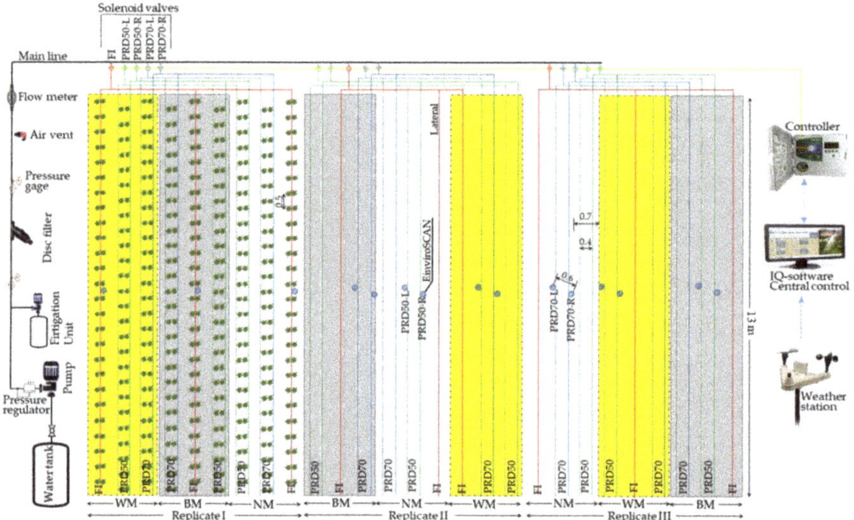

Figure 3. Schematic diagram of the experimental fields under mulch treatments (black mulch-BM, transparent mulch-WM and non-mulch- NM) and irrigation treatments (full irrigation-FI, partial root drying with 50% of evapotranspiration-PRD50, partial root drying with 70% of evapotranspiration-PRD70).

Soil physical and chemical analyses were conducted by taking soil samples every 0.1 m down to a depth of 0.5 m, as shown in Table 1. Soil physical parameters were determined, including the field capacity (FC), wilting point (WP), saturated hydraulic conductivity (ks), bulk density (ρ_b), and soil saturation (S). The experiment was conducted in a split-plot design (Figure 3). Treatments were allocated three levels of irrigation and three mulching treatments. The mulching treatments, transparent mulch (WM), black mulch (BM), and without mulch (NM), were assigned as main plots, and the irrigation treatments, FI with 100% of crop evapotranspiration (ET_c), irrigation with 70% of ET_c (PRD70), and irrigation with 50% of ET_c (PRD50%) were allocated in subplots. The experimental plot area was 13 m in length by 0.70 m in row width (9.1 m^2). A total of 27 plots were made by replicating each treatment three times.

Table 1. Soil physical and chemical properties.

Depth (cm)	Particle Size (%)			Texture	FC %	WP %	k_s (mm/h)	S %	ρ_b (g cm^{-3})
	Sand	Silt	Clay						
0–10	82.90	8.80	8.30	sandy loam	22.11	5.53	48.06	38.15	1.40
10–30	74.35	16.85	8.80	sandy loam	21.30	4.72	18.10	35.00	1.51
30–50	70.32	20.8 8	8.80	sandy loam	22.44	4.46	11.39	33.17	1.57

Depth (cm)	pH	Cation (meq L^{-1})				Anions (meq L^{-1})			
		Ca^{2+}	Mg^{2+}	Na^+	K^+	$HCO_3{}^-$	$CO_3{}^{2-}$	Cl^-	$SO_4{}^{2-}$
0–10	7.56	2.95	0.95	1.98	0.39	1.25	0.00	2.45	2.35
10–30	7.47	3.73	0.59	3.85	0.44	1.28	0.00	3.10	3.45
30–50	7.35	4.40	0.98	4.78	0.73	1.78	0.00	4.00	4.48

FC: field capacity; WP: wilting point; k_s: saturated hydraulic conductivity; S: soil saturation; ρ_b: bulk density.

2.2. Applied Irrigation Water

Drip pipes were buried 15 cm below the soil surface and had 26 inline emitters, which were spaced at intervals of 0.5 m, and had a discharge rate of 8 L h^{-1} at an operating pressure of 100 kPa. In the FI experimental plot, one lateral was installed adjacent to the crop rows, while in the PRD treatments, two laterals with two control valves were installed 0.4 m apart in each crop row. Irrigation in the PRD treatment was shifted between the two sides of plants every five days.

A weather station (WS-PRO LT Weather Station, Rain Bird) was launched in the experiment field. Daily reference evapotranspiration (ET_o) was calculated from daily climate data according to Allen et al. [39] using Equation (1),

$$ET_O = \frac{0.408\Delta(R_n - G) + \gamma\frac{900}{T+273}u_2\left(e_S - e_a\right)}{\Delta + \gamma(1 + 0.34u_2)} \qquad (1)$$

where ET_o is reference crop evapotranspiration (mm day^{-1}), R_n is net radiation at the crop surface (MJ m^{-2} day^{-1}), G is soil heat flux density (MJ m^{-2} day^{-1}), T is mean daily temperature at 2 m height (°C), u_2 is wind speed at 2 m height (m s^{-1}), e_s is saturation vapor pressure (kPa), e_a is actual vapor pressure (kPa), Δ is the slope of the vapor pressure curve (kPa °C^{-1}), and γ is a psychrometric constant (kPa °C^{-1}).

Irrigation was conducted every day using an automatic controller (ESP-LXME controllers, Rain Bird Corporation, Tucson, AZ, USA), which was connected with a central control (IQ v2.0, Rain Bird Corporation, Azusa, CA, USA). The IQ-software monitored and adjusted watering schedules for the controller and site from a compatible Windows PC, which was connected with the weather station to schedule irrigation automatically based on ET_c. The crop water requirements (ET_c) were estimated using Equation (2),

$$ET_c = ET_o \times K_c \qquad (2)$$

where ET_c is the crop water requirement (crop evapotranspiration; mm day^{-1}), and K_c is the crop coefficient. The crop growth stages, initial, development, mid, and late stage, were 20, 30, 25, and 15 days, respectively, and K_c of 0.6, 1.0, and 0.75 were used for the initial, mid, and late stage, respectively [39]. Moreover, the values of K_c were adjusted according to Allen et al. [39] based on the relative humidity, wind speed at 2 m, percentage of wetted soil surface in the experimental field using Equations (3) and (4),

$$K_{c\ ini} = f_w\, K_{c\ ini(Table)} \qquad (3)$$

where $K_{c\,ini}$ is the adjusted value of initial K_c, f_w is the fraction of surfaced wetted by irrigation, and $K_{c\,ini(Table)}$ is the value of initial K_c from Allen et al. [39].

$$K_{c\,mid\,OR\,end} = K_{c\,mid\,OR\,end(Table)} + [0.04(u_2 - 2) - 0.004(RH_{min} - 45)]\left(\frac{h}{3}\right)^{0.3} \tag{4}$$

where $K_{c\,mid\,OR\,end}$ is the adjusted values of mid K_c or end K_c, $K_{c\,mid\,OR\,end(Table)}$ is the value of of mid K_c or end K_c from Allen et al. [39], RH_{min} is the mean value for daily minimum relative humidity during the mid-season growth stage or the end-season stage [%], and h is mean plant height during the mid-season stage or the end-season stage [m].

At 20 days after sowing (DAS), PRD70 and PRD50 were applied until harvesting.

2.3. Plant Management

Two seeds of zucchini squash, *Cucurbita pepo* L., were hand-sown 10 cm apart on both sides of the central line of the planting rows, and there was 0.5 m between plants within a row. Seeds were planted on November 18, 2018 and terminated on 15 February 2019 in the WS, and in the SS, seeds were planted on 23 March 2019 and terminated on 20 June 2019. Chemical fertilization was applied at the recommended rate for squash production in this area: 5.1 g N/plant, 5.1 g P_2O_5/plant, 16.8 g K_2O/plant, 37.5 g $Ca(NO_3)_2$/plant, 28.5 mL H_3PO_4/plant, 14.52 mL HNO_3/plant, and 1.41 g humic acid/plant. Pest management and disease control were conducted based on local squash protection procedures.

2.4. Soil Moisture Measurements

Capacitance probes (EnviroSCAN®, Sentek Sensor Technologies, Stepney, Australia) were used to measure soil moisture. Enviroscan probes were used to continuously monitor volumetric soil water content (θ_v) down to a depth of 0.5 m in the root zone of each irrigation treatment. Probes were installed vertically at a distance of 0.10 m from laterals and had five sensors at 0.10 m intervals. Soil frequencies (F_s) were converted into scaled frequencies (S_f) according to Equation (5) following Buss [40],

$$S_f = \frac{F_A - F_S}{F_A - F_W} \tag{5}$$

where F_A is the sensor reading in the air, F_S is the sensor reading in the soil, and F_W is the sensor reading in non-saline water. According to Vera et al. [41], θ_v can be calculated using Equation (6),

$$\theta_v = \left(\frac{S_f - C}{A}\right)^{\frac{1}{b}} \tag{6}$$

where $A = 0.1957$, $b = 0.404$, and $C = 0.02852$. The determination coefficient value provided using standard default calibration was 0.97. One EnviroScan device per plot was installed in single lateral treatments (FI), while two EnviroScan devices were installed 0.6 m apart in the diagonal direction in the two lateral treatments: PRD70 and PRD50 (Figure 3).

2.5. Physiological and Agronomic Measurements

Chlorophyll index (soil-plant analysis development (SPAD) value) and gas exchange measurements, including stomatal conductance (g_s), photosynthesis (P_n), and transpiration rate (T_r), were measured at three different growth stages: development (35 DAS), mid (63 DAS), and late stage (83 DAS). One leaf (of the same age) was selected per plant from five plants per plot. A total of 15 measurements per treatment were made at every growth stage.

The chlorophyll index (SPAD value) was measured using a SPAD 502 Plus Chlorophyll Meter (Minolta Co. Ltd., Osaka, Japan). Using a chlorophyll meter is a non-destructive method that quickly and precisely approximates the chlorophyll concentration of leaves by measuring the red (650 nm) and infrared (940 nm) radiation of leaves [42]. The sample readings were made for every plot using the center section of the selected leaf at all measured growth stages.

The gas exchange measurements g_s, P_n, and T_r were measured using an LI-6400XT portable photosynthesis system (LiCor Inc., Lincoln, NE, USA). The samples were measured for each treatment from functional leaves on a cloudless day from 08h00 to 10h00 local time.

Total fresh squash yields (Mg ha^{-1}) were determined by manually collecting and weighing fruits from each line for all harvested squash fruits. The irrigation water use efficiency (IWUE) was calculated by dividing the total weight of harvested squash fresh fruits (kg ha^{-1}) by the volume of water applied to the crop (m^3 ha^{-1}) [9,10].

The fruit quality parameters, total soluble solids (*TSS*, %), vitamin C (*V_C*, mg 100 g^{-1} fruit fresh weight-FW), and titratable acidity (*TA*, % citric acid), were assessed by choosing samples of three mature fruits in the third, fifth, and seventh harvestings per treatment in each growing season. A squash extract was taken by blending and filtering the flesh of each fruit. A digital refractometer (PR-101 model, ATAGO, Tokyo, Japan) was used to determine the TSS using standard methods of analysis [43], while TA was determined using the procedure, described by Caruso et al. [44]. 2,6-dichlorophenol-indophenol-dye was used to measured Vc in the extracted juice [45].

2.6. Statistical Analysis

Statistical analysis was conducted using analysis of variance procedures using CoStat version 6.451 [46]. The difference between means was compared using a least significant difference test (LSD) at the 5% level ($p \leq 0.05$).

3. Results and Discussion

3.1. Evapotranspiration and Applied Irrigation

There was variation in the weather parameters of the WS and SS (Figure 1). Air temperature, solar radiation intensity, and wind speed were higher in the SS than the WS. However, in the WS, the relative humidity was higher than in the SS. This caused a 73% increase in the seasonal reference evapotranspiration (ET$_o$) in the SS, compared with the WS (Figure 2). As irrigation scheduling was based on ET$_c$, more water was consumed in the SS than the WS (Figure 4).

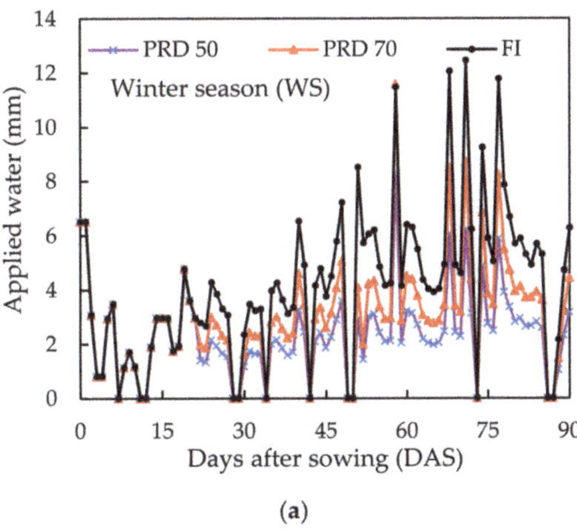

(a)

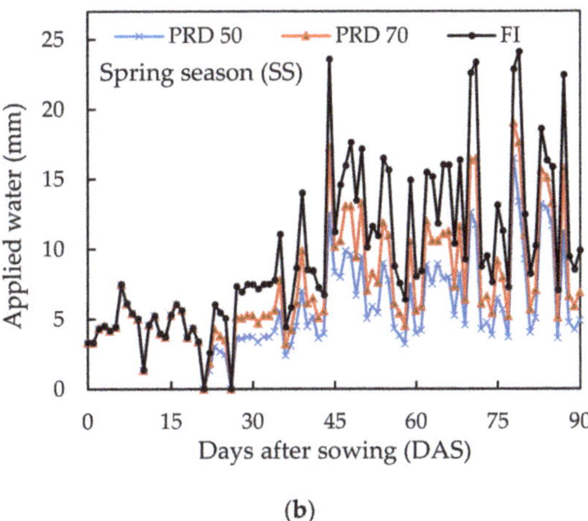

(b)

Figure 4. Seasonal water application to zucchini squash crops for the two growing seasons: (a) Winter season, and (b) spring season.

3.2. Soil Moisture Content

Figure 5 shows the different patterns of soil moisture distribution in their response to FI, PRD70, and PRD50, combined with WM, BM, and NM during the WS and SS. The values presented for volumetric soil moisture content (θ_v) are an average of 0.1, 0.2, 0.3, 0.4, and 0.5 m soil depths. Before irrigation treatments (20 DAS), the θ_v for all treatments was almost the same for each growing season. Irrigation was scheduled based on ET, and this caused a variation in θ_v between the two seasons. In the WS, the θ_v for FI was below the FC compared with the SS when the θ_v of FI was almost near the FC. The average θ_v in the SS was higher by 16%, 22%, and 32% for BM, WM, and NM, respectively, than the corresponding values in WS. This was primarily due to the applied water in the SS, which was higher than in the WS (Figure 4). For the mulch treatments, WM and BM showed higher θ_v than NM. The increased moisture retention capacity of the mulched treatments in the two growing seasons could be attributed to less evaporation from the soil, as shown in Figure 5. Besides, vapor accumulation from irrigated water trapped within the mulches cause the formation of fog, which precipitates back into the soil. These findings are in agreement with Yaghi et al. [32] and Rashid et al. [47], who found that mulched treatments showed higher soil moisture content compared to non-mulched treatments. The θ_v values of the PRD treatments showed alternately an increase in the wet side (right) of the root zone, while the dry side (left) showed a reduction in soil moisture content, as shown in Figure 5. The wet side of the root zone delivers water to the plant, while the dry side improves root ventilation. In PRD70, θ_v was between the FC and WP. However, in PRD50, θ_v was below the WP, and this had a negative impact on plant growth. The patterns of soil water dynamics in PRD-treated plants in this study were similar to those described by Barideh et al. [16] and Rashid et al. [47], who found that the soil water content in PRD treatments increased and decreased interchangeably.

3.3. Stomatal Conductance (g_s), Photosynthesis (P_n), and Transpiration (T_r)

Data in Table 2 show that irrigation quantity significantly ($p < 0.001$) affected the values of g_s at all growth stages. At 83 DAS, PRD70 and PRD50 reduced the value of g_s by 10% and 37% compared with FI, respectively. This is due to plant age, which reduces its activity. FI showed a higher P_n rate compared with the PRD treatments. At 63 DAS, P_n values under the FI plot were 10.685 μmol m^{-2} s^{-1}. This is a 7% and 16% increase compared with PRD70, and PRD50, respectively. T_r was significantly affected ($p < 0.001$) by irrigation quantity at all measured days. The lowest T_r values were observed under PRD50 treatments at 63 DAS.

However, FI treatments showed the highest T_r (4.046 mmol m^{-2} s^{-1}) at 35 DAS, which was not statistically different from PRD70. This finding indicates that the water deficit in PRD70 did not affect transpiration efficiency. At 83 DAS, PRD50 and PRD70 reduced the T_r values by 20% and 17.6%, respectively, compared with the FI treatments. In this study, the irrigation treatment significantly affected the g_s, P_n, and T_r values, indicating that both P_n and T_r are controlled by g_s, and they mutually affect each other [48,49]. Liu et al. [50] stated that under water-stressed conditions, g_s decreases due to the closure of stomata to maintain leaf water status. However, there are opposing reports on the mechanism behind stomatal closure [51]. Although some studies suggest that chemical signals, such as abscisic acid (ABA) and pH are behind stomatal closure [50,52], others endorse that hydraulic signals, such as soil, root, and shoot resistances, are responsible for stomatal closure [53]. Many questions still arise related to the mechanism behind stomatal closure, even though many studies have been conducted [51]. Farooq et al. [54] stated that stomatal closure reduces the amount of carbon dioxide going into the parenchyma cells, which causes inhabitation of CO_2 and light that ultimately affects plant photosynthesis efficiency.

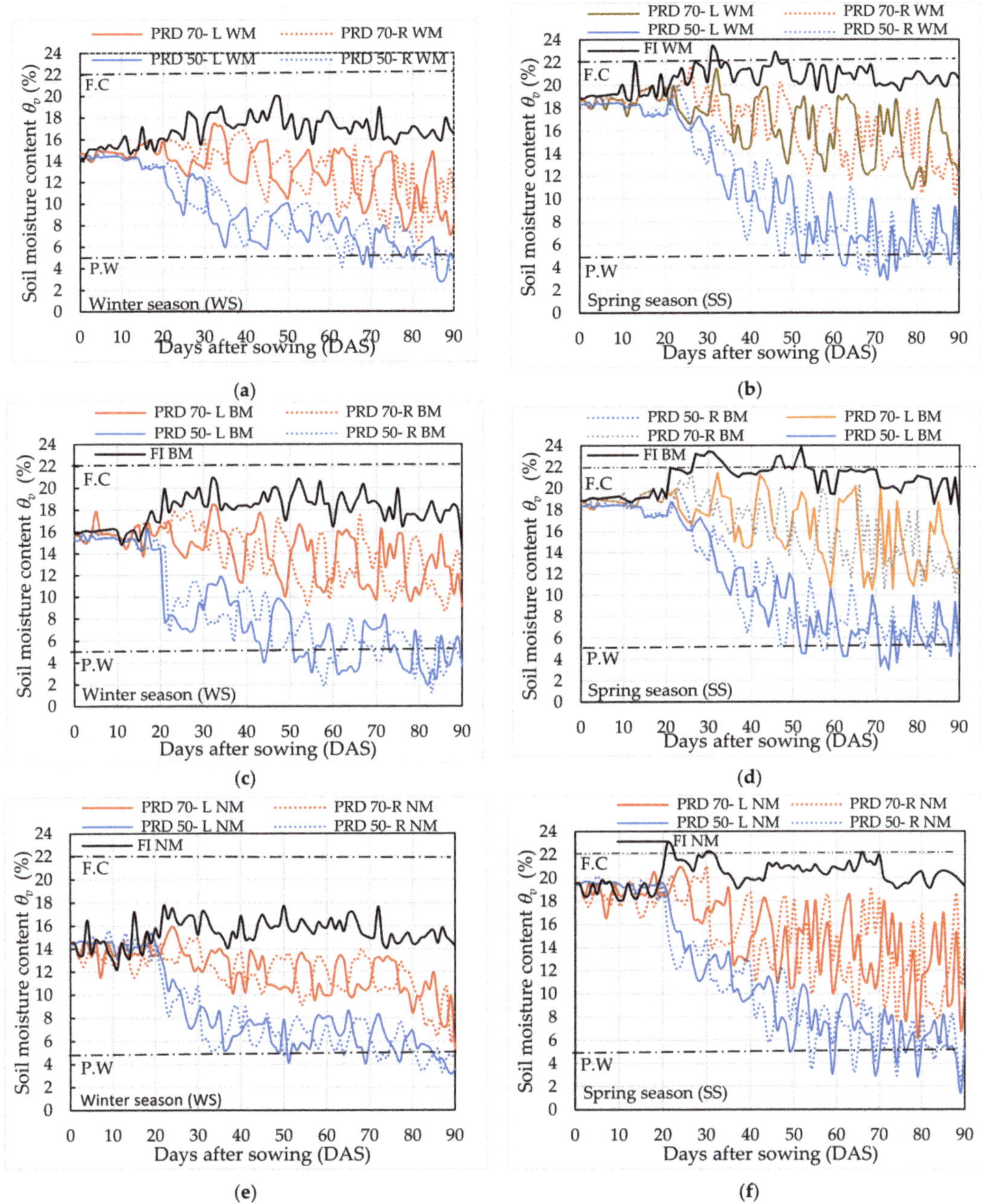

Figure 5. Soil moisture for full irrigation (FI) and deficit partial rootzone drying under 70% and 50% of evapotranspiration (PRD70 and PRD50, respectively) at the left (L) and right (R) rootzone sides combined with: (**a**) Transparent mulch (WM) during the winter season (WS), (**b**) WM during the spring season (SS), (**c**) black mulch (BM) during the WS, (**d**) BM during the SS, (**e**) no mulch (NM) during the WS, and (**f**) NM during SS. FC, field capacity, and WP, wilting point.

Table 2. Analysis of variance of stomatal conductance (g_s), photosynthesis (P_n), and transpiration (T_r) at the development stage (35 DAS), mid stage (63 DAS), and late stage (83 DAS) of squash growth during the winter (WS) and spring (SS) growing seasons.

Treatments		g_s (mol H_2O m^{-2} s^{-1})			P_n (μmol CO_2 m^{-2} s^{-1})			T_r (mmol H_2O m^{-2} s^{-1})		
		35 DAS	63 DAS	83 DAS	35 DAS	63 DAS	83 DAS	35 DAS	63 DAS	83 DAS
Season (S)	**WS**	0.2912 [b]	0.2192 [b]	0.1722 [b]	8.647 [b]	9.335 [b]	9.1668 [b]	3.115 [b]	2.804 [b]	2.887 [b]
	SS	0.3881 [a]	0.3182 [a]	0.2122 [a]	9.657 [a]	10.414 [a]	10.381 [a]	4.285 [a]	3.952 [a]	3.836 [a]
	p-value	0.0059 **	0.0006 **	0.0126 *	0.023 *	0.0465 *	0.0172 *	0.014 *	0.0002 **	0.0039 **
	LSD0.05	0.032	0.0107	0.019	0.678	1.037	0.694	0.602	0.072	0.254
Mulch (M)	**WM**	0.3445	0.3004 [a]	0.2257 [a]	10.176 [a]	9.961 [b]	10.256 [a]	4.046 [a]	3.812 [a]	3.852 [a]
	BM	0.3401	0.2660 [b]	0.1818 [b]	9.778 [a]	10.461 [a]	10.088 [a]	3.901 [a]	3.359 [b]	3.173 [b]
	NM	0.3262	0.2397 [c]	0.1690 [c]	7.502 [b]	9.200 [c]	8.976 [b]	3.150 [b]	2.963 [c]	3.061 [b]
	p-value	0.0527 [ns]	0.00001 **	0.0001 **	0.00001 **	0.00001 **	0.0001 **	0.00001 **	0.00001 **	0.0001 **
	LSD0.05	–	0.014	0.0095	0.451	0.151	0.273	0.169	0.093	0.132
Irrigation (I)	**FI**	0.375 [a]	0.3222 [a]	0.2284 [a]	10.562 [a]	10.685 [a]	10.333 [a]	4.044 [a]	3.826 [a]	3.603 [a]
	PRD70	0.333 [b]	0.2557 [b]	0.2051 [a]	9.0196 [b]	9.972 [b]	9.594 [b]	3.792 [a]	3.272 [b]	3.414 [b]
	PRD50	0.310 [c]	0.2283 [c]	0.1431 [b]	7.875 [c]	8.965 [c]	9.394 [b]	3.265 [b]	3.036 [c]	3.069 [c]
	p-value	0.0001 **	0.00001 **	0.00001 **	0.00001 **	0.00001 **	0.0005 **	0.00001 **	0.00001 **	0.00001 **
	LSD0.05	0.0149	0.016	0.028	0.416	0.313	0.44	0.261	0.092	0.142
S $\times$ M	*p*-value	0.9732 [ns]	0.68 [ns]	0.013 *	0.99 [ns]	0.971 [ns]	0.936 [ns]	0.664 [ns]	0.404 [ns]	0.0251 *
S $\times$ I	*p*-value	0.9948 [ns]	0.99 [ns]	0.85 [ns]	0.99 [ns]	0.989 [ns]	0.909 [ns]	0.924 [ns]	0.438 [ns]	0.0066 **
M $\times$ I	*p*-value	0.4528 [ns]	0.15 [ns]	0.23 [ns]	0.0002 **	0.011 *	0.171 [ns]	0.851 [ns]	0.0005 **	0.00001 **
S $\times$ M $\times$ I	*p*-value	0.9980 [ns]	0.99 [ns]	0.91 [ns]	1 [ns]	0.999 [ns]	0.999 [ns]	0.994 [ns]	0.788 [ns]	0.00001 **

WS: winter season; SS: spring season; WM: transparent mulch; BM: black mulch; NM: no mulch FI: full irrigation; PRD70 and PRD50: deficit partial rootzone drying under 70 and 50 of evapotranspiration, respectively; S, M and I: season, mulch and irrigation treatments, respectively; S $\times$ M $\times$ I: interaction between season, mulch and irrigation treatments; [ns]: not statistically significant; **: significant at the 1% level ($p < 0.01$); *: significant at the 5% level ($p < 0.05$); different letters indicate significant difference between treatments; bold letters and words indicate treatments names.

Our results indicated that P_n decreased with a decrease in g_s at the same stage of plant growth, but in different growth stages, a decrease in g_s did not cause a decrease in P_n. For instance, at the mid stage (63 DAS), the P_n values increased despite g_s reduction in both the mulch and irrigation treatments for the two growing seasons (Table 2). This could be explained by the squash leaves having reached their maximum area at this stage, when plants reach their peak values of most photosynthetic parameters [55]. The leaf g_s, P_n, and T_r in the PRD treatments were significantly lower than that of FI at all measured days. In the PRD treatment, two sides of the root were alternately irrigated. The side of the root that undergoes a water deficit for a period induces ABA, which reduces g_s, affecting both transpiration and photosynthesis efficiencies. However, the watered side of the root keeps the plant in a preferable situation [13]. In the current study, due to water stress under PRD treatment, plants induced ABA from the root to the leaves, resulting in the accumulation of ABA in the leaves causing stomatal closure [50,52]. Several studies showed that plants under PRD could enhance leaf T_r [56] and improve the P_n rate [57] compared to FI. These results are in agreement with other studies [5,58], which indicated that g_s decreases with increasing water stress levels.

The g_s, P_n, and T_r were significantly affected by mulching treatments. However, at 35 DAS, g_s showed no significant difference ($p > 0.05$) between mulch and non-mulched treatments. At 63 DAS, compared with NM, WM and BM increased g_s by 20% and 10%, respectively, indicating that plants under mulched treatments were healthier at the mid growth stage. At 83 DAS, the NM treatment reduced the P_n value by 11% and 12.5% compared with the BM and WM treatments, respectively (Table 2). At 35 DAS, the non-mulched treatments reduced the T_r by 19% and 22% compared with the BM and WM treatments, respectively. Our results agree with the findings described by Ibarra-Jiménez et al. [59] and Lira-Saldivar et al. [36], who found that plastic mulch significantly increased photosynthetic activity in zucchini plants compared with non-mulched treatments. This finding is due to the advantage of plastic mulch, which can control soil temperature, enhance soil moisture, and elevate crop photosynthesis [60]. Yang et al. [61] and Zhang et al. [62] emphasized that soil hydrothermal state is an essential element in photosynthesis. The proper soil moisture and temperature situation under mulched treatments boost the movement of water from the deep soil to the surface soil by capillary and steam action, increasing the intercellular CO_2 concentration in the ear-leaf [62]. These activities help increase carbon sources for leaf photosynthesis, thereby decreasing the limitations of stomatal factors [63] and leading to consistently higher P_n in mulched than non-mulched treatments.

Data in Table 2 indicate that g_s, P_n, and T_r were significantly affected by growing season for all measured days. The highest g_s was 0.3881 and 0.2912 mmol m^{-2} s^{-1} in the SS, and WS at 35 DAS, respectively. The Tr in SS and WS followed the same trend as g_s; the highest T_r was observed at 35 DAS. At 63 DAS, the P_n value in the SS increased by 10%, compared with in the WS. Urban et al. [64] revealed that high temperatures affect all physiological processes in plants. Furthermore, Jones et al. [65] and Scherrer et al. [66] asserted that environmental factors, such as radiation, air temperature, and wind, affect the size of the stomata aperture. In this study, the physiological trend (g_s, P_n, and T_r) could be explained by the environmental differences between the two growing seasons, where the SS had higher air temperature, radiation, and wind speed than the WS (Figure 1).

The effects of the growing season, mulch treatment, and irrigation quantities on g_s, P_n, and T_r were significant at all squash stages (Table 2). This finding indicates that sowing squash during a suitable growing season and choosing a suitable combination of irrigation volume and plastic mulch could enhance squash physiological response, which would ultimately increase the yield and IWUE.

The g_s was not significantly affected by interactions between growing season, mulch, and irrigation quantities, as shown in Table 2. However, at 83 DAS, the interaction between season and mulch showed a significance difference ($p < 0.05$). No interaction effect on P_n was observed, except for interaction between irrigation and mulch, which significantly affected ($p < 0.05$) P_n values at 35 DAS and 63 DAS. In 35 DAS, comparing with same irrigation strategies FI, BM increased P_n 3% and 20%, respectively compared with WM and NM. In PRD70, P_n values under BM and WM were not different, while BM and WM enhanced P_n values 37%, and 36%, respectively, compared with NM. in PRD50, BM increased P_n 21%, 39%, compared with WM and NM, respectively. Data in Table 2 indicate that there was no significant interaction between growing season, mulch, and irrigation on Tr, except after 83 DAS. This indicates that squash plants were not able to withstand environmental changes at a late stage of growth, and there was a water deficit due to the age of the plants. Tr values were reduced under PRD strategies compared with FI for both mulched and non-mulched treatments. At 63 DAS, WM increased Tr 6% and 14% compared with BM, and NM, respectively under FI strategy. Using PRD70 and PRD50 Tr values under WM was higher 18% and 38% compared with BM, and NM, respectively. At 83DAS, under FI, Tr under BM was higher 10%, and 17%, respectively, compared with WM and NM. In PRD 70, WM increased Tr 22%, and 40%, respectively, compared with BM and NM. in PRD50, Tr was reduced dramatically due to water stress. However, WM increased Tr by 9% while BM increased by 7%, compared NM.

3.4. Chlorophyll Index (SPAD Value)

The chlorophyll index (SPAD value) was statistically analyzed, as shown in Table 3. High chlorophyll content is a desired attribute, as it implies a low degree of photoinhibition of the photosynthetic apparatus [67]. Li et al. [42] suggested that SPAD values could perfectly trace the variations in chlorophyll content of plants. At 35 and 83 DAS, squash plants sown in the SS showed high chlorophyll content (SPAD value) compared with those sown in the WS. This could be due to the higher photosynthesis rate (P_n), observed in squash plants sown in the SS, compared with those sown in the WS (Table 2). Li et al. [68] and Peiguo and Mingqi [69] emphasized that the relative chlorophyll and photosynthetic rate interact positively with each other, as chlorophyll represents the primary chloroplast component of photosynthesis.

Table 3. Analysis of variance of the chlorophyll index (SPAD value) at the development stage (35 DAS), mid stage (63 DAS), and late stage (83 DAS) of squash growth during winter (WS) and spring (SS) growing seasons.

Treatments		Chlorophyll Index (SPAD Value)		
		35 DAS	**63 DAS**	**83 DAS**
Season (S)	**WS**	43.88 [b]	42.99 [b]	41.66 [b]
	SS	48.31 [a]	43.88 [a]	45.85 [a]
	p-value	0.0083 **	0.196 [ns]	0.010 *
	LSD 0.05	1.746	–	1.81
Mulch (M)	**WM**	48.09 [a]	45.61 [a]	46.28 [a]
	BM	46.75 [b]	46.21 [a]	45.00 [b]
	NM	43.46 [c]	38.42 [b]	39.98 [c]
	p-value	0.00001 **	0.00001 **	0.0001 **
	LSD 0.05	1.12	1.27	0.99
Irrigation (I)	**FI**	48.35 [a]	45.81 [a]	46.07 [a]
	PRD70	45.73 [b]	43.05 [b]	43.53 [b]
	PRD50	44.22 [b]	41.38 [c]	41.66 [c]
	p-value	0.00001 **	0.00001 **	0.0005 **
	LSD 0.05	1.66	1.09	1.41
S × M	*p*-value	0.0016 **	0.035 *	0.023 *
S × I	*p*-value	0.806 [ns]	0.101 [ns]	0.440 [ns]
M × I	*p*-value	0.831 [ns]	0.0265 *	0.265 [ns]
S × M × I	*p*-value	0.718 [ns]	0.122 [ns]	0.063 [ns]

[ns]: not statistically significant, **: significant at the 1% level ($p < 0.01$), *: significant at the 5% level ($p < 0.05$); different letters indicate significant difference between treatments; bold letters and words indicate treatments names.

Mulched treatments significantly affected ($p < 0.001$) chlorophyll index values (Table 3). Our study showed that the SPAD value of mulch treatments was significantly higher than non-mulched treatments. The primary reason for the high SPAD value with mulch treatment could be that the film mulch changed the soil water content (Figure 5) and the heat environment in the root area of the squash, causing a change in the physical and chemical properties of the soil, which accelerated root system growth. Kante et al. [70] showed that a reduction in the chlorophyll content of plant leaves was directly associated with root growth. This result follows the same trend as the findings of Hugar et al. [71], Nasrullah et al. [72], and Iqbal et al. [73], who found that soil mulch enhances chlorophyll content compared with non-mulched treatments.

Drought stress reduced the chlorophyll index at all growth stages. PRD70 and PRD50 reduced the chlorophyll content. Under conditions of water stress, chlorophyll content declines as a result of damage to chloroplast membranes and structure and photo-oxidation of chlorophyll [74–76]. The reduction of leaf chlorophyll values due to a water deficit has been reported for squash [23], cabbage [58], cotton [73], and wheat [67] crops.

Chlorophyll index values were not significantly affected by the interactions between S×M×I. However, the interaction between S × M was significant ($p < 0.05$) at all measured days. At 63 DAS, the interaction effect between mulch and irrigation treatments on SPAD value was significance. In FI treatments, BM increased SPAD value 6% and 23% compared with NM. In PRD70, the SPAD values under BM and WM were not different. BM and WM both increased SPAD values 17% compared with NM. Under PRD50, WM increased SPAD values 3% and 24 %, respectively compared with BM and NM. Overall, FI and BM improved Pn, Tr and SPAD value.

3.5. Fruit Quality

Table 4 shows the statistical analysis of squash fruit quality, total soluble solids (TSS), total acidity (TA), and vitamin C (V_C) under mulch and irrigation treatments for the WS and SS. The fruit qualities of the FI treatment were significantly different ($p < 0.001$) to those of the PRD treatments. Squash plants under the PRD50 treatments reduced TSS, TA, and V_C by 17%, 25%, and 19%, respectively, compared with the FI treatment. The severe water stress treatment (PRD50) negatively affected the squash fruit quality. This finding could be explained by the water deficit causing a reduction in fruit water potential [25]. These results are in agreement with the findings of Al-Ghobari and Dewidar [24], Abd El-Mageed et al. [34], Kuslu et al. [77] and Zhang et al. [25], who found that water-stressed treatments reduced fruit qualities compared with non-stressed water. Fruit quality under PRD can be affected by many factors, including plant type, developmental stage, soil type, and environmental conditions [62].

Table 4. Analysis of variance of squash fruit quality, total soluble solids (TSS), total acidity (TA), and vitamin C (V_C) for winter and spring growing seasons.

Treatments		TSS (%)	TA (% Citric Acid)	V_C (mg/100 g FW)
Season (S)	WS	4.98 [b]	0.311	0.727
	SS	5.52 [a]	0.334	0.746
	p-value	0.036 *	0.1785 [ns]	0.602 [ns]
	LSD 0.05	0.149	–	–
Mulch (M)	WM	5.47 [b]	0.340 [b]	0.760 [b]
	BM	5.63 [a]	0.342 [a]	0.775 [a]
	NM	4.71 [c]	0.287 [c]	0.675 [c]
	p-value	0.00001 **	0.0002 **	0.0018 **
	LSD 0.05	0.052	0.018	0.045
Irrigation (I)	FI	5.85 [a]	0.373 [a]	0.813 [a]
	PRD70	5.63 [b]	0.313 [b]	0.733 [b]
	PRD50	4.86 [c]	0.281 [c]	0.663 [c]
	p-value	0.0001 **	0.00001 **	0.00001 **
	LSD 0.05	0.048	0.017	0.043
S × M	*p*-value	0.0003 **	0.0457 *	0.036 *
S × I	*p*-value	0.00001 **	0.0047 **	0.182 [ns]
M × I	*p*-value	0.357 [ns]	0.958 [ns]	0.908 [ns]
S × M × I	*p*-value	0.635 [ns]	0.917 [ns]	0.906 [ns]

[ns]: not statistically significant, **: significant at the 1% level ($p < 0.01$), *: significant at the 5% level ($p < 0.05$); different letters indicate significant difference between treatments; bold letters and words indicate treatments names.

Mulching significantly affected ($p < 0.0001$) all fruit quality attributes. Mulch treatments increased the TSS, TA, and V_C by 16%, 16%, and 13%, respectively, compared with non-mulched treatments. This result is consistent with those of Lira-Saldivar et al. [36] and Li et al. [78], who found that soil mulching enhances fruit quality, compared with non-mulching. Abd El-Mageed et al. [34] indicated that mulch could reduce the influence of water stress on squash fruit quality, as mulch reduces soil evaporation, while preserving soil moisture content near the root zone.

Growing seasons did not significantly ($p > 0.05$) affect fruit quality, except for TSS. The interaction effect between S×I on TSS and TA was significant ($p < 0.001$). However, there was no significant ($p > 0.05$) difference in the value of V_C. Squash fruit qualities were not significantly affected by the interactions of S × M × I. In contrast, the effect of the interaction of S × M showed a significant difference ($p < 0.05$) between all fruit qualities.

3.6. Yield and Irrigation Water Use Efficiency (IWUE)

Statistical analysis of squash yield and IWUE are shown in Table 5. Squash yield was significantly ($p < 0.05$) affected by growing season. The squash yield obtained in the SS was higher (19%) than that in the WS. The reduction of squash yield in the WS could be due to extreme lower temperatures and solar radiations during the WS than SS (Figure 1). Similar results were obtained for cucumber by Wan et al. [79] and for squash by Amer [37], who reported that the different yields, obtained in different growing seasons, were due to non-favorable weather conditions. Similarly, the higher yield recorded during the SS was due to an increase in physiological properties (g_s, P_n, and T_r) and the chlorophyll index, compared with WS (Tables 2 and 3).

Table 5. Analysis of variance of squash fresh fruit yield and irrigation water use efficiency (IWUE) for winter (WS) and spring (SS) growing seasons.

Treatments		Fresh Fruit Yield (Mg ha^{-1})	IWUE (kg m^{-3})
Season (S)	**WS**	72.12 [b]	26.71 [a]
	SS	85.88 [a]	12.92 [b]
	p-value	0.0118 *	0.0005 **
	LSD 0.05	6.49	1.35
Mulch (M)	**WM**	87.46 [a]	22.51 [a]
	BM	85.30 [a]	21.74 [b]
	NM	64.23 [c]	15.20 [c]
	p-value	0.00001 **	0.0001 **
	LSD 0.05	3.41	0.45
Irrigation (I)	**FI**	80.62 [a]	15.07 [c]
	PRD70	82.53 [a]	20.48 [b]
	PRD50	73.85 [c]	23.90 [a]
	p-value	0.0001 **	0.0001 **
	LSD 0.05	2.51	0.52
S × M	*p*-value	0.0001 **	0.0001 **
S × I	*p*-value	0.0003 **	0.0001 **
M × I	*p*-value	0.474 [ns]	0.0001 **
S × M × I	*p*-value	0.773 [ns]	0.0001 **

[ns]: not statistically significant, **: significant at the 1% level ($p < 0.01$), *: significant at the 5% level ($p < 0.05$); different letters indicate significant difference between treatments; Bold letters and words indicate treatments names.

The mulching treatments showed a significant difference ($p < 0.0001$) in squash yield compared with the non-mulched treatments (Table 5). Mulched treatments increased squash yield by 36% compared with non-mulched treatments. However, no statistical difference was observed between mulched treatments (BM and WM). The yield increase observed in the plastic mulch treatment could be attributed to its ability to reduce evaporation, fertilizer leaching, weed accumulation, and soil compaction and increase soil temperature, which enhances root growth [30,31]. These properties led to higher soil moisture and nutrient holding in the root zone, which eventually enhanced squash yield, compared with NM. Many studies have reported that mulch enhances crop yield in squash [34], cucumber [59], chili [80] and broccoli [35].

Squash yield was significantly ($p < 0.001$) affected by irrigation treatments. The highest squash yield was obtained under the PRD70 treatment (82.53 Mg ha^{-1}). Although this yield was not significantly different from that of the FI treatment (80.62 Mg ha^{-1}). This suggests that reducing the irrigation volume perfectly could improve fruit yield. The higher squash yield in the PRD70 treatment than the FI treatment could be partially explained by the PRD having parallel drip lines that irrigate the root zone of the plant interchangeably. This could reduce water losses due to deep percolation in sandy soil, resulting in nutrient availability near the root zone in plants under PRD treatments. Another possible reason for the PRD70 plot having a higher yield than the FI plot is that plastic mulch could prevent soil evaporation to some degree. Therefore, plots under the FI treatment might be over irrigated, and irrigation of 70% of crop water requirement supplies sufficient water for crop growth without stress [81]. Hakim et al. [17] indicate that plants receiving FI could encounter higher soil moisture in the root zone, which reduces root activity, delaying maturity, and lowering yield compared with plants under PRD treatments. This result is consistent with the findings of Qin et al. [20] and Hooshmand et al. [19], who found that the yield of the FI treatment was lower than the deficit treatments, but not significantly different. Howerver, the squash yield obtained in this study was more than three times higher than the squash yield obtained by Al-Omran et al. [82] under the same environmental conditions. This finding could be attributed to the higher plant density and good fertilization program used in this experiment, resulting in a higher squash yield compared with the mentioned study.

The interaction effects between S × M × I were not statistically significant ($p > 0.05$) for squash yield, while there were strong significant ($p < 0.001$) interactions between S × M and S × I (Table 5). It is worth mension that sowing squash in the SS under non-mulched treatment was almost doubled the squash yield compared to sowing in the WS. The variation of squash yield under mulched treatments in the SS and WS was not considerable. This shows that WM and BM were effective during both growing seasons (Figure 6). The highest squash yield was recorded under SS-BM-PRD70 treatment (95.84 Mg ha^{-1}), while the lowest yield (46.06 Mg ha^{-1}) was obtained under WS-NM-PRD50. In the WS, the highest squash yield obtained was 87.9 Mg ha^{-1} in WM PRD70, while in the SS, the lowest squash yield obtained was 75.33 Mg ha^{-1} in NM PRD50. The Squash yield obtained under SS-NM-FI and WS-NM-FI were 77.4, and 53.1 Mg ha^{-1}, respectively, while in SS-WM-PRD50 and WS-WM-PRD50 were 85.29, and 79.34 Mg ha^{-1}, respectively (Figure 6). This shows that using the PRD strategy and soil mulching technique reduces 50% of applied water, while increasing squash yield in both growing seasons. These results suggest that in arid and semi-arid regions where there are water scarcity problems, soil mulch with PRD50 could be used as a water-saving strategy to maintain the squash yield.

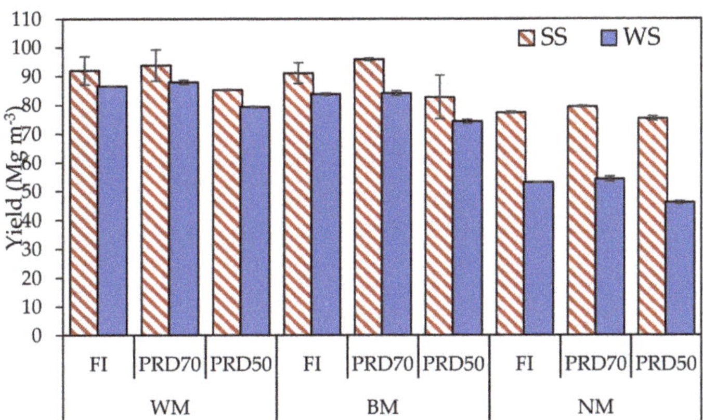

Figure 6. Squash yield under irrigation and mulch treatments during the winter and spring seasons.

Data presented in Table 5 and Figure 7 show that the effect of S, M, and I on IWUE was significant ($p < 0.001$). The IWUE in the WS was two times higher than in the SS. This result could be due to the water applied to squash in the SS, which was higher than that applied in the WS. This finding is in line with those recorded by Rouphael and Colla [83], Abd El-Mageed and Semida [9], Abd El-Mageed et al. [34] and Silva et al. [38], who worked on squash and observed that the IWUE was affected by environmental factors under different growing seasons.

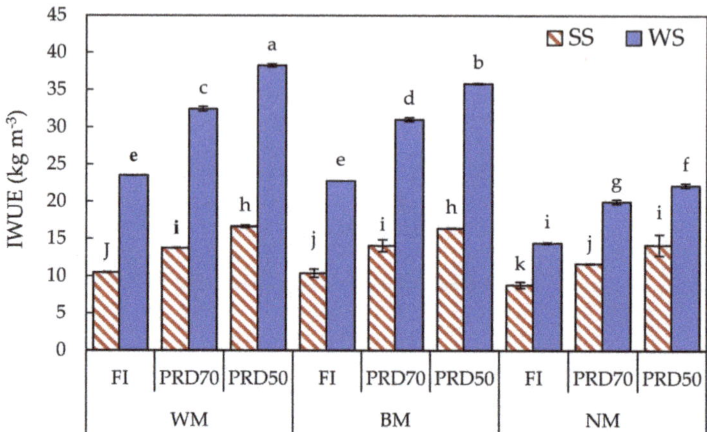

Figure 7. Squash irrigation water use efficiency (IWUE) during winter (WS) and spring (SS) under mulch treatments (black mulch-BM, transparent mulch-WM and non-mulch- NM) and irrigation treatments (full irrigation-FI, partial root drying with 50% of evapotranspiration- PRD50, partial root drying with 70% of evapotranspiration-PRD70); the data is the mean value ± standard error; different letters indicate significant difference between treatments.

In terms of the mulching treatments, WM increased the IWUE by 48% compared with the NM treatments (Table 5). Soil mulching decreased evaporation and increased the soil moisture content near the root zone, which positively affected the squash yield, and finally, it contributed to higher IWUE. This result is consistent with the findings of Zhang et al. [81], Chen et al. [84], and Yang et al. [61], who found that mulching treatments had higher IWUE than the control treatment (NM).

In terms of irrigation quantities, the highest IWUE was observed under PRD50 treatments (23.90 kg m^{-3}). The corresponding value for the FI treatments was 15.07 kg m^{-3} (Table 5). PRD50 and PR70 increased the IWUE by 59%, and 36%, respectively, compared with the FI treatment. These results are in agreement with Amer [37], Abd El-Mageed et al. [34], and Zhang et al. [81], who found that water-stressed treatments increase the IWUE, compared with FI.

Data presented in Table 5 show that the interaction effects of the S×M, S×I, M×I, and S×M×I on IWUE were significant ($p < 0.001$). The highest IWUE (38.24 kg m^{-3}) was recorded under WS-WM-PRD50, while the lowest value was 8.82 kg m^{-3} under SS-NM-FI (Figure 7). The IWUE in the WS were doubled compared with SS for mulch treatments under same irrigation treatments. It can be seen from Figure 7 that PRD50 obtained higher IWUE in mulch and non-mulched treatments, compared with PRD70 and FI. Overall, Sowing squash in WS, FI-NM obtained squash yield of 53.1 Mg ha^{-1} with IWUE 14.42 kg m^{-3}, while PRD50-WM obtained 79.34 Mg ha^{-1} with IWUE 38.24 kg m^{-3}. This result led to conculde that sowing squash in WS using PRD50-WM saves 50% of applied water while increases squash yield by 49%, compared with FI-NM.

4. Conclusions

The effect of growing season DI integrated with PRD, and soil mulching on the yield and IWUE of squash plants, was studied. The results indicated that plant density postively affected squash yield in both growing seasons for all treatments. The spring growing season positively affected squash yield. In contrast, the SS negatively affected the IWUE, compared with the WS. Moreover, soil mulching enhanced the physiological properties of the squash plants (g$_s$, P$_n$, and T$_r$), fruit quality (TSS, TA, and Vc), increasing the squash yield, and IWUE, compared with non-mulched treatments. g$_s$, P$_n$ and T$_r$ were significantly affected by growing season for all measured days. Furthermore, PRD70 and PRD50 reduced the chlorophyll index at all growth stages. Mulch treatments increased the TSS, TA, and V_C, compared with non-mulched treatments. However, growing seasons did not significantly affect fruit quality. In addition, PRD strategy improved both squash yield and IWUE in both growing seasons. This emphasizes that sowing squash plants in the winter season, using PRD50 and plastic mulch as water-saving strategies, could increase the yield and IWUE in arid and semi-arid regions.

Author Contributions: Conceptualization, H.M.A.-G., T.K.Z.E.-A. and A.A.E.-S.; methodology, A.A.E.-S., A.H.F. and M.S.A.; software, A.A.E.-S., A.H.F. and M.S.A.; validation, H.M.A.-G., T.K.Z.E.-A. and A.A.E.-S.; formal analysis, A.H.F. and M.S.A.; investigation, A.H.F. and M.S.A.; resources, H.M.A.-G. and T.K.Z.E.-A.; data curation, A.A.E.-S., A.H.F. and M.S.A.; writing—original draft preparation, A.A.E.-S. and A.H.F.; writing—review and editing, H.M.A.-G., T.K.Z.E.-A., A.A.E.-S. and A.H.F.; visualization, T.K.Z.E.-A. and A.A.E.-S.; supervision, H.M.A.-G. and T.K.Z.E.-A.; project administration, H.M.A.-G. and T.K.Z.E.-A.; funding acquisition, H.M.A.-G., T.K.Z.E.-A. and M.S.A. All authors have read and agreed to the published version of the manuscript.

Funding: This research was funded by the Deanship of Scientific Research at King Saud University, through research group No. RG-1441-321.

Institutional Review Board Statement: Not applicable.

Informed Consent Statement: Not applicable.

Data Availability Statement: The data presented in this study are available on request from the corresponding author.

Acknowledgments: The authors extend their appreciation to the Deanship of Scientific Research at King Saud University for funding this work through research group No. RG-1441-321.

Conflicts of Interest: The authors declare no conflict of interest. The funders had no role in the design of the study; in the collection, analyses, or interpretation of data; in the writing of the manuscript, or in the decision to publish the results.

References

1. Wang, X.; Huo, Z.; Guan, H.; Guo, P.; Qu, Z. Drip irrigation enhances shallow groundwater contribution to crop water consumption in an arid area. *Hydrol. Process.* **2018**, *32*, 747–758. [CrossRef]
2. Biswas, S.K.; Akanda, A.R.; Rahman, M.S.; Hossain, M.A. Effect of drip irrigation and mulching on yield, water-use efficiency and economics of tomato. *Plant Soil Environ.* **2015**, *61*, 97–102. [CrossRef]
3. Woodrow, J.E.; Seiber, J.N.; Lenoir, J.S.; Krieger, R.I. Determination of methyl isothiocyanate in air downwind of fields treated with metam-sodium by subsurface drip irrigation. *J. Agric. Food Chem.* **2008**, *56*, 7373–7378. [CrossRef]
4. Ayars, J.E.; Fulton, A.; Taylor, B. Subsurface drip irrigation in California—Here to stay? *Agric. Water Manag.* **2015**, *157*, 39–47. [CrossRef]
5. Hashem, M.S.; Zin El-Abedin, T.; Al-Ghobari, H.M. Assessing effects of deficit irrigation techniques on water productivity of tomato for subsurface drip irrigation system. *Int. J. Agric. Biol. Eng.* **2018**, *11*, 156–167. [CrossRef]
6. Abdalhi, M.A.M.; Jia, Z. Crop yield and water saving potential for AquaCrop model under full and deficit irrigation managements. *Ital. J. Agron.* **2018**, *13*, 267–278. [CrossRef]
7. Khapte, P.S.; Kumar, P.; Burman, U.; Kumar, P. Deficit irrigation in tomato: Agronomical and physio-biochemical implications. *Sci. Hortic.* **2019**, *248*, 256–264. [CrossRef]
8. Puértolas, J.; Albacete, A.; Dodd, I.C. Irrigation frequency transiently alters whole plant gas exchange, water and hormone status, but irrigation volume determines cumulative growth in two herbaceous crops. *Environ. Exp. Bot.* **2020**, *176*, 104101. [CrossRef]
9. Abd El-Mageed, T.A.; Semida, W.M. Effect of deficit irrigation and growing seasons on plant water status, fruit yield and water use efficiency of squash under saline soil. *Sci. Hortic.* **2015**, *186*, 89–100. [CrossRef]
10. Jensen, M.E. *Design and Operation of Farm Irrigation Systems*; ASAE: Detroit, MI, USA, 1983; pp. 108–118.
11. Bacon, M.A. Water use efficiency in plant biology. In *Water Use Efficiency in Plant Biology*; Bacon, M.A., Ed.; CRC Press: Boca Raton, FL, USA, 2004; pp. 1–26.
12. Velasco-Muñoz, J.F.; Aznar-Sánchez, J.A.; Belmonte-Ureña, L.J.; López-Serrano, M.J. Advances in Water Use Efficiency in Agriculture: A Bibliometric Analysis. *Water* **2018**, *10*, 377. [CrossRef]
13. Kang, S.; Zhang, J. Controlled alternate partial root-zone irrigation: Its physiological consequences and impact on water use efficiency. *J. Exp. Bot.* **2004**, *55*, 2437–2446. [CrossRef]

14. Jensen, C.R.; Battilani, A.; Plauborg, F.; Psarras, G.; Chartzoulakis, K.; Janowiak, F.; Stikic, R.; Jovanovic, Z.; Li, G.; Qi, X.; et al. Deficit irrigation based on drought tolerance and root signalling in potatoes and tomatoes. *Agric. Water Manag.* **2010**, *98*, 403–413. [CrossRef]

15. Jovanovic, Z.; Stikic, R. Partial root-zone drying technique: From water saving to the improvement of a fruit quality. *Front. Sustain. Food Syst.* **2018**, *1*, 1–9. [CrossRef]

16. Barideh, R.; Besharat, S.; Morteza, M.; Rezaverdinejad, V. Effects of partial root-zone irrigation on the water use efficiency and root water and nitrate uptake of corn. *Water* **2018**, *10*, 526. [CrossRef]

17. Hakim, A.; Qinyan, Z.; Khatoon, M.; Gullo, S. Impact of partial root-zone drying on growth, yield and quality of tomatoes produced in green house condition. *Adv. Hortic. Sci.* **2019**, *33*, 133–138. [CrossRef]

18. Mattar, M.A.; Zin El-Abedin, T.K.; Alazba, A.A.; Al-Ghobari, H.M. Soil water status and growth of tomato with partial root-zone drying and deficit drip irrigation techniques. *Irrig. Sci.* **2020**, *38*, 163–176. [CrossRef]

19. Hooshmand, M.; Albaji, M.; Boroomand nasab, S.; Alam zadeh Ansari, N. The effect of deficit irrigation on yield and yield components of greenhouse tomato (Solanum lycopersicum)in hydroponic culture in Ahvaz region, Iran. *Sci. Hortic.* **2019**, *254*, 84–90. [CrossRef]

20. Qin, J.; Ramírez, D.A.; Xie, K.; Li, W.; Yactayo, W.; Jin, L.; Quiroz, R. Is Partial root-zone drying more appropriate than drip irrigation to save water in China? A preliminary comparative analysis for potato cultivation. *Potato Res.* **2018**, *61*, 391–406. [CrossRef]

21. Jovanovic, Z.; Stikic, R.; Vucelic-Radovic, B.; Paukovic, M.; Brocic, Z.; Matovic, G.; Rovcanin, S.; Mojevic, M. Partial root-zone drying increases WUE, N and antioxidant content in field potatoes. *Eur. J. Agron.* **2010**, *33*, 124–131. [CrossRef]

22. Xie, K.; Wang, X.X.; Zhang, R.; Gong, X.; Zhang, S.; Mares, V.; Gavilán, C.; Posadas, A.; Quiroz, R. Partial root-zone drying irrigation and water utilization efficiency by the potato crop in semi-arid regions in China. *Sci. Hortic.* **2012**, *134*, 20–25. [CrossRef]

23. Ors, S.; Ekinci, M.; Yildirim, E.; Sahin, U. Changes in gas exchange capacity and selected physiological properties of squash seedlings (Cucurbita pepo L.) under well-watered and drought stress conditions. *Arch. Agron. Soil Sci.* **2016**, *62*, 1700–1710. [CrossRef]

24. Al-Ghobari, H.M.; Dewidar, A.Z. Integrating deficit irrigation into surface and subsurface drip irrigation as a strategy to save water in arid regions. *Agric. Water Manag.* **2018**, *209*, 55–61. [CrossRef]

25. Zhang, K.; Dai, Z.; Wang, W.; Dou, Z.; Wei, L.; Mao, W.; Chen, Y.; Zhao, Y.; Li, T.; Zeng, B.; et al. Effects of partial root drying on strawberry fruit. *Eur. J. Hortic. Sci.* **2019**, *84*, 39–47. [CrossRef]

26. Guang-Cheng, S.; Rui-Qi, G.; Na, L.; Shuang-En, Y.; Weng-Gang, X. Photosynthetic, chlorophyll fluorescence and growth changes in hot pepper under deficit irrigation and partial root zone drying. *Afr. J. Agric. Res.* **2011**, *6*, 4671–4679. [CrossRef]

27. Chakraborty, D.; Nagarajan, S.; Aggarwal, P.; Gupta, V.K.; Tomar, R.K.; Garg, R.N.; Sahoo, R.N.; Sarkar, A.; Chopra, U.K.; Sarma, K.S.S.; et al. Effect of mulching on soil and plant water status, and the growth and yield of wheat (*Triticum aestivum* L.) in a semi-arid environment. *Agric. Water Manag.* **2008**, *95*, 1323–1334. [CrossRef]

28. Sharma, R.; Bhardwaj, S. Effect of mulching on soil and water conservation: A review. *Agric. Rev.* **2017**, *38*, 311–315. [CrossRef]

29. Zhao, H.; Wang, R.Y.; Ma, B.L.; Xiong, Y.C.; Qiang, S.C.; Wang, C.L.; Liu, C.A.; Li, F.M. Ridge-furrow with full plastic film mulching improves water use efficiency and tuber yields of potato in a semiarid rainfed ecosystem. *Field Crops Res.* **2014**, *161*, 137–148. [CrossRef]

30. Mutetwa, M.; Mtaita, T. Effects of mulching and fertilizer sources on growth and yield of onion. *J. Glob. Innov. Agric. Soc. Sci.* **2014**, *2*, 102–106. [CrossRef]

31. Kader, M.A.; Senge, M.; Mojid, M.A.; Ito, K. Recent advances in mulching materials and methods for modifying soil environment. *Soil Tillage Res.* **2017**, *168*, 155–166. [CrossRef]

32. Yaghi, T.; Arslan, A.; Naoum, F. Cucumber (*Cucumis sativus*, L.) water use efficiency (WUE) under plastic mulch and drip irrigation. *Agric. Water Manag.* **2013**, *128*, 149–157. [CrossRef]

33. Kumari, P.; Ojha, R.K.; Job, M. Effect of plastic mulches on soil temperature and tomatoyield inside and outside the polyhouse. *Agric. Sci. Digest* **2016**, *36*, 333–336. [CrossRef]

34. Abd El-Mageed, T.A.; Semida, W.M.; Abd El-wahed, M.H. Effect of mulching on plant water status, soil salinity and yield of squash under summer-fall deficit irrigation in salt affected soil. *Agric. Water Manag.* **2016**, *173*, 1–12. [CrossRef]

35. Verma, V.K.; Jha, A.K.; Verma, B.C.; Nonglait, D.; Chaudhuri, P. Effect of Mulching materials on soil health, yield and quality attributes of broccoli grown under the mid-hill conditions. *Proc. Natl. Acad. Sci. India Sect. B Biol. Sci.* **2018**, *88*, 1589–1596. [CrossRef]

36. Lira-Saldivar, R.H.; Méndez-Argüello, B.; Felipe-Victoriano, M.; Vera-Reyes, I.; Cardenas-Flores, A.; Méndez-Argüello, B.; Felipe-Victoriano, M.; Vera-Reyes, I.; Cardenas-Flores, A.; Méndez-Argüello, B.; et al. Gas exchange, yield and fruit quality of Cucurbita pepo cultivated with zeolite and plastic mulch. *Agrochimica* **2017**, *61*, 123–139. [CrossRef]

37. Amer, K.H. Effect of irrigation method and quantity on squash yield and quality. *Agric. Water Manag.* **2011**, *98*, 1197–1206. [CrossRef]

38. Silva, G.H.; Cunha, F.F.; Morais, C.V.; Freitas, A.R.J.; Silva, D.J.H.; Souza, C.M.D. Mulching materials and wetted soil percentages on zucchini cultivation. *Ciênc. Agrotecnol.* **2020**, *44*, e006720. [CrossRef]

39. Allen, R.G.; Pereira, L.S.; Raes, D.; Smith, M. *Crop Evapotranspiration–Guidelines for Computing Crop Water Requirements*; FAO Irrigation and Drainage Paper 56; FAO: Rome, Italy, 1998.

40. Buss, P. The use of capacitance based measurement of real time soil water profile dynamics for irrigation scheduling. In Proceedings of the National Conference Irrigation Association, Australia and National Committee Irrigation Drainage, Launceston, TAS, Australia, 17–19 May 1993.

41. Vera, J.; Mounzer, O.; Ruiz-Sánchez, M.C.; Abrisqueta, I.; Tapia, L.M.; Abrisqueta, J.M. Soil water balance experiments utilizing capacitance and neutron probe measurements in irrigation scheduling. In *Transactions of the Second International Symposium on Soil Water Measurement Using Capacitance Impedance and Time Domain Transmission (TDT)*; Paltineanu, I.C., Ed.; Paltin International Incorporated: Beltsville, MD, USA, 2007; p. 180.

42. Li, Y.; Song, H.; Zhou, L.; Xu, Z.; Zhou, G. Tracking chlorophyll fluorescence as an indicator of drought and rewatering across the entire leaf lifespan in a maize field. *Agric. Water Manag.* **2019**, *211*, 190–201. [CrossRef]

43. Helrich, K. *Official Methods of Analysis of the Association of Official Analytical Chemists*; Association of Official Analytical Chemists: Arlington, VA, USA, 1990.

44. Caruso, G.; Conti, S.; Villari, G.; Borrelli, C.; Melchionna, G.; Minutolo, M.; Russo, G.; Amalfitano, C. Effects of transplanting time and plant density on yield, quality and antioxidant content of onion (*Allium cepa* L.) in southern Italy. *Sci. Hortic.* **2014**, *166*, 111–120. [CrossRef]

45. Patanè, C.; Tringali, S.; Sortino, O. Effects of deficit irrigation on biomass, yield, water productivity and fruit quality of processing tomato under semi-arid Mediterranean climate conditions. *Sci. Hortic.* **2011**, *129*, 590–596. [CrossRef]

46. CoStat. *CoStat Version 6.451—Statistics Software*; CoHort Software: Monterey, CA, USA, 2018.

47. Rashid, M.A.; Zhang, X.; Andersen, M.N.; Olesen, J.E. Can mulching of maize straw complement deficit irrigation to improve water use efficiency and productivity of winter wheat in North China plain? *Agric. Water Manag.* **2019**, *213*, 1–11. [CrossRef]

48. Wong, S.C.; Cowan, I.R.; Farquhar, G.D. Stomatal conductance correlates with photosynthetic capacity. *Nature* **1979**, *282*, 424–426. [CrossRef]

49. Tuzet, A.; Perrier, A.; Leuning, R. A coupled model of stomatal conductance, photosynthesis and transpiration. *Plant Cell Environ.* **2003**, *26*, 1097–1116. [CrossRef]

50. Liu, F.; Jensen, C.R.; Andersen, M.N. Hydraulic and chemical signals in the control of leaf expansion and stomatal conductance in soybean exposed to drought stress. *Funct. Plant Biol.* **2003**, *30*, 65–73. [CrossRef] [PubMed]

51. Parkash, V.; Singh, S. A Review on potential plant-based water stress indicators for vegetable crops. *Sustainability* **2020**, *12*, 3945. [CrossRef]

52. Raghavendra, A.S.; Gonugunta, V.K.; Christmann, A.; Grill, E. ABA Perception and Signalling. *Trends Plant Sci.* **2010**, *15*, 395–401. [CrossRef]

53. Sperry, J.S.; Alder, N.N.; Eastlack, S.E. The effect of reduced hydraulic conductance on stomatal conductance and xylem cavitation. *J. Exp. Bot.* **1993**, *44*, 1075–1082. [CrossRef]

54. Farooq, M.; Wahid, A.; Kobayashi, N.; Fujita, D.; Basra, S.M.A. Plant Drought Stress: Effects, Mechanisms and Management. In *Sustainable Agriculture*; Lichtfouse, E., Navarrete, M., Debaeke, P., Véronique, S., Alberola, C., Eds.; Springer: Dordrecht, The Netherlands, 2009; pp. 153–188.

55. Shirke, P.A. Leaf photosynthesis, dark respiration and fluorescence as influenced by leaf age in an evergreen tree. *Prosop. Juliflor.* **2001**, *39*, 305–311. [CrossRef]

56. Du, T.; Kang, S.; Zhang, J.; Li, F.; Hu, X. Yield and physiological responses of cotton to partial root-zone irrigation in the oasis field of northwest China. *Agric. Water Manag.* **2006**, *84*, 41–52. [CrossRef]

57. Romero, P.; Dodd, I.C.; Martinez-Cutillas, A. Contrasting physiological effects of partial root zone drying in field-grown grapevine (*Vitis vinifera* L. cv. Monastrell) according to total soil water availability. *J. Exp. Bot.* **2012**, *63*, 4071–4083. [CrossRef]

58. Sahin, U.; Ekinci, M.; Ors, S.; Turan, M.; Yildiz, S.; Yildirim, E. Effects of individual and combined effects of salinity and drought on physiological, nutritional and biochemical properties of cabbage (*Brassica oleracea* var. capitata). *Sci. Hortic.* **2018**, *240*, 196–204. [CrossRef]

59. Ibarra-Jiménez, L.; Zermeño-González, A.; Munguía-López, J.; Rosario Quezada-Martín, M.A.; De La Rosa-Ibarra, M. Photosynthesis, soil temperature and yield of cucumber as affected by colored plastic mulch. *Acta Agric. Scandinav. Sect. B Plant Soil Sci.* **2008**, *58*, 372–378. [CrossRef]

60. Wang, F.; Wang, Z.; Zhang, J.; Li, W. Combined effect of different amounts of irrigation and mulch films on physiological indexes and yield of drip-irrigated maize (*Zea mays* L.). *Water* **2019**, *11*, 472. [CrossRef]

61. Yang, Y.; Ding, J.; Zhang, Y.; Wu, J.; Zhang, J.; Pan, X.; Gao, C.; Wang, Y.; He, F. Effects of tillage and mulching measures on soil moisture and temperature, photosynthetic characteristics and yield of winter wheat. *Agric. Water Manag.* **2018**, *201*, 299–308. [CrossRef]

62. Zhang, X.; Yang, L.; Xue, X.; Kamran, M.; Ahmad, I.; Dong, Z.; Liu, T.; Jia, Z.; Zhang, P.; Han, Q. Plastic film mulching stimulates soil wet-dry alternation and stomatal behavior to improve maize yield and resource use efficiency in a semi-arid region. *Field Crops Res.* **2019**, *233*, 101–113. [CrossRef]

63. Li, Q.; Shen, J.; Zhao, D. Effect of irrigation frequency on yield and leaf water use efficiency of winter wheat. *Trans. Chin. Soc. Agric. Eng.* **2011**, *27*, 33–36.

64. Urban, J.; Ingwers, M.; McGuire, M.A.; Teskey, R.O. Stomatal conductance increases with rising temperature. *Plant Signal. Behav.* **2017**, *12*, e1356534. [CrossRef] [PubMed]

65. Jones, H.G.; Serraj, R.; Loveys, B.R.; Xiong, L.; Wheaton, A.; Price, A.H. Thermal infrared imaging of crop canopies for the remote diagnosis and quantification of plant responses to water stress in the field. *Funct. Plant Biol.* **2009**, *36*, 978–989. [CrossRef]
66. Scherrer, D.; Bader, M.K.-F.; Körner, C. Drought-sensitivity ranking of deciduous tree species based on thermal imaging of forest canopies. *Agric. Forest Meteorol.* **2011**, *151*, 1632–1640. [CrossRef]
67. Talebi, R. Evaluation of chlorophyll content and canopy temperature as indicators for drought tolerance in durum wheat (*Triticum durum* Desf.). *Aust. J. Basic Appl. Sci.* **2011**, *5*, 1457–1462.
68. Li, R.H.; Guo, P.G.; Michael, B.; Stefania, G.; Salvatore, C. Evaluation of Chlorophyll Content and Fluorescence Parameters as Indicators of Drought Tolerance in Barley. *Agric. Sci. China* **2006**, *5*, 751–757. [CrossRef]
69. Peiguo, G.; Mingqi, L. Studies on photosynthetic characteristics in rice hybrid progenies and their parents I. Chlorophyll content, chlorophyll-Protein complex and chlorophyll fluorescence kinetics. *J. Trop. Subtrop. Bot.* **1996**, *4*, 60–65.
70. Kante, M.; Revilla, P.; De La Fuente, M.; Caicedo, M.; Ordás, B. Stay-green QTLs in temperate elite maize. *Euphytica* **2016**, *207*, 463–473. [CrossRef]
71. Hugar, A.Y.; Halemani, H.L.; Aladakatti, Y.R.; Nandagavi, R.A.; Hallikeri, S.S. Studies on the effect of polyethylene mulching on rainfed cotton genotypes: II. Influence on status of soil moisture, microbial population in soil and uptake of nutrients. *Karnataka J. Agric. Sci.* **2009**, *22*, 284–288.
72. Nasrullah, M.; Khan, M.B.; Ahmad, R.; Ahmad, S.; Hanif, M.; Nazeer, W. Sustainable cotton production and water economy through different planting methods and mulching techniques. *Pak. J. Bot.* **2011**, *43*, 1971–1983.
73. Iqbal, R.; Raza, M.A.S.; Saleem, M.F.; Khan, I.H.; Ahmad, S.; Zaheer, M.S.; Aslam, M.U.; Haider, I. Physiological and biochemical appraisal for mulching and partial rhizosphere drying of cotton. *J. Arid Land* **2019**, *11*, 785–794. [CrossRef]
74. Kingston-Smith, A.H.; Foyer, C.H. Bundle sheath proteins are more sensitive to oxidative damage than those of the mesophyll in maize leaves exposed to paraquat or low temperatures. *J. Exp. Bot.* **2000**, *51*, 123–130. [CrossRef] [PubMed]
75. Amirjani, M.R.; Mahdiyeh, M. Antioxidative and biochemical responses of wheat to drought stress. *J. Agric. Biol. Sci.* **2013**, *8*, 291–301.
76. Kabiri, R.; Nasibi, F.; Farahbakhsh, H. Effect of exogenous salicylic acid on some physiological parameters and alleviation of drought stress in Nigella sativa plant under hyroponic culture. *Plant Protect. Sci.* **2014**, *50*, 43–51. [CrossRef]
77. Kuslu, Y.; Sahin, U.; Kiziloglu, F.M.; Memis, S. Fruit yield and quality, and irrigation water use efficiency of summer squash drip-irrigated with different irrigation quantities in a semi-arid agricultural area. *J. Integrat. Agric.* **2014**, *13*, 2518–2526. [CrossRef]
78. Li, H.; Yang, X.; Chen, H.; Cui, Q.; Yuan, G.; Han, X.; Wei, C.; Zhang, Y.; Ma, J.; Zhang, X. Water requirement characteristics and the optimal irrigation schedule for the growth, yield, and fruit quality of watermelon under plastic film mulching. *Sci. Hortic.* **2018**, *241*, 74–82. [CrossRef]
79. Wan, S.; Kang, Y.; Wang, D.; Liu, S.P. Effect of saline water on cucumber (*Cucumis sativus* L.) yield and water use under drip irrigation in North China. *Agric. Water Manag.* **2010**, *98*, 105–113. [CrossRef]
80. Khan, M.N.; Ayub, G.; Ilyas, M.; Haq, F.U.; Ali, J.; Alam, A. Effect of different mulching materials on weeds and yield of Chili cultivars. *Pure Appl. Biol.* **2016**, *5*, 1160–1170. [CrossRef]
81. Zhang, H.; Xiong, Y.; Huang, G.; Xu, X.; Huang, Q. Effects of water stress on processing tomatoes yield, quality and water use efficiency with plastic mulched drip irrigation in sandy soil of the Hetao Irrigation District. *Agric. Water Manag.* **2017**, *179*, 205–214. [CrossRef]
82. Al-Omran, A.M.; Sheta, A.S.; Falatah, A.M.; Al-Harbi, A.R. Effect of drip irrigation on squash (*Cucurbita pepo*) yield and water-use efficiency in sandy calcareous soils amended with clay deposits. *Agric. Water Manag.* **2005**, *73*, 43–55. [CrossRef]
83. Rouphael, Y.; Colla, G. Growth, yield, fruit quality and nutrient uptake of hydroponically cultivated zucchini squash as affected by irrigation systems and growing seasons. *Sci. Hortic.* **2005**, *105*, 177–195. [CrossRef]
84. Chen, Z.; Sun, S.; Zhu, Z.; Jiang, H.; Zhang, X. Assessing the effects of plant density and plastic film mulch on maize evaporation and transpiration using dual crop coefficient approach. *Agric. Water Manag.* **2019**, *225*, 105765. [CrossRef]

Article

Pearl Millet Forage Water Use Efficiency

Bradley Crookston [1],*, Brock Blaser [2],*, Murali Darapuneni [3] and Marty Rhoades [2]

[1]   Department of Plants, Soils and Climate, Utah State University, Logan, UT 84322, USA
[2]   Department of Agricultural Sciences, West Texas A&M University, Canyon, TX 79016, USA;
      mrhoades@wtamu.edu
[3]   Department of Plant and Environmental Sciences, New Mexico State University, Tucumcari, NM 88401, USA;
      dmk07@nmsu.edu
*   Correspondence: bradley.crookston@aggiemail.usu.edu (B.C.); bblaser@wtamu.edu (B.B.)

Received: 9 October 2020; Accepted: 27 October 2020; Published: 29 October 2020

Abstract: Pearl millet (*Pennisitum glaucum* L.) is a warm season C_4 grass well adapted to semiarid climates where concerns over scarce and depleting water resources continually prompt the search for water efficient crop management to improve water use efficiency (WUE). A two-year study was conducted in the Southern Great Plains, USA, semi-arid region, to determine optimum levels of irrigation, row spacing, and tillage to maximize WUE and maintain forage production in pearl millet. Pearl millet was planted in a strip-split-plot factorial design at two row widths, 76 and 19 cm, in tilled and no-till soil under three irrigation levels (high, moderate, and limited). The results were consistent between production years. Both WUE and forage yield were impacted by tillage; however, irrigation level had the greatest effect on forage production. Row spacing had no effect on either WUE or forage yield. The pearl millet water use-yield production function was y = 6.68 × x (mm) − 837 kg ha^{-1}; however, a low coefficient of determination (r^2 = 0.31) suggests that factors other than water use (WU), such as a low leaf area index (LAI), had greater influence on dry matter (DM) production. Highest WUE (6.13 Mg ha^{-1} mm^{-1}) was achieved in tilled soil due to greater LAI and DM production than in no-till.

Keywords: optimum water use; forage

1. Introduction

Crop yield loss occurs under water deficit, however, many studies have found that higher crop water use efficiency (WUE) is often achieved under water stress conditions [1], albeit, with reduced yield. It is therefore, imperative to identify crop management strategies that optimize WUE without sacrificing attainable yield under limited water availability. The ratio of crop fodder (forage) or grain biomass produced per unit water used (transpiration and losses to soil evaporation) is considered crop water use efficiency [2]. Although there are many location characteristics and environmental conditions that influence WUE (e.g., climate regime, soil type) that cannot easily be manipulated by land managers, there are various crop management schemes known to increase water availability and promote greater WUE [1,2]. It is up to researchers to identify useful combinations of practices fit for particular regions based on the best science available from around the globe.

Pearl millet is a C_4 warm season grass predominately in production for grain for human consumption and forage for livestock feed throughout Africa and India and is noted for its tolerance to semiarid conditions where there is low rainfall and limited levels of soil nutrients and organic matter [3,4]. It is typically grown on rainfed (dryland) areas in systems with grain sorghum (*Sorghum bicolor* L.), maize (*Zea mays* L.), or often integrated with legume crops such as cowpea (*Vigna unguiculata* (L.) Walp) in Africa [4,5]. Pearl millet is also gaining recognition in the Southern Great Plains, USA, semi-arid region, as a potential forage crop to be integrated into grain sorghum-winter wheat (*Triticum aestivum*, L.)

livestock feed cropping systems by replacing summer fallow [6–8]. The United Nations Food and Agriculture Organization Statistical Database (FAO) reported that worldwide millet production in 2018 was approximately 31 million metric tons, with Africa and Asia accounting for 51 and 46% of total production, respectively; the Americas make up only 1% of global millet production [9]. Pearl and other millets (e.g., foxtail (*Setaria italica*, (L.) P. Beauvois), proso (*Panicum miliaceum*, L.), and Japanese (*Echinochloa esculenta*, A.Braun)) are common in semi-arid regions for livestock feed because of their rich nutrient value and fit in cropping rotations with short growing seasons that have limited water supply [6,10].

Water use efficiency can be improved using soil management, crop husbandry, water management, genetic selection, and crop competition management [1]. Many of these methods to improve WUE have been reported in studies of pearl and other millet species in locations around the world over the past 25 years (Table 1). Husbandry studies have tested various planting arrangement methods and population densities in hill type sowing [11–13]; forage harvest intervals have also been tested [8]. Studies of irrigation technology have investigated sprinkler, surface drip, and subsurface drip methods or irrigation scheduling throughout the season [10,14]. Other researchers explored increasing water stress and nutrient management; many of those studies utilized the factorial combination of the treatments to benefit from the interaction of improved WUE with nutrient management under limited water supply [15–20]. However, throughout all these studies, the range of millet WUE is 1–92 kg ha^{-1} mm^{-1} with a mean and median of 20.6 and 13.4 kg ha^{-1} mm^{-1}, respectively (Table 1). In a global meta-analysis of WUE in various crops, [2], reported many environmental factors impact WUE variability within and among crop species. For example, greater WUE was observed in tropical versus desert climates, and greater organic matter content was strongly correlated with positive WUE, whereas clay content was negatively correlated with WUE [2]. Ultimately, [2] and co-authors admonished researchers to investigate the mechanisms that might further explain WUE variability within a species.

To our knowledge, no studies with pearl millet have investigated the combination of irrigation levels to limit water availability, plant arrangement in narrow versus wide row spacing and soil management with conventional rotary tillage versus no-till. Goals of this research were to determine the optimum levels irrigation, row spacing, and tillage to maximize WUE while maintaining stable forage production [21] in pearl millet in the semi-arid region of the Southern Great Plains, USA. The underlying hypotheses guiding the objectives were (a) WUE will increase with decreased water availability [22]; (b) no-till soil management techniques can promote more efficient WU [23]; (c) pearl millet crop canopy can be modified through row spacing to increase incidence of leaf area index and light interception [24].

Table 1. Location, soil type, millet crop, experimental treatment, soil tillage management, crop row spacing, water source, crop water use (WU) and water use efficiency (WUE) of pearl and other millet species.

Reference	Location	Soil Type [a]	Crop [b]	Treatment [c]	Tillage	Row Spacing (cm)	Water Source [d]	WU (mm)	WUE (kg ha^{-1} mm^{-1})	Comments
[15]	Texas, USA	CL	Pearl, F	Irr levels, cultivars	Disk	100	Subdrip	193–480	87–92	Max WUE in limited Irr only 1 yr.
[8]	Texas, USA	CL	Pearl, F	Harvest intervals	Rotary till	19	Surfdrip	594	16.2	WUE not different across harvest intervals
[12]	Tanzania	Snd	Pearl, G	Cultivation, micro-dosing	various	na	Rf	481	1.0–4.0	Max WUE in tide-ridges with micro-dose
[25]	China	SndL	Foxtail, Japanese, G	Species, sowing dates	Rotary till	42	Rf	356	10.2–22.8	Max WUE in early sowing
[17]	Niger	Snd	Pearl, G	Manure, micro-dosing	na	100	Rf	582	4.9–12.8	Manure and fertilizer increased WUE
[11]	India	SndL	Pearl, G	Planting arrangement	na	100	Rf	263	5.8–7.7	Tide-ridge recommended
[16]	Jordan	CL	Pearl, F	Species, irr levels	Disk	50	Surfdrip	325–516	21.3	Deficit increased millet WUE
[10]	Saudi Arabia	SndL	Pearl, F	Irr type	na	20	Sprinkler, Surfdrip, subdrip	100–1300	8.4–13.4	Max WUE in subdrip
[19]	Australia	CL	Pearl, F	Irr levels	na	na	Sprinkler	496	22.7	Forage species, high WUE
[20]	Iran	CL	Pearl, F	Irr levels, N levels	na	60	Flood	321–755	28.4–41	Max WUE in deficit with N applied
[18]	Niger	na	Pearl, G	Animal integrated cropping	na	na	Rf	372–393	0.9–6.2	Max WUE with animal integrated cropping
[26]	Colorado, USA	SltL	Proso, F	Cropping systems	No-till	na	Rf	100–250	3.0–22.0	WUE improved in forage-based systems
[14]	Nebraska, USA	SltL	Pearl, G	Irr scheduling	No-till	76	Flood	353	13.4–28.5	Sorghum had higher WUE than millet
[27]	Texas, USA	CL	Pearl, G/F	Cropping systems	No-till	25	Rf	268	4–31.0	Forage crops improved system WUE
[13]	Niger	Snd	Pearl, G	Population, varieties	na	100	Rf	366	7.6–9.7	Max WUE in high populations with nutrients applied
[22]	India	SndL	Pearl, F	Irr scheduling, cultivars	na	30	Flood	223–568	13.8–17.9	Max WUE under water stress

Note: na, not available. [a] Cl, clay loam; SltL, silt loam; Snd, sand; SndL; sandy loam. [b] F, forage; G, grain. [c] Irr, irrigation. [d] Rf, rainfed; Subdrip, subsurface drip; Surfdrip, surface drip.

2. Materials and Methods

2.1. Location and Field Preparation

This forage pearl millet study was conducted during the 2016 and 2017 growing seasons at West Texas A&M University Nance Ranch near Canyon, TX (34°58′6″ N, 101°47′16″ W; 1097 m above sea level elevation) located on Olton clay loam (fine, mixed, superactive, thermic, Aridic Paleustoll). Treatments were arranged as strip-split-plots with four replications with irrigation as main plots, row spacing as subplots, and tillage as sub-subplots.

The plot area was prepared for planting in 2016 by first a using a rotary mower to reduce native perennial grasses, which had established in an unused portion of the field, and application of Roundup© PowerMax™ (Monsanto, St. Louis, MO, USA) herbicide at 9.5 L ha on 31 May 2016. The field was further prepared in 2016 by applying 88 mm of preplant irrigation through drip tape (system described below) to increase the ease of mechanical tillage. In 2017, pearl millet was planted following winter wheat sown in October 2016 terminated in March using Roundup© PowerMax™ herbicide at the same 2016 application rate. In both years, tilled plots were cultivated 20 cm depth using a CountyLine rotary tiller (King Kutter, Inc., Winfield, AL 35594, USA) the day of planting. 'Graze King' BMR pearl millet (176,405 seeds kg^{-1}, 85% germination, 98% purity) was planted on 17 June 2016. Seed used in 2017 was procured from Winfield United (Shoreview, MN 55126, USA) (116,280 seeds kg^{-1}, 85% germination, 98% purity). Sowing utilized a Great Plains 3P500 grain drill (Great Plains Mfg., Inc., Salina, KS, USA). The 3.05 × 6.1 m plots were seeded at 125 seeds m^{-2} and 95 seeds m^{-2} in 2016 and 2017, respectively. Plot width being 3.05 m, we accommodated 16 rows and 4 rows in 19 and 76 cm row spacing treatments, respectively. Different seeding rates were the result of different seed sizes and planter limitation to accommodate the small seed size at the same rate for both years.

2.2. Irrigation Management

Irrigation was administered with a metered surface drip line system with two lines 152 cm apart in each plot with emitters spaced 61 cm, each applying 7.5 L hour^{-1} emitter^{-1}. In-line control valves were used to regulate water flow to the irrigation treatment levels so that when irrigation was applied to high (H) only, flow was restricted to moderate (M) and limited (L). However, when L was irrigated, the other levels were irrigated as well, and when M was irrigated, H also received water (Figure 1).

Irrigation management was preplanned to simulate grower conditions where a specific amount of irrigation allowable by aquifer pumping would be provided during the season. Approximately 225 mm was selected as the upper quantity for H. The M and L levels were to be approximately 60 and 30% of the high level, respectively. Weekly applications of 25 mm were planned for H, biweekly in M, and only early season irrigation in L. Ultimately, irrigation management decisions were influenced by growing season climatic conditions, therefore, irrigation events took place when necessary.

Crop WU was defined as the sum of growing season precipitation, irrigation, and plant available soil water (PAW) at planting, minus PAW remaining at harvest. Water lost to runoff or drainage was not measured. Water use efficiency is DM divided by total WU.

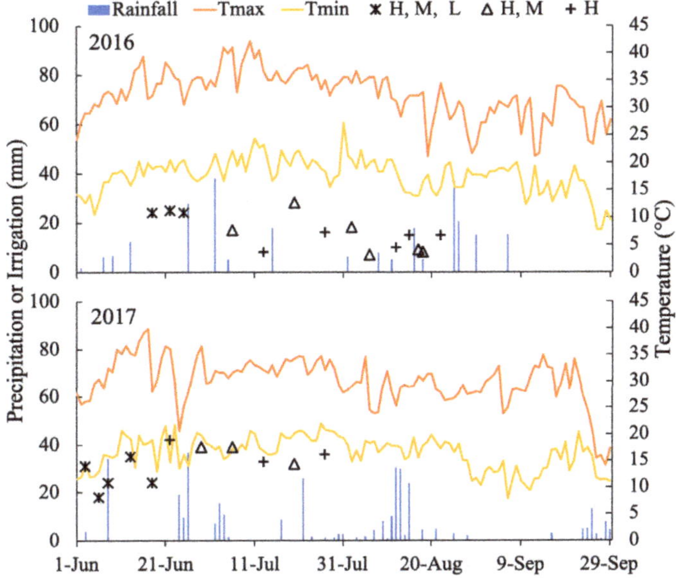

Figure 1. Precipitation, irrigation by water level, and maximum and minimum daily temperature from 1 June to 30 September for 2016 and 2017 near Canyon, TX. Tmax, maximum daily temperature; Tmin, minimum daily temperature; H, high, M, moderate, L, limited irrigation levels.

2.3. Soil Moisture

Soil cores used to calculate average soil volumetric water content at planting were obtained by sampling four random locations throughout the plot area using a tractor mount Giddings hydraulic press (Giddings Machine Company, Inc., Windsor, CO, USA). Soil samples were taken the day prior to planting to a depth of 60 cm in 2016 and to a depth of 75 cm in 2017. Each core sample was divided into three incremental depths: 0–15, 15–30, 30–60 cm in 2016 and 0–15, 15–45, 45–75 cm in 2017. The soil cores were weighed, and oven dried at 104 °C for 72 h until a constant dry weight was attained. Soil volumetric water content at harvest in both years was determined using one core sample from each plot divided into segments of 0–15, 15–45, 45–75 cm.

Soil characteristics and properties utilized data obtained from the United States Department of Agriculture Natural Resources Conservation Service Web Soil Survey [28] (Table 2). Plant available soil water (PAW) was calculated by subtracting volumetric water content at permanent wilting point (−1500 kPa) from volumetric water content at −33 kPa and multiplied by the soil depth.

Table 2. Soil moisture characteristics for Olton clay loam [a].

		Water Content (kPa) [b]			
Depth	ρ_b [c]	**−33**	**−1500**	**Plant Available Water**	
cm	g cm^{-3}	% by Volume		cm cm^{-3}	cm
0–15	1.54	34.8	23.5	0.13	
15–30	1.49	35.3	24.5	0.16	
30–60	1.46	35.6	25	0.17	
60–90	1.45	35.2	24.3	0.16	
Mean	1.49	35.23	24.33	0.16	
Profile					13.95

[a] NRCS, Web Soil Survey, Randall County, Texas (TX381), Olton Clay Loam, 0–1% slopes. [b] Water content at field capacity (−33 kPa) and permanent wilting point (−1500 kPa). [c] ρ_b, soil bulk density.

2.4. Weather Data

Precipitation during the 2016 growing season was collected with a graduated plastic rain gauge at the study site and cross referenced with National Weather Service in Amarillo, TX (approximately 32 km from the study location). Temperature values were taken from the National Weather Service in Amarillo. Weather data for 2017 was collected on site utilizing a Campbell Scientific (Logan, UT, USA) weather station approximately 100 m from the plot location. The American Society of Civil Engineers reference evapotranspiration (ET_0) for a well-watered grass crop was calculated using the REF-ET [29] macro software for Microsoft Excel (Microsoft Corp., Redmond, WA, USA). Growing degree-days (GDD) were calculated using Equation (1):

$$GDD = crop\ max\left(\frac{T_{max} + T_{min}}{2} - T_{base},\ 0\right) \tag{1}$$

where crop maximum is 35 °C and base temperature is 10 °C [30]. Growing degree-days were calculated for the time period between planting date and harvest.

2.5. Crop Growth and Forage Yield Measurements

Plant emergence and density were observed and counted at 308 GDD (19 days after planting; DAP) and 255 GDD (18 DAP) in 2016 and 2017, respectively. Crop canopy height was determined using a 0.25 m^2 circular clear plastic disk and the method described by [31] and used by [32]. Harvest DM was the mean of two 1 m^2 quadrats from each plot, arranged in the 76 cm row spacing to include two rows, and cut at 15 cm above the soil surface [33].

Photosynthetically active solar radiation (PAR) intercepted by the crop canopy was measured every 7 to 12 d beginning 7 July 2016 and 19 June 2017 using the AccuPAR Linear PAR ceptometer, model 80 light measuring instrument (Decagon Devices, Pullman, WA, USA). Light interception measurements were collected by placing the instrument diagonally across three 19 cm spaced rows and perpendicular across two 76 cm spaced rows. Percent light intercepted by the canopy was determined by calculating the difference of one above (incident) measurement from the mean of two below canopy measurements, divided by the above measurement.

Leaf area index (LAI) was measured and calculated utilizing the LAI-2200 Plant Canopy Analyzer (Li-Cor, Inc., Lincoln, NE, USA). Measurements were obtained for row crops of narrow and wide row spacing per the Li-Cor instruction manual to calculate a mean from two sequences of one above canopy and four below canopy readings. Measurements were obtained at sunrise or sunset every 10 to 14 d beginning 12 July 2016 and 19 June 2017.

2.6. Statistical Design and Analysis

Year, irrigation, row spacing, and tillage were treated as fixed effects. Analysis of variance of main effects utilized Proc MIXED and GLM of the Statistical Analysis System, Version 9.4 (SAS Institute, Cary, NC, USA). Linear regression and analysis of covariance utilized Proc REG and Proc GLM with contrast statements to determine differences among slopes and intercepts of the regression equations [26]. Significance was determined at $p < 0.05$ for all means separation tests and used the P diff statement with Tukey's adjustment for all main effects.

Interactions for all main effects were initially tested. When no interactions were detected, only main effect results are presented. Years were analyzed and are presented separately for analysis of variance as a consequence of management differences between the seasons: planting in 2016 was into native perennial grass, into wheat stubble in 2017; different seed used in each year. Years were combined for regression analysis after observing similar trends for both years.

3. Results

3.1. Weather, Irrigation, and Soil Moisture

3.1.1. Seasonal Weather

Total growing season precipitation was 217 and 309 mm in 2016 and 2017, respectively (Figure 1; growing seasons: 17 June to 30 September 2016; 1 June to 30 August 2017). The 30-year precipitation average for the Amarillo, TX area for June through September is 275 mm. Precipitation received was 89% and 137% of normal for the respective years (Table 3). Higher temperatures in July 2016 resulted in 10% greater cumulative degree days than in July 2017. Lower temperatures and greater precipitation in August 2017 generated 13% lower GDD accumulation for that month compared to 2016 (Table 3). The local 30-year average GDD accumulation for a 90-d pearl millet growing season is 1676 GDD for a 1 June planting date (Table 3).

Table 3. Average maximum daily temperature, total precipitation, and estimated cumulative thermal time by month for 2016 and 2017 growing seasons near Canyon, TX from planting to harvest. [a]

Month	Avg. [b] Max Daily Temp.			Precipitation			Heat Units		
	°C			mm			GDD		
	2016	2017	Avg.	2016	2017	Avg.	2016	2017	Avg.
June	33	32	30	57	103	80	437	411	405
July	36	32	33	61	79	72	528	478	486
August	31	28	31	113	127	74	440	389	461
September	29		28	15		49	362		324
Total				246	309	275	1767	1279	1676
Percent of normal				89%	137% [c]		105%	94%	

[a] Growing Season, 17 June to 30 September 2016; 1 Jun to 30 Aug 2017. [b] Averages for temperature and precipitation are averages for Amarillo, TX, USA. 1981–2010. [c] 2017 percent of normal for precipitation and heat units is taken from historical avg. of June–August. The percent of normal is indicated for each season's precipitation.

3.1.2. Irrigation

Total irrigation applied in 2016 for each irrigation level was 220, 157, and 72 mm for the H, M, and L levels, respectively (Table 4). In 2017, 356, 242, and 132 mm were applied to the H, M, and L irrigation treatments, respectively (Table 4). The L level received three irrigation events in 2016, and five events in 2017; greater water was applied in 2017 early in the season due lower soil moisture at planting and to prevent soil surface crusting. The M level received nine events in 2016 and eight events in 2017. The H level received 14 events in 2016 and 11 events during the 2017 season. As a result of more frequent rainfall during the second half of July and in August of 2017, there were fewer irrigation events needed than in 2016.

Table 4. Millimeters of irrigation applied to pearl millet by month for the 2016 and 2017 growing seasons [a] near Canyon, TX.

	2016			2017		
	High	Moderate	Limited	High	Moderate	Limited
June	72	72	72	214	171	132
July	66	43		142	71	
August	82	42				
September						
Total	220	157	72	356	242	132

[a] Growing Season, 17 June to 30 September 2016; 1 June to 30 August 2017.

Applied irrigation accounted for approximately 24–53% of seasonal total water (irrigation and precipitation) for the crop at the respective irrigation levels over both growing seasons (Table 5). During the growing seasons, irrigation accounted for an average range of 50–53%, 41–43%, and 24–29% of the total water received by the crop (excluding stored soil moisture) in H, M, and L levels, respectively (Table 5). The total water is compared to ET [29] to evaluate the efficacy of irrigation management to meet climatic demand on the crop (Table 5). The ET_o climatic demand was estimated at 772 and 902 mm in 2016 and 2017, respectively. The range of satisfied climatic demand across water regimes was 48–73% in 2017 vs. 37–56% in 2016 (Table 5). Although 2017 had fewer irrigation events for H and M levels, additional rainfall in 2017 was available to meet the climatic demand.

Table 5. Precipitation and irrigation by month and irrigation level for 2016 and 2017 near Canyon, TX. Percent of total water as irrigation and total as water percent of ET_o are also included.

	2016 [a]				2017			
	mm							
	High	Moderate	Limited	ET_o [b]	High	Moderate	Limited	ET_o
Total	437	374	289	772	665	551	441	902
Percent of total as irrigation	50%	41%	24%		53%	43%	29%	
Total water as percent of ET_o	56%	48%	37%		73%	61%	48%	

[a] Growing Season, 17 June to 30 September 2016; 1 June to 30 August 2017. [b] ET_o, ASCE reference evapotranspiration (ET) for a well-watered grass crop, calculated using REF-ET [29].

The authors of [14] reported applying an average 353 and 350 mm in two study years on pearl millet and grain sorghum (*Sorghum bicolor* L.), respectively, in Nebraska where annual precipitation was 285 and 480 mm in the study two locations. In studies of rainfed millet production, [27] at Bushland, TX reported pearl millet forage WU range was 236 to 289 mm (precipitation and PAW from the soil). In a study testing the response of sorghum, maize, and pearl millet to four irrigation levels in India, [22], reported total WU ranged from 242 to 568 mm during the field study. In Akron, CO, during the five-year study of opportunity cropping systems [34], millet growing season precipitation ranged from 57 to 298 mm, similar to other semiarid locations.

3.1.3. Soil Moisture

Plant available water at planting was 3.35 cm for the 60 cm profile in 2016 and was 1.9 cm for the 75 cm profile in 2017. Mean PAW (soil depth 75 cm) at harvest across all treatments in 2016 (0.63 cm) was 20% of mean PAW in 2017 (3.09 cm). Harvest PAW in 2016 was influenced by irrigation, there was no influence of row spacing or tillage treatments, averaging 0.64 cm. High and M irrigation levels averaged 0.86 cm, which was different from L at 0.19 cm of PAW. Differences in PAW were found in 2017 among irrigation levels and in an irrigation × tillage interaction. Plant available water was highest in M (3.52 cm) and lowest in L (2.49 cm), a difference of 41%; H (3.1 cm) was not different from the M and L levels. The interaction resulted from L irrigation × no-till, 1.8 cm PAW, being 47% lower than the average of all other treatments (3.42 cm; SE = 0.42 cm), ranging from 2.7 to 3.8 cm. The primary reason harvest PAW in 2016 was 80% less than 2017 is the additional 31 d of growing season in 2016 and the rainfall prior to harvest in 2017 (Figure 1).

3.2. Treatment Effects: Irrigation, Row Spacing and Tillage

3.2.1. Plant Growth and Forage Dry Matter

Average pearl millet plant population was 14.6 and 30.9 plants m^{-2} in 2016 and 2017, respectively, a difference of 52%. A year x till interaction was observed, most likely due to low plant population in no-till in 2016 compared to no-till in 2017 (7.9 vs. 45.8 plants m^{-2}, respectively). Additionally, in 2016, average population in no-till was 67% less than till (7.9 vs. 21.2 plants m^{-2}, respectively). However, in 2017, average till plant population was 65% less than no-till (16.1 vs. 45.7 plants m^{-2}). The no-till planting in 2016 was into herbicide killed native perennial grasses which prevented adequate seed to soil contact for proper germination. In 2017, an afternoon rainstorm following a morning irrigation event three days after sowing caused runoff that was observed in the till treatment. In contrast, no-till in 2017 was protected by the winter wheat stubble remaining on the soil surface after planting.

Forage DM was influenced by year, irrigation level, and tillage ($p < 0.05$) as main effects in both years; row spacing had no effect on forage production (Table 6). Average pearl millet DM in 2016 was approximately 38% less than in 2017 (1712 and 2759 kg DM ha^{-1}, respectively). Average WU was 523 and 395 mm in 2017 and 2016, respectively. Average WUE was 5.28 and 4.26 kg ha^{-1} mm^{-1} in 2017 and 2016, respectively (Figure 2). Forage DM in H irrigation was 46% greater than L, ranging from 1219 to 2213 kg ha^{-1} in 2016. The H level was 54% greater than L in 2017, which ranged from 2213 to 3450 kg ha^{-1} (Figure 2). Dry matter in M was not different from H in either year but was different from L ($p < 0.05$; Figure 2). Water use in H was 15% greater than M and 43% greater than L in 2016. In 2017, H received 59% more water than L and 21% more than M (Figure 2). Water use efficiency was not different among irrigation levels ranging from 3.78 to 4.79 kg ha^{-1} mm^{-1}. In contrast, a C$_3$ bioenergy crop, hybrid poplar (*Populus generosa* Henry $\times$ *P. nigra* L.), achieved differing WUE values of 5.6 and 8.9 kg ha^{-1} mm^{-1} grown under 115 mm and 240 mm of applied irrigation, respectively [35]. Row spacing did not influence pearl millet forage DM, WU, or WUE in 2016 or 2017 (Figure 2, Table 6).

Table 6. The p-values from ANOVA of pearl millet dry matter (DM), water use (WU), and water use efficiency (WUE).

Treatment [a]	2016		
	DM	**WU**	**WUE**
I	0.001	<0.0001	ns
R	ns	ns	ns
I × R	ns	ns	ns
T	0.0001	ns	<0.0001
I × T	ns	ns	ns
R × T	ns	ns	ns
I × R × T	ns	ns	ns
	2017		
	DM	**WU**	**WUE**
I	0.0097	<0.0001	ns
R	ns	ns	ns
I × R	ns	ns	ns
T	0.0064	ns	0.0095
I × T	ns	ns	ns
R × T	ns	ns	ns
I × R × T	ns	ns	ns

[a] I, irrigation; R, row spacing; T, tillage; ns, not significant at $p = 0.05$.

Pearl millet DM and WUE in the tilled plots were greater than in no-till in both years, although, WU was not affected by tillage management in either year (Figure 2). Forage DM in till was 66 and 39% greater than no-till, in 2016 and 2017, respectively. Pearl millet WUE in tilled soil was 68% greater than no-till in 2016 and 39% greater in 2017 (Figure 2).

Forage DM in 2016 till was greater as a consequence of the low plant population in no-till due to poor seed to soil contact. The low forage yields in 2017 are attributed to N deficiency observed in the no-till plants. This deficiency in no-till may be a result of higher plant population, compared to the till treatment, leading to the deficiency observed mid-season and no deficiency observed in the till plants.

Crop canopy height in 2016 was not influenced by irrigation, row spacing, or tillage; the end of season canopy height was 50 cm. For the 2017 season, canopy height was affected by irrigation over time. Height at harvest was 72, 67, and 59 cm for H, M, and L, respectively. In addition, there was an irrigation × tillage interaction in 2017, where L × no-till was 38 cm and H × till was 49 cm at harvest. No difference for crop canopy height was measured in the tillage treatment for either year.

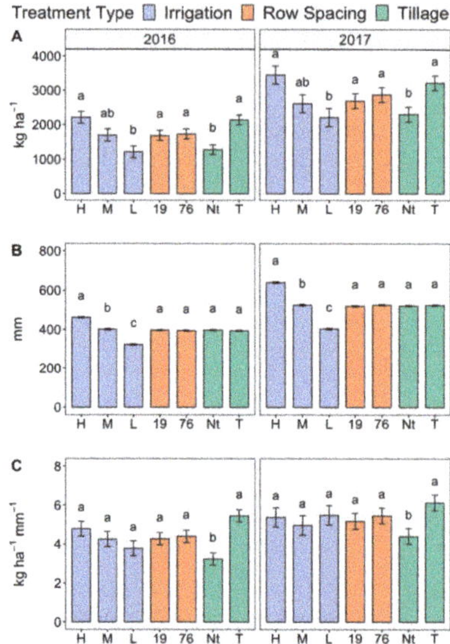

Figure 2. (**A**) pearl millet dry matter; (**B**) water use = (rainfall + irrigation + plant available soil moisture at planting)–plant available soil moisture at harvest; (**C**) water use efficiency = dry matter/water use. Lowercase letters indicate significance at $p < 0.05$ within treatment and year; bars indicate standard error within treatment. H, high irrigation level, M, moderate irrigation level, L, limited irrigation level; 19 cm and 76 cm row spacing; N-t, no-till, T, tilled.

3.2.2. Light Interception and Leaf Area Index

Crop LAI during the 2016 season was impacted by irrigation, which reached a maximum of 2.47 at 1262 GDD (Figure 3). In 2017, LAI was similarly affected by irrigation and reached a maximum of 3.07 at 1203 GDD (Figure 3). Incidence of crop canopy light interception responded to changes in irrigation through time in 2016 only, reaching 74.6% interception at 967 GDD (Figure 3).

Light interception was influenced by row spacing though the 2016 season, which reached 74% in 19 cm and 65% in 76 cm at 1262 GDD (Figure 4). During 2017, LAI reached 2.87 in 19 cm and 2.59 in 76 cm at 1203 GDD (Figure 4). In both years, light interception and LAI were lower in wide row spacing. However, other work with pearl millet and row spacing has shown that in years where rainfall was less than the local average, wide row spacing outperformed narrow rows [36]. Weeds were not controlled either year of this field study and in 2016 rainfall was 11% less than the 30 year normal (Table 3), yet, row spacing had no effect on pearl millet DM.

During 2016, no differences for LAI nor light interception were observed between tillage treatments (Figure 5). Through 2017, mid-season LAI was different for the tillage treatments but there was no difference by harvest time, reaching a maximum value of 2.8 and 2.6 in till and no-till, respectively (Figure 5). Light interception was greater in no-till through the 2017 season, reaching 77 vs. 70% in till at end of season (Figure 5). For the 2017 season, light interception interactions were found for irrigation × tillage and row spacing × tillage. In the till treatment, there was an increasing trend following increasing irrigation and narrow row spacing, however, that trend was interrupted in the no-till where M had 6% less light interception than L, and 76 cm row spacing had 20% less than 19 cm.

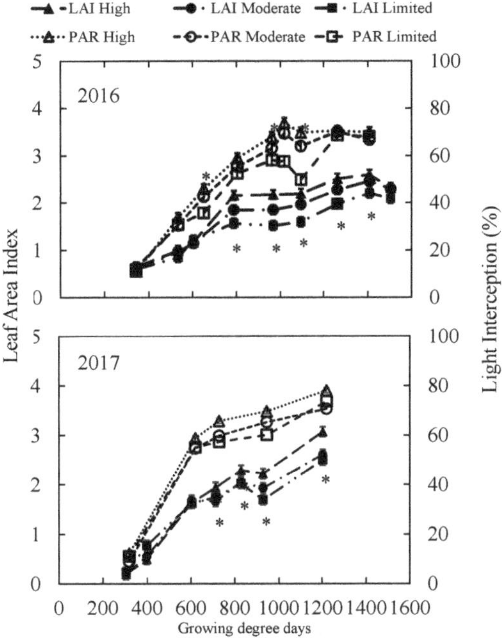

Figure 3. Leaf area index (LAI) and percent light interception (PAR) for irrigation levels in 2016 and 2017 in pearl millet at Canyon, TX. Asterisks indicate treatment $p < 0.05$ on measurement date.

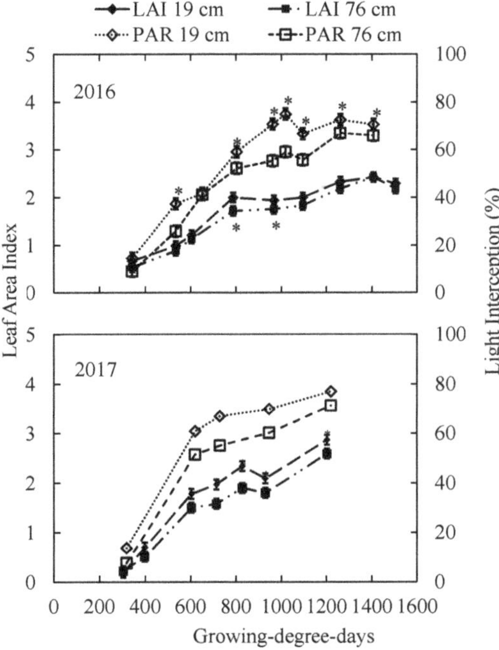

Figure 4. Leaf area index (LAI) and percent light interception (PAR) for row spacing in 2016 and 2017 in pearl millet at Canyon, TX. Asterisk indicate treatment $p < 0.05$ on measurement date.

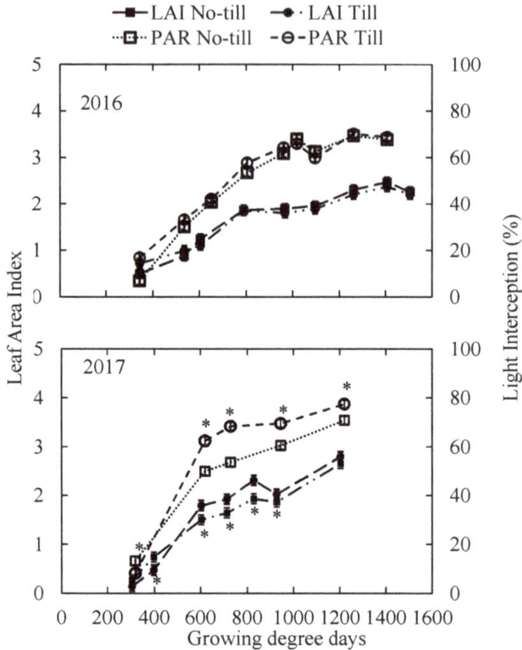

Figure 5. Leaf area index (LAI) and percent light interception (PAR) for tillage levels in 2016 and 2017 in pearl millet at Canyon, TX. Asterisk indicate treatment $p < 0.05$ on measurement date.

4. Discussion

4.1. Explaining Water Use Efficiency

4.1.1. Water Use Efficiency and Soil Nutrient Supply

Although nutrient management was not a treatment in this study, field observations of possible nutrient deficiencies prompted investigation to the relationship between WUE and nutrient management. Proper soil nutrient supply improves plant growth, photosynthetic rate, transpiration, root growth and ultimately yield, all result in increased WUE [23]. For example, pearl millet, being tested under four levels of increasing N and irrigation supply, achieved 41 kg ha^{-1} mm^{-1} WUE in the most water stressed treatment with the highest level of N application, 225 kg N ha^{-1} [20]. The authors of [37] also illustrated the effect that irrigation and N application can have on WUE in an experiment with irrigated and rainfed wheat when "adequate" N was applied with irrigation, mean WUE was 78% greater and grain production was 16% greater in irrigated than in rainfed wheat without N application. The 2016 and 2017 of this pearl millet study, DM was influenced by observed nutrient deficiency, especially in the no-till treatment of 2017. The authors of [8] conducted an experiment with forage pearl millet at the same time just adjacent to the plot where this forage pearl millet experiment took place. Agronomic management similar to the H irrigation, 19 cm row spacing, and tilled treatment was similar in both studies except that in [8] N was applied in both study years at rates of 84 kg ha^{-1} and 78 kg ha^{-1} in 2016 and 2017, respectively. Pearl millet DM reported by [8] used approximately 514 and 602 mm of water in 2016 and 2017, respectively, which produced 6287 and 9874 kg ha^{-1}, with WUE of 12.2 and 16.4 kg ha^{-1} mm^{-1} in those years, respectively. The DM yields from [8] were approximately three times greater than the mean across all treatments of this study in 2016 and 2017 having used a maximum of 462 and 641 mm in those years without N application. The lack of N, demonstrated by comparison to [8], lead to lack of nutrients reduced plant productivity and negatively impacted WUE [38].

4.1.2. Weed Influence on Pearl Millet Water Use Efficiency

Native grass/weed DM may have impacted pearl millet WUE due to no weed control in either year of this study. In 2016 grass/weed DM was not influenced by row spacing, however grass/weed DM was different among irrigation and tillage levels (Table 7). In 2017, only row spacing influenced grass/weed DM. The authors of [39] demonstrated in forage sorghum, when weeds are present, there was no difference in DM yields among wide (76 cm) or narrow (25 cm) row spacing. Although weed control was not a treatment in this study, regression analysis showed that grass/weed DM did influence pearl millet WUE across treatments and years (Figure 6). As weed proliferation was not checked by high leaf area index and resulting light interception correlated with crop competitive ability [40], weed/grass DM accumulated, thus reducing pearl millet WUE.

Table 7. Grass/weed dry matter (DM) at pearl millet harvest by treatment near Canyon, TX in 2016 and 2017.

Treatment	DM	
	kg ha^{-1}	
	2016	2017
Irrigation		
High	710a [a]	324a
Moderate	561b	363a
Limited	484b	297a
SE	37.8	44.7
Row Spacing		
19 cm	576a	218b
76 cm	594a	438a
SE	30.9	36.5
Tillage		
No-till	792a	318a
Till	379b	338a
SE	30.9	36.5

[a] Lowercase letters indicate mean separation within effect, $p < 0.05$.

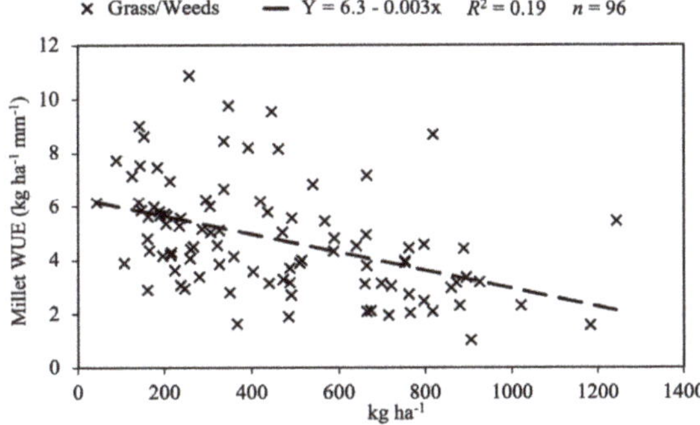

Figure 6. Pearl millet water use efficiency (WUE) as a function of grass/weed dry matter (DM) across treatments and years.

4.1.3. Water Use Efficiency Response to Agronomics

Pearl millet WUE reported in the literature has been consistent yet wide ranging. During a five-year rotation study by [27], mean pearl millet production from two seasons was 3670 kg ha^{-1}. Although weather conditions were not considerably different between the two years, [27] reported WUE of 31 kg ha^{-1} mm^{-1} in 1995 and 4.5 kg ha^{-1} mm^{-1} 1996, most likely due to one month earlier planting date in 1995, allowing the crop to better utilize soil water prior to harvest. The authors of [13] found the WUE of pearl millet, in a study of various cultivars and plant populations in Niger, ranged from 7.6 to 8.7 kg ha^{-1} mm^{-1}. In Tunisia, [41] reported a pearl millet WUE range of 6.4 to 7.6 kg ha^{-1} mm^{-1} among four water stress treatments. The authors of [22] reported pearl millet DM values of 4000 to 8300 kg ha^{-1} and WUE from 13.8 to 17.9 kg ha^{-1} mm^{-1}, respectively, in India.

Whereas, in Akron, CO, among forage and grain, foxtail and proso millet cropping systems described by [33], DM production ranged from 562 to 4545 kg ha^{-1} with WUE ranging from 4.8 to 27.5 kg ha^{-1} mm^{-1} during the five-year study. Additionally, [20] also reported high WUE ranging between 35.7 and 41 kg ha^{-1} mm^{-1} under four levels of irrigation. Grain sorghum WUE has been reported between 11 to 49.2 kg ha^{-1} mm^{-1} among locations throughout the Central and Southern Great Plains U.S. [42,43].

In the studies of pearl millet and sorghum cited above, minimum or no-till was used by [27] and [43], however, tillage information was absent from the other studies. In this study, a WUE response to agronomic management was only observed from tillage practices (Figure 2), whereas WUE was not impacted in the same way in this study by the increasing DM and WU from irrigation treatments as previously observed in other studies [20,22,41]. In a review of soil management effects on WUE, [23] reported that in no-till, where plant residue is maintained on the soil surface, higher WUE is typically found due to reduced evaporative water loss (E) from soil, lower soil surface temperatures, and improved water infiltration. The authors of [23] analyzed nineteen studies of crop and soil management practices and synthesized their results into a graphical depiction of crop WUE responses to soil management and seasonal effects. They demonstrated that crop biomass and yield increased or decreased due to changes in soil management. In this study, pearl millet DM response to soil tillage reflects differences in biomass due to soil management changes as described by [23] (Figure 7). However, when, soil surface E is roughly approximated at the x-intercept of the DM/WU regression line (Figure 7) [38,44], the two-season average estimated E from tilled and no-till soil was approximately 114 and 131 mm, respectively. While these estimates are similar, the result is contrary to reports of less E from no-till soil [23]. Additionally, WU was not different between tillage or row spacing treatments in either year. Furthermore, the spread of DM responses observed, lack-of-fit to the DM/WU predicted line, and no differences in DM or WU detected between row space levels and WU between tillage levels indicates that WU was insensitive to changes in the management practices other than when water was added by irrigation (Figures 2, 7 and 8). The authors of [24] reported that pearl millet yield is largely independent of WU when LAI is low, as was observed in both seasons of this pear millet forage study, due to the majority of light being intercepted by the soil surface allowing water loss to E in the sparse canopied crop with LAI values < 2 [44]. In 2016, LAI was not different between the tillage levels, however, in 2017, LAI in the tilled treatment reached approximately 2 nearly 100 GDDs before no-till reached a LAI of approximately 2. As mentioned above, the N deficiency observed predominately in no-till explains slower growth, lower DM and WEU for both years even though till had lower early season plant count than in no-till during 2017. Thus, higher WUE in tilled soil was achieved as a result of earlier canopy development, more light interception and greater DM production.

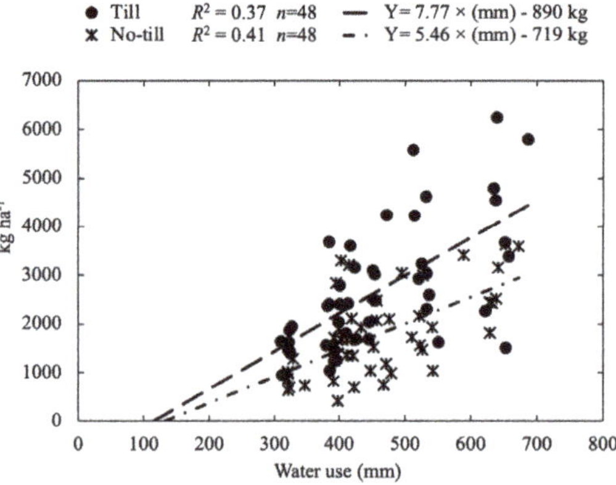

Figure 7. Pearl millet forage dry matter (DM)/water use (WU) relationship with tillage levels indicating management impact on DM production for 2016 and 2017 at Canyon, TX.

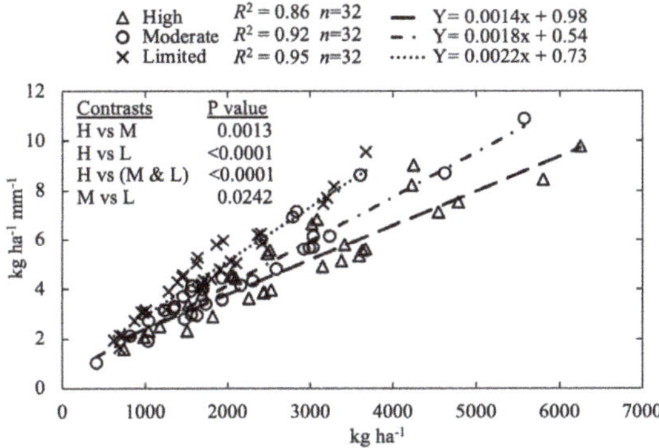

Figure 8. Direct relationship between water use efficiency (WUE) and pearl millet dry matter (DM). Slopes show trends of increasing WUE as DM increases for each irrigation level.

The proceeding discussion explaining WUE differences in the tilled treatment also helps explain the absence of main effect interactions and lack of WUE response to decreasing irrigation. Poor soil nutrient supply and high weed competition reduced pearl millet crop growth thereby obscuring any possible interactions among the main effects and responses typically reported from no-till or water stress studies. The authors of [20,22,41] found that pearl millet WUE increased as water stress also increased, which was not observed in this study. However, the relationship between WUE and DM [26,37,38] affords direct comparison of WUE at each irrigation level (Figure 8). Statistical contrast analysis of the slopes and intercepts for the respective irrigation levels [26] show that L had greater positive gain in WUE than was achieved in the H or M. This analysis suggests that WUE at the L level had a higher rate of return for accumulated DM than the other water levels. The authors of [22] demonstrated that in water stressed pearl millet WUE increased 23% from well-watered conditions. Analysis of their results demonstrates that pearl millet maintains physiological functions even when constrained by water stress. For example, the ratio of net photosynthesis (g CO_2 m^{-2} h^{-1}) to WU (mm) was greater in the limited water treatment than in the full water treatment (0.029, 0.002, respectively) [22]. This result can be explained by the biochemical pathway in warm season C_4 plants used to continue photosynthesis when vapor pressure deficit is high in low density canopies [45,46]. Furthermore, [44] explained that increased WUE this is the result of a greater proportional increase in WUE for low LAI values than when LAI is >2. Thus, the direct relationship between WUE and DM illustrates an ecophysiological response from pearl millet interacting with the crop environment.

Although irrigation applied lead to increased DM, WU did not well predict the behavior of pearl millet forage production across both years and all treatments. The production function y = 6.68 × x (mm) − 837 kg ha^{-1} (R^2 = 0.31, n = 96) cannot be readily utilized in other production situations that might have better weed and sol nutrient management. In contrast, for most DM/WU linear regressions, the coefficient of determination is much higher (>0.7) [33,42] and more confidence is given to estimating crop yield from WU given a set of WUE optimizing management strategies.

5. Conclusions

These results demonstrate the challenges of field study management to optimize WUE. Future research in the study region might explore agroecological strategies to mitigate poor soil nutrient supply and weed competition for pearl millet by practicing intercropping with legumes or agroforestry where possible [24]. Despite the factors that obscured the main effects and interactions, greatest average forage production was achieved with the highest irrigation level, however, highest WUE was attained in tilled soil due to greater LAI, light interception, and plant growth than in no-till. While the application of water increases forage production, low LAI values will increase estimated E and reduce WUE, especially without adequate nutrient application. Differences in DM were associated with changes in soil management, which ultimately resulted in higher WUE for tilled soil. The DM/WU pearl millet forage production function lacked strong correlation because of weeds and LAI < 2. An agronomic WUE optimization plan for pearl millet forage production should include conventional till, weed control, and proper nutrient management. Narrow is preferable to wide row spacing for greater LAI and light interception earlier in the growing season. If irrigation is available, water application should range from 400–600 mm of total water available for ET to maximize forage production, however, climatic demand will cause greater water loss from transpiration than if limited water is applied. Future research has an opportunity to conduct a more comprehensive review and meta-analysis of global pearl millet management practices used to improve water use efficiency and the effective use of water [47].

Author Contributions: Conceptualization, B.C., B.B., M.R.; Methodology, B.C., B.B., and M.R.; Formal analysis, B.C., M.D., and M.R.; Funding acquisition, B.B.; Investigation, B.C., and B.B.; Project administration, B.C., and B.B.; Resources, B.B.; Supervision, B.B., M.D., and M.R.; Validation, B.B., M.D., and M.R.; Visualization, B.C.; Writing—original draft, B.C.; Writing—review and editing, B.C., B.B., M.D., and M.R. All authors have read and agreed to the published version of the manuscript.

Funding: This research received no external funding.

Acknowledgments: Research funding was provided by the Dryland Agriculture Institute, West Texas A&M University, Canyon, TX, USA. The authors also wish to express their gratitude for the time and efforts of two anonymous reviewers whose thoughts contributed to the improvement of the manuscript.

Conflicts of Interest: The authors declare no conflict of interest.

References

1. Farooq, M.; Hauuain, M.; Ul-Allah, S.; Siddique, K.H.M. Physiological and Agronomic Approaches for Improving Water-Use Efficiency in Crop Plants. *Agric. Water Manag.* **2019**, *219*, 95–108. [CrossRef]

2. Mbava, N.; Mutema, M.; Zengeni, R.; Shimelis, H.; Chaplot, V. Factors Affecting Crop Water Use Efficiency: A Worldwide Meta-Analysis. *Agric. Water Manag.* **2020**, *228*. [CrossRef]

3. Andrews, D.L.; Kumar, K.A. Pearl millet for food, feed, and forage. In *Advances in Agronomy*; Sparks, D.L., Ed.; Academic Press, Inc.: Cambridge, MA, USA, 1992; Volume 48, pp. 89–139.

4. Mason, S.C.; Maman, N.; Pale, S. Pearl Millet Production Practices in Semi-Arid West Africa: A Review. *Exp. Agric.* **2015**, *15*, 501–541. [CrossRef]

5. Samake, O.; Stomph, T.J.; Kropff, M.J. Integrated Pearl Millet Management in the Sahel: Effects of Legume Rotation and Follow Management on Productivity and Striga hermonthica infestation. *Plant Soil* **2006**, *286*, 245–257. [CrossRef]

6. Baumhardt, R.L.; Salinas-Garcia, J. Dryland agriculture in Mexico and the U.S. Southern Great Plains. In *Dryland Agriculture*, 2nd ed.; Agronomy Monograph No. 23; American Society of Agronomy, Crop Science Society of America, Soil Science Society of America: Madison, WI, USA, 2006.

7. Bhattarai, B.; Singh, S.; West, C.P.; Saini, R. Forage Potential of Pearl Millet and Forage Sorghum Alternatives to Corn under the Water-Limiting Conditions of the Texas High Plains: A Review. *Crop Forage Turfgrass Manag.* **2019**, *5*, 1–12. [CrossRef]

8. Machicek, J.A.; Blaser, B.C.; Darapuneni, M.; Rhoades, M.B. Harvesting regimes affect brown midrib sorghum-sudangrass and brown midrib pearl millet forage production and quality. *Agronomy* **2019**, *9*, 416. [CrossRef]

9. FAO. Production Quantities of Millet by Country. FAOSTAT, Crops. 2020. Available online: http://www.fao.org/faostat/en/#data/QC/visualize (accessed on 22 October 2020).

10. Ismail, S.M. Optimizing Productivity and Irrigation Water Use Efficiency of Pearl Millet as a Forage Crop in Arid Regions Under Different Irrigation Methods and Stress. *Afr. J. Agric. Res.* **2012**, *7*, 2509–2518. [CrossRef]

11. Sharma, B.; Kumari, R.; Kumari, P.; Meena, S.K.; Singh, R.M. Effect of Planting Pattern of Productivity and Water Use Efficiency of Pearl Millet in the Indian Semi-Arid Region. *J. Indian Soc. Soil Sci.* **2015**, *6*, 259–265. [CrossRef]

12. Silungwe, F.R.; Graef, F.; Bellingrath-Kimura, S.D.; Tumbo, S.D.; Kahimba, F.C.; Lana, M.A. The Management Strategies of Pearl Millet Farmers to Cope with Seasonal Rainfall Variability in a Semi-Arid Agroclimate. *Agronomy* **2019**, *9*, 400. [CrossRef]

13. Payne, W.A. Managing yield and water use of pearl millet in the Sahel. *Agron. J.* **1997**, *89*, 481–490. [CrossRef]

14. Maman, N.; Lyon, D.J.; Mason, S.C.; Galusha, T.D.; Higgins, R. Pearl millet and grain sorghum yield responses to water supply in Nebraska. *Agron. J.* **2003**, *95*, 1618–1624. [CrossRef]

15. Bhattarai, B.; Singh, S.; West, C.P.; Ritchie, G.L.; Trostle, C.L. Water Depletion Pattern and Water Use Efficiency of Forage Sorghum, Pearl Millet, and Corn Under Water Limiting Conditions. *Agric. Water Manag.* **2020**, *238*, 106206. [CrossRef]

16. Jahansouz, M.R.; Afshar, R.K.; Heidari, H.; Hashemi, M. Evaluation of Yield and Quality of Sorghum and Millet as Alternative Forage Crops to Corn under Normal and Deficit Irrigation Regimes. *Jordan J. Agric. Sci.* **2014**, *10*, 699–715. Available online: https://platform.almanhal.com/Files/2/94696 (accessed on 20 October 2020).

17. Ibrahim, A.; Abaidoo, R.C.; Fatandji, D.; Opoku, A. Hill Placement of Manure and Fertilizer Micro-Dosing Improves Yield and Water Use Efficiency in the Sahelian Low Input Millet-Based Cropping System. *Field Crops Res.* **2015**, *180*, 29–36. [CrossRef]

18. Manyame, C. On-Farm Yield and Water Use Response of Pearl Millet to Different Management Practices in Niger. Ph.D. Thesis, Texas A&M University, College Station, TX, USA, 2006. Available online: https://core.ac.uk/download/pdf/147132537.pdf (accessed on 20 October 2020).

19. Neal, J.S.; Fulkerson, W.J.; Hacker, R.B. Differences in Water Use Efficiency among Annual Forages Used by the Dairy Industry Under Optimum and Deficit Irrigation. *Agric. Water Manag.* **2011**, *98*, 759–774. [CrossRef]

20. Rostamza, M.; Chaichi, M.; Jahansouz, M.; Alimadadi, A. Forage Quality, Water Use and Nitrigen Utilization Efficiencies of Pearl Millet (*Pennisetum americanum* L.) Grown Under Different Soil Moisture and Nitrogen Levels. *Agric. Water Manag.* **2011**, *98*, 1607–1614. [CrossRef]

21. Geerts, S.; Raes, D. Deficit Irrigation as an on-farm Strategy to Maximize Crop Water Productivity in Dry Areas. *Agric. Water Manag.* **2009**, *96*, 1275–1284. [CrossRef]

22. Singh, B.R.; Singh, D.P. Agronomic and physiological responses of sorghum, maize, pearl millet, to irrigation. *Field Crops Res.* **1995**, *42*, 57–67. [CrossRef]

23. Hatfield, J.L.; Sauer, T.J.; Prueger, J.H. Managing Soils to Achieve Greater Water Use Efficiency: A review. *Agron. J.* **2001**, *93*, 271–280. [CrossRef]

24. Payne, W.A. Optimizing crop water use in sparse stands of pearl millet. *Agron. J.* **2000**, *92*, 808–814. [CrossRef]

25. Zhang, Z.; Whish, J.P.M.; Bell, L.W.; Nan, Z. Forage Production, Quality and Water-Use-Efficiency of Four Warm-Season Annual Crops at Three Sowing Times in the Loess Plateau Region of China. *Eur. J. Agron.* **2017**, *84*, 84–94. [CrossRef]

26. Nielsen, D.C.; Vigil, M.F.; Benjamin, J.G. Forage response to water use for dryland corn, millet, and triticale in the Central Great Plains. *Agron. J.* **2006**, *98*, 992–998. [CrossRef]

27. Unger, P.W. Alternative and opportunity dryland crops and related soil conditions in the Southern Great Plains. *Agron. J.* **2001**, *93*, 216–226. [CrossRef]

28. Web Soil Survey. 2019. Available online: https://websoilsurvey.sc.egov.usda.gov/App/HomePage.htm.

29. Allen, R.G. *REF-ET, Reference Evapotranspiration Calculator, Version 4.1*; Research and Extension Center, University of Idaho: Kimberly, ID, USA, 2016.

30. Ritchie, J.T.; Alagarswamy, G. Simulation of sorghum and pearl millet phenology. In *Modeling the Growth and Development of Sorghum and Pearl Millet*; Virmani, S.M., Tandon, H.L.S., Alagarswamy, G., Eds.; Research Bulletin No. 12; ICRISAT: Patancheru, Andhra Pradesh, India, 1989.

31. Olsen, J.E.; Hansen, P.K.; Berntsen, J.; Christensen, S. Simulation of above-ground suppression of competing species and competition tolerance in winter wheat varieties. *Field Crops Res.* **2004**, *89*, 263–280. [CrossRef]

32. Blaser, B.C.; Singer, J.W.; Gibson, L.R. Winter cereal canopy effect on cereal and interseeded legume productivity. *Agron. J.* **2011**, *103*, 1080–1185. [CrossRef]

33. Stephenson, R.J.; Posler, G.L. Forage yield and regrowth of pearl millet. *Trans. Kans. Acad. Sci.* **1984**, *87*, 91–97. [CrossRef]

34. Nielsen, D.C.; Vigil, M.F.; Benjamin, J.G. Evaluating decision rules for dryland rotation crop selection. *Field Crops Res.* **2011**, *120*, 254–261. [CrossRef]

35. Paris, P.; Di Matteo, G.; Tarchi, M.; Tosi, L.; Spaccino, L.; Lauteri, M. Precision Subsurface Drip Irrigation Increases Yield While Sustaining Water-Use Efficiency in Mediterranean Poplar Bioenergy Plantations. *For. Ecol. Manag.* **2018**, *409*, 749–756. [CrossRef]

36. Bouchard, A.; Vanasse, A.; Seguin, P.; Belanger, G. Yield and Composition of Sweet Pearl Millet as Affected by Row Spacing and Seeding Rate. *Agron. J.* **2011**, *103*, 995–1001. [CrossRef]

37. Musick, J.T.; Jones, O.R.; Stewart, B.A.; Dusek, D.A. Water-yield relationships for irrigated and dryland wheat in the U.S. Southern Plains. *Agron. J.* **1994**, *86*, 980–986. [CrossRef]

38. Ehlers, W.; Goss, M. *Water Dynamics and Plant Production*, 2nd ed.; CAB International: Boston, MA, USA, 2016.

39. Myers, R.J.K.; Foale, M.A. Row spacing and population density in grain sorghum—A simple analysis. *Field Crops Res.* **1981**, *4*, 147–154. [CrossRef]

40. Mohler, C.L. Enhancing the Competitive Ability of Crops. In *Ecological Management of Agricultural Weeds*; Liebman, M., Mohler, C.L., Staver, C.P., Eds.; Cambridge University Press: Cambridge, UK, 2004; pp. 269–305.

41. Nagaz, K.; Toumi, I.; Mahjoub, I.; Masmoudi, M.M.; Mechlia, N.B. Yield and Water-Use Efficiency of Pearl Millet (*Pennisetum glaucum* (L.) R. Br.) Under Deficit Irrigation with Saline Water in Arid Conditions of Southern Tunisia. *Res. J. Agron.* **2009**, *3*, 9–17.

42. Nielsen, D.C.; Vigil, M.F. Defining a Dryland Grain Sorghum Production Function for the Central Great Plains. *Agron. J.* **2017**, *109*, 1582–1590. [CrossRef]

43. Hao, B.; Xue, Q.; Bean, B.W.; Rooney, W.L.; Becker, J.D. Biomass Production, Water and Nitrogen Use Efficiency in Photoperiod-Sensitive Sorghum in the Texas High Plains. *Biomass Bioenergy* **2014**, *62*, 108–116. [CrossRef]

44. Ritchie, J.T. Efficient water use in crop production: Discussion on the generality of relationships between biomass production and evapotranspiration. In *Limitations to Efficient Water Use in Crop Production*; Tanner, H.M., Jordan, W.R., Sinclair, T.R., Eds.; ASA-CSSA-SSSA: Madison, WI, USA, 1983; pp. 29–44.

45. Sinclair, T.R.; Weiss, A. *Principles of Ecology in Plant Production*, 2nd ed.; CABI: Cambridge, MA, USA, 2010.

46. Hardwick, S.R.; Toumi, R.; Pfeifer, M.; Turner, E.C.; Nilus, R.; Ewers, R.M. The Relationship between Leaf Area Index and Microclimate in Tropical Forest and Oil Palm Plantation: Forest Disturbance Drives Changes in Microclimate. *Agric. For. Meteorol.* **2015**, *201*, 187–195. [CrossRef]

47. Blum, A. Effective Use of Water (EWU) and not Water-Use Efficiency (WUE) is the Target of Crop Yield Improvement under Drought Stress. *Field Crops Res.* **2009**, *112*, 119–123. [CrossRef]

Publisher's Note: MDPI stays neutral with regard to jurisdictional claims in published maps and institutional affiliations.

Article

Feeding Emitters for Microirrigation with a Digestate Liquid Fraction up to 25% Dilution Did Not Reduce Their Performance

Simone Bergonzoli [1], Massimo Brambilla [1], Elio Romano [1], Sergio Saia [2,*], Paola Cetera [2], Maurizio Cutini [1], Pietro Toscano [1], Carlo Bisaglia [1] and Luigi Pari [2]

[1] Council for Agricultural Research and Economics-Research Centre for Engineering and Agro-Food Processing, CREA-IT, Treviglio, 24047 Bergamo, Italy; simone.bergonzoli@crea.gov.it (S.B.); massimo.brambilla@crea.gov.it (M.B.); elio.romano@crea.gov.it (E.R.); maurizio.cutini@crea.gov.it (M.C.); pietro.toscano@crea.gov.it (P.T.); carlo.bisaglia@crea.gov.it (C.B.)

[2] Council for Agricultural Research and Economics—Research Centre for Engineering and Agro-Food Processing (CREA-IT), Via della Pascolare, 16, Monterotondo, 00015 Roma, Italy; paola.cetera@crea.gov.it (P.C.); luigi.pari@crea.gov.it (L.P.)

* Correspondence: sergio.saia@crea.gov.it

Received: 11 July 2020; Accepted: 5 August 2020; Published: 6 August 2020

Abstract: Irrigation with wastewater can strongly contribute to the reduction of water abstraction in agriculture with an especial interest in arid and semiarid areas. However, its use can have drawbacks to both soil and micro-irrigation systems, especially when the total solids in the wastewater are high, such as in digestate liquid fractions (DLF) from plant material. The aim of this study was thus to evaluate the performances of a serpentine shaped micro-emitter injected with a hydrocyclone filtered DLF (HF-DLF) from corn + barley biomass and evaluate the traits of the liquid released within a 8-h irrigation cycle. HF-DLF was injected at 10%, 25%, and 50% dilution compared to tap water (at pH = 7.84) and the system performances were measured. No clogging was found, which likely depended on both the shape of the emitter and the high-pressure head (200 kPa). HF-DLF dilution at 10%, 25%, and 50% consisted in +1.9%, +3.5, and −4.9% amount of liquid released compared to the control. Fluid temperature during irrigation (from 9:00 to 17:00) did not explain the difference in the released amounts of liquid. In 10% HF-DLF % and 25% HF-DLF, a pH difference of + 0.321 ± 0.014 pH units compared to the control was found, and such difference was constant for both dilutions and at increasing the time. In contrast, 50% HF-DLF increased pH by around a half point and such difference increased with time. Similar differences among treatments were found for the total solids in the liquid. These results indicate that 50% HF-DLF was accumulating materials in the serpentine. These results suggest that a low diluted HF-DLF could directly be injected in irrigation systems with few drawbacks for the irrigation system and contribute to water conservation since such wastewater are available from the late spring to the early fall, when water requirements are high.

Keywords: clogging; drip irrigation; emitter; hydrocyclone; digestate liquid fraction; wastewater

---

1. Introduction

Water availability for crops in various areas of the world is reducing because of climate change and the use of fresh water for other human uses. Climate change is increasing the demand for water in agriculture through both a general increase of temperatures, and thus of the evapotranspiration demand, and the increase of their variability [1,2].

Irrigation with low-quality water, and especially wastewater, was thus proposed a long time ago as a suitable measure to mitigate the shortage of high quality water [3,4]. The use of wastewater or filtrate of the liquid fractions of various wastes can increase the availability of water for agriculture. However, its use may result in a wealth of problems following the effects wastewater has on soil and irrigation systems. These include salinization or pH variation [5], a reduction of other soil fertility properties [6], and an increase of soil hydrophobicity [7], the clogging of the emitters [3], as well as the deposition of solid materials in the tanks or other parts of the irrigation system [8]. Besides, wastewaters may contain pollutants and pathogens which harm plants and animals, albeit the treatments they undergo are meant to prevent health risk following their use or disposal [9,10].

Drip systems allow the achievement of high irrigation efficiency in areas with high water demand and low water availability. In these systems, water pressure is usually below 200 kPa, and emitters have internal serpentine to compensate pressure loss along the line and potential fluctuations in the water pressure.

The success of the use of wastewater in the irrigation depends on a wealth of factors. These include the amount of solids in the wastewater or its filtrate, the ability of the suspended material to form biofilms, the pressure of the water in the system, the type of filters and emitters, and age of the systems [8,11–14]. In case of low pressure (60 kPa) and low rate emitters (0.9 and 1.4 L h^{-1} emitter^{-1}), high quality drip tapes showed a reduction of uniformity of distribution by 5.2% on average depending on the activated sludge used as secondary effluent [15]. Similar results were found by Puig-Bargués et al. [14], who also reported that pressure compensating emitters performed better than non-pressure compensating emitters. Chlorination or flushing of the pipes at the end of each irrigation cycle proved to reduce the impact on clogging [14,16]. However, such treatments imply an additional cost, and application of chlorine to the soil may have harmful effects both on the soil and on plants. In the latter work [16], application of compressed air cleaning at a pressure of 1.96 kPa did not mitigate the incidence of drippers clogging.

Digestate from crop biomass and manure is increasingly being used, and its liquid fraction was indicated as a potential source for a wastewater irrigation [17]. When used for irrigation purposes, information on the solid particles fractions, mostly salts, of these liquids in the irrigation systems are scarce. Such salts are likely to precipitate and, together with other suspended solids, can easily clog the emitters of a drip irrigation system by a fouling accumulation [18]. In turn, digestate filtrates used for irrigation can increase plant yield [6], even when compared to an irrigated + fertilized treatment [19]. However, little information is available about the efficiency of many emitters when subjected to wastewater, especially when using the liquid fraction of the biomass-based digestate, as the solid fraction can contain high amounts of organic material [20].

The aim of the present study was thus to test the efficiency of a commercial emitter when injected with the liquid fraction of the effluent from an agricultural biogas unit previously treated with a hydrocyclone filtration system.

2. Materials and Methods

2.1. Experimental Setup

The study was conducted in February 2020 at the CREA-IT institute (45°31′21.9″ N 9°33′54.9″ E, Treviglio, Bergamo, Italy). The liquid digestate used was gathered from an anaerobic digestion plant, stored in a tank and filtered using a hydrocyclone filter (Alfaturbo Hydrocyclone Sand separator 2″). The hydrocyclone filter was placed between a storage tank and the operating tank. Both tanks had a 1 m^3 maximum volume. After the filtration, the filtered liquid was shaken once per day before the beginning of the tests (see below) by gently shaking the tank using a forklift truck. The storage of the liquid digestate before using it at the irrigation setup lasted for 8 days.

The characteristic of the liquid digestate before the filtration are depicted in Table 1. The digestate liquid fraction was kindly provided by the Società Agricola Pallavicina S.R.L. (Via Fara—24047 Treviglio, IT), which also undergoes quality analyses, and did not display the presence of any pathogen nor pollutants according to the Italian laws.

Table 1. Composition of the digestate liquid fraction used for the test. Values are means ± s.d. (3 subsamples) provided by the management office of the digester. The analyses were made on the digestate liquid fraction hydrocyclone filtering.

Trait	Value	Unit	Method
Dry matter content at 105 °C	4.91 ± 0.03	%	IRSA CNR Q 64 Vol 2 1985
Ashes (dry matter content at 600 °C)	1.42 ± 0.02	%	
Chemical Oxigen demand (COD)	51.0 ± 1.8	$g\ O_2\ kg^{-1}$	APAT IRSA CNR 5130 Man 29 2003
Total N	3.55 ± 0.02	$g\ N\ kg^{-1}$	IRSA CNR Q 64 Vol 3 1985
NH_4^+-N	1.848 ± 0.025	$g\ N\ kg^{-1}$	
Phosphorus	0.843 ± 0.051	$g\ P\ kg^{-1}$	UNI EN 16174:2012 and 16170:2015
Potassium	4.96 ± 0.25	$g\ K\ kg^{-1}$	

A water pump of 0.75 kW power (Pedrollo company, model: JSWm 2CX, San Bonifacio, Verona, Italy) was used to pump the digestate liquid fraction in the irrigation system. The operating pressure was set to 0.2 MPa. Samples were taken from the non-filtered digestate liquid fraction (DLF), from the filter outlet and from the filtered DLF to determine dry matter content and pH.

The experimental design consisted of two factors: filtrate dilution (FD) × time of the sampling within the irrigation cycle (TIME). The FD factor had four treatments: One control using freshwater and three filtrate dilutions (10%, 25%, and 50% of filtrate in freshwater). Each irrigation cycle lasted 8 h and samples were taken once per hour.

One irrigation cycle per day was performed, the irrigation cycles began with the tap water and continued with each increasing concentration of the digestate to avoid contamination. Each water-HF-DLF mixture was prepared mixing the relevant amount of tap water and HF-DLF in an operating tank and reflushing it several times with the pump. The water pump used was set at 0.2 MPa operating pressure, and the irrigation tank was filled with 400 L of DLF. Before starting each test, three samples were collected from the irrigation tank to measure the dry matter and the pH of the solution, following the methodology described above. In addition, flushing with tap water was performed by 15 min at the end of each cycle to allow for the cleaning the system. A pre-flushing was also made before the beginning of the first experiment with tap water. Within each irrigation cycle, the pump recycled part of the water or diluted HF-DLF into the tank to keep it mixed and avoid particle deposition. The irrigation system was organized by three polyethylene 1-m long dripper tubes (Stocker company N°26085, Bozen, Italy), as replicates, with three emitters each. The dripper tubes used had 0.016 m of diameter (maximum design pressure 0.4 MPa) and were spaced 0.33 m each other. Water flow declared by the manufacturer was 2 L h^{-1}. Emitters were not changed from each cycle to the following one.

2.2. Measurements and Analyses

During the tests, the water or filtrate dropping from the tubes was collected in plastic flagons placed underneath each emitter (Figure 1). The flagons were weighted once per hour with a portable scale (RADWAG WLC6/C1/R, Radom, Poland, used with 0.1 g sensitivity) to calculate the water flow (g h^{-1}). At the time of weighting the turbidity and the temperature of the liquid were measured. The temperature of the water or DLF were measured in °C using the DS18B20 digital thermometer (Maxim IC, San Jose, CA, USA). The turbidity of water or DLF were measured by using the turbidity sensor SKU:SEN0189 (Arduino, Ivrea, Italia), which was used as an indirect measurement of filtrate and water quality. The turbidity sensor SKU:SEN0189 uses light to detect suspended particles in water by measuring the light transmittance and scattering rate, which depends on the concentration of the total suspended solids (TSS) in the solution/dispersion. In particular, the sensor provide an output expressed in mV, which should be calibrated to the corresponding Nephelometric Turbidity Units (NTU). The output slightly and linearly decreases at increasing temperatures. In addition, the relationship between output and NTU value is not linear (for a brief description see https://wiki.dfrobot.com/Turbidity_sensor_SKU__SEN0189). In particular, the higher the sensor output, the lower the liquid NTU value. The manufacturer provide an output for pure water of 4.1 ± 0.3*V* when temperature span from 10 °C to 50 °C. Integration of the temperature and turbidity systems was made according to [21]. In the present work, the output of sensor was provided along with a direct measurement of the total suspended solids.

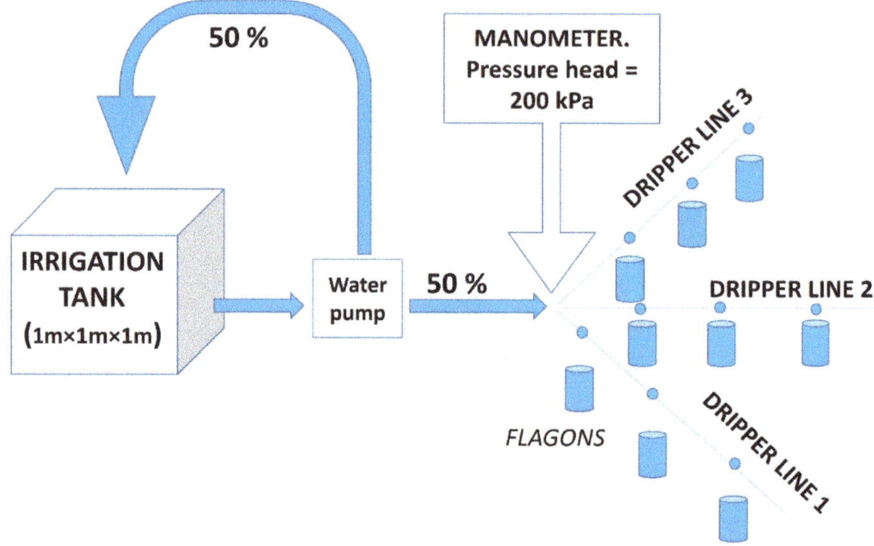

Figure 1. Design of the irrigation tests.

Dry matter and moisture content of the samples were assessed by oven drying at 105 °C until constant weight [22]. The pH of the samples was measured with no dilution by using a CRISON GLP21 pH-meter (Hach Lange Spain, S.L.U., Barcelona, Spain).

Then, the samples of each emitter per line were mixed and a random composite subsample of 500 mL of liquid was taken. In total, 24 sample per irrigation test were obtained (8 sampling moments × 3 irrigation lines). Each irrigation test consisted of the injection of a DLF dilution in an 8-h irrigation cycle. Thus 72 total samples of DLF released by each line were obtained. Dry matter of each subsample and its pH, following the methodology described above, were measured. For the control test (100% water), turbidity and pH were measured only before starting the test and no samples were collected during the test. This was done since these variables did not change by the time in the control from previous tests (data not shown). To monitor the air temperature and humidity of the indoor environment where the test was performed a sensor DHT22 (Guangzhou Aosong Electronics Co., Ltd., Guangzhou, China) was used. The Waterproof DS18B20 Digital Temperature Sensor was used to read the liquid temperature.

2.3. Computations and Statistical Analysis

The amounts of OH$^-$ ions per ton of solution released by the emitters were computed by using the pH and used as a proxy of the potential of the irrigation with the DLF to increase the pH of a soil compared to the tap water used as a control. The analysis of variance was performed according to the statistical design by means of a general linear mixed model (Glimmix procedure in SAS/STAT 9.2 statistical package; SAS Institute Inc., Cary, NC, USA). The model used was similar to that shown in Saia et al. [23] (see the supplementary material in [23] for both a description of the procedure and the SAS package model applied) in which TIME was modelled as a repeated measurement [24]. Unbiased estimates of variance and covariance parameters were estimated by restricted maximum likelihood (REML). Repeated measurement analysis was modelled by applying a random statement with a first-order autoregressive covariance structure. In particular, the subject of reference was the emitter for the data related to the amount of liquid released, its turbidity and temperature, and it was the line for date on the pH. Denominator degrees of freedom of each error were estimated by Kenward–Roger approximation and least square means (LSmeans, see below for a definition) of the treatment distributions were computed. Data were provided both as LSmeans in figures and arithmetic means in supplemental materials, along with each standard error estimation or computation, respectively. Differences among means were compared by applying t-grouping at the 5% probability level to the LSMEANS *p*-differences. Time-sliced significance was also computed.

When the effect of time was significant, variation per unit time was modelled. Variation by time per each variable and treatment significantly varying by time was fitted to a linear distribution function and significance of the regression models were computed using the Slide Write Plus for Windows version 7.01 (Advanced Graphics Software, Inc., Encinitas, CA, USA).

3. Results

3.1. Effects of the Hydrocyclone Filtration and Resting Time on the Traits of the Digestate Liquid Fraction

Filtration of the digestate liquid fraction (DLF) influenced the pH of the different resulting fractions (Figure 2) pointing out that the native fraction before filtering had a pH value higher than 8.2 and not statistically appreciable differences were found after the filtration (hereafter referred as hydrocyclone filtered DLF, or HF-DLF). When the HF-DLF was allowed to rest for eight days before the beginning of the experiment, the solution at 50% dilution showed a lower pH compared to both the freshly made native DLF and the freshly made HF-DLF soon after the filtration. However, such latter difference was not statistically appreciable according to the conservative post-hoc test used. The pH value of the native DLF was higher than the fraction discarded from the filter and the tap water.

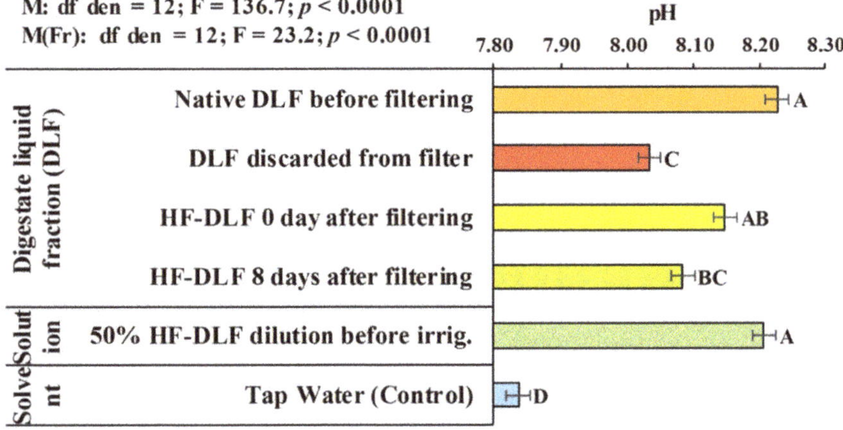

Figure 2. Values of pH of the digestate liquid fraction (DLF), before and after hydrocyclone filtering (HF), the 50% dilution of the HF-DLF, and the tap water used for the experiment. Bares are least square means (LSmeans) ± Lsmeans standard error estimates. Bars with a letter in common cannot be considered different according to a conservative Tukey-grouping applied to the *p*-differences of the LSmeans. Results of the statistical analysis are embedded.

The output of the turbidity sensor did not change by the filtration among the DLFs (Figure 3), which recorded value closed to 14.2 mV (please mind that, in theory, the lower the conductibility, the higher the turbidity). The tap water showed a turbidity sensor output close to 800 mV, 56-fold than any of DLFs or HF-DLF, on average.

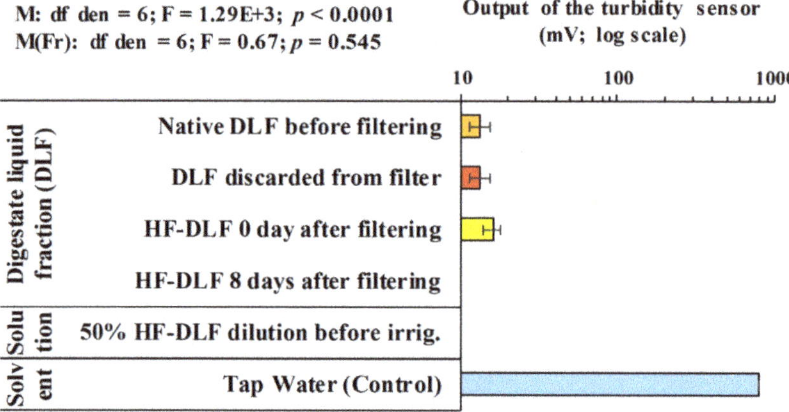

Figure 3. Output of the turbidity sensor in the digestate liquid fraction (DLF), before and after hydrocyclone filtering (HF), the 50% dilution of the HF-DLF, and the tap water used for the experiment. Bars are LSmeans ± Lsmeans standard error estimates. Bars with a letter in common cannot be considered different according to a conservative Tukey-grouping applied to the *p*-differences of the LSmeans. Results of the statistical analysis are embedded.

The dry matter concentration varied among the DLFs (Figure 4), with the fraction discarded from the filter showing a relative concentration compared to the native DLF 45% higher. The native DLF also showed a marginally, albeit significantly, lower dry matter concentration than the HF-DLF (−1.1% relative difference, corresponding to −0.02%).

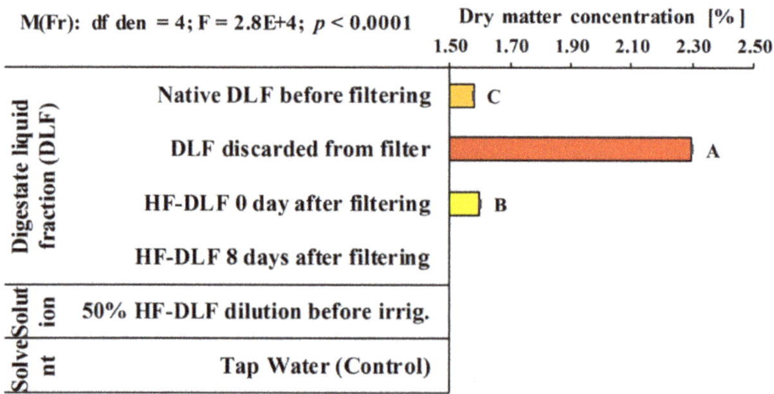

Figure 4. Dry matter concentration of the digestate liquid fraction (DLF), before and after hydrocyclone filtering (HF), the 50% dilution of the HF-DLF, and the tap water used for the experiment. Bars are LSmeans ± Lsmeans standard error estimates. Bars with a letter in common cannot be considered different according to a conservative Tukey-grouping applied to the *p*-differences of the LSmeans. Results of the statistical analysis are embedded.

3.2. Effects of the Digestate Liquid Fraction Dilution on the Emitter Performances and Solution Traits

Results of the statistical analyses of the irrigation test are shown in Table 2. Both the amount of water in the control and HF-DLF and its turbidity varied by the treatment at increasing time, with differences more marked among treatments in the early stages of the irrigation. The dilution of the DLF influenced the quantity of the solution released by the emitters during the test depending on the percentage of the dilution (Figure 5, Supplementary Material Table S1). In particular, water release increased almost constantly, whereas 10% and 25% dilution during the first 3 h. The 50% showed milder increases, and a total amount of water released slightly lower than the other treatments. The coefficient of variation of the system was in general lower than 5%, with some outlier only in the HF-DLF 25% dilution (Table 3).

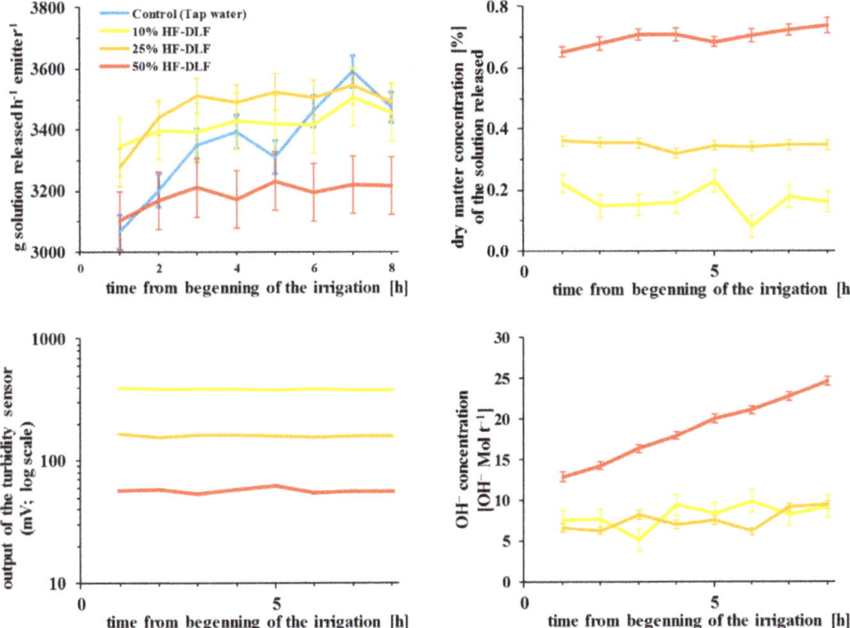

Figure 5. Amount of solution (tap water in the control and digestate liquid fractions (HF-DLF) diluted at the 10%, 25%, and 50%) released by the emitter each hour (upper left panel); dry matter concentration of the DLF released (upper right panel); turbidity of the HF-DLF released (lower left panel); and amount of OH$^-$ ions released per ton of water released each hour. Data are LSmeans ± LSmeans standard error estimates. For post-hoc comparisons and raw data see Supplementary Material Table S3.

Table 2. Results of the statistical analysis (degrees of freedom estimate of the error (DF den); F statistics; and *p* values) of the general linear mixed model applied to the amount of solution released each hour, its turbidity, dry matter concentration, pH difference compared to the control, and the amount of OH^- released with the solution per hour (expressed as mol ton^{-1}). Factors were the solution dilution (FD) and variation in time (1-h step of a 8-h irrigation cycle). Data were analyzed with a repeated treatment option and after which differences by each time-step were sliced. When the *p*-values were lower than 0.05, F and *p* were shown in bold. See the Supplementary Material for the post-hoc test.

Effect	Amount of Solution Released Each Hour			Turbidity			Dry Matter Concentration			[OH$^-$] Concentration			pH Difference Compared to Control		
	Den DF	F	*p*	Den DF	F	*p*	Den DF	F	*p*	Den DF	F	*p*	Den DF	F	*p*
FD	101.8	3.2	0.0284	52.03	3.6×10^4	<0.0001	7.395	540.3	<0.0001	11.12	912.2	<0.0001	15.5	188.2	<0.0001
Time	83.06	47.3	<0.0001	139.6	9.1	<0.0001	30.7	1.2	0.3452	25.62	24.0	<0.0001	20.06	12.7	<0.0001
FD × Time	149.6	52.9	<0.0001	149.8	7.4	<0.0001	28.01	2.2	0.0389	29.78	12.9	<0.0001	21.15	7.4	<0.0001
Sliced By Time [h]															
1	97.0	3.9	0.011	64.9	9.7×10^3	<0.0001	23.6	121.5	<0.0001	18.5	34.4	<0.0001	13.0	41.4	<0.0001
2	109.4	4.7	0.004	91.6	9.7×10^3	<0.0001	11.4	97.8	<0.0001	24.5	72.8	<0.0001	20.0	64.2	<0.0001
3	111.0	3.2	0.025	95.0	1.0×10^4	<0.0001	24.1	165.7	<0.0001	25.5	74.4	<0.0001	20.5	65.4	<0.0001
4	111.4	3.1	0.031	95.4	9.8×10^3	<0.0001	12.7	128.3	<0.0001	25.5	111.4	<0.0001	20.5	85.9	<0.0001
5	111.5	4.1	0.008	95.4	9.1×10^3	<0.0001	22.3	129.1	<0.0001	25.5	149.7	<0.0001	20.5	101.7	<0.0001
6	111.6	3.0	0.034	95.4	9.8×10^3	<0.0001	13.2	142.3	<0.0001	25.5	205.9	<0.0001	20.5	140.0	<0.0001
7	111.7	4.4	0.006	95.4	9.5×10^3	<0.0001	20.3	159.0	<0.0001	25.5	183.6	<0.0001	20.5	110.3	<0.0001
8	111.8	2.5	0.065	95.4	9.5×10^3	<0.0001	15.5	112.9	<0.0001	25.5	226.2	<0.0001	20.5	119.0	<0.0001

Table 3. Coefficient of variation of each treatment (n = 9, consisting of 3 lines with 3 emitters each) and relative distribution percentiles at 0.025, 0.25, 0.5 (median), 0.75, and 0.975. Data of the three dilutions of the hydrocyclone filtered digestate liquid fraction (HF-DLF) were showed singly and bulked.

	Control (Tap Water)	10% Dil. HF-DLF	25% Dil. HF-DLF	50% Dil. HF-DLF	10% + 25% + 50% Dil. HF-DLF Bulked Data
Mean of coefficient of variation	1.65	3.41	3.69	3.66	3.59
Percentile 0.025	1.24	3.26	2.65	3.55	2.71
Percentile 0.250	1.45	3.35	2.79	3.59	2.89
Percentile 0.500 (median)	1.66	3.41	2.82	3.65	3.43
Percentile 0.750	1.86	3.46	2.86	3.68	3.61
Percentile 0.975	2.07	3.49	8.86	3.86	6.44

The dry matter content of the solutions (Figure 5) showed that the concentration strongly depended on the dilution rate, and to a lesser extent on time. In fact, higher values >0.6%) were recorded in the DLF 50%, compared to those in DLF 10% < 0.2%) and DLF 25% showed values ranging from 0.4% to 0.3%. The value of dry matter concentration (%) was quite stable all along the 8 h of irrigation test for all the treatments, except for DLF50% that showed a slight increase with time (Supplementary Material Table S1; Supplementary Material Table S2).

A similar trend, but more pronounced by the time, was found regarding turbidity (Figure 5; please mind that the higher is the turbidity value, the lower the liquid turbidity). For all the treatments the values recorded were stable over time and around 60 mV, 160 mV, and 380 mV for DLF 50%, 25% and 10% respectively. The analysis of the concentration of ions OH^- (Figure 5), calculated using the pH values, showed that even if the trend of treatments DLF 10% and 25% was slightly variable during the irrigation test, the values were included between 5 and 10 OH^- mol t^{-1} solution, with scarce differences by the time. Instead, treatment DLF 50% showed an increasing trend during the test with an initial value of 13 OH^- mol t^{-1} and a final value of 25 OH^- mol t^{-1}. Finally, we inspected a serpentine from the control and the DLF 50% (Figure 6) and found that no clogging occurred.

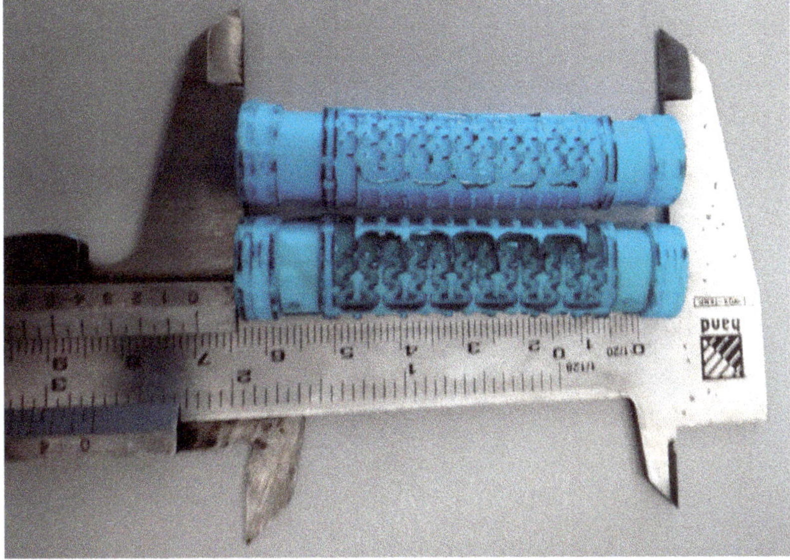

Figure 6. Serpentine inspection of the control (tap water; above) vs. 50% diluted hydrocyclone filtered digestate liquid fraction (below) emitters.

4. Discussion

In the present work, we studied the role of an increasing ratio between a hydrocyclone-filtered digestate liquid fraction (referred to as HF-DLF) and tap water on the performance of an irrigation system and water quality. Treatments included 3 HF-DLF ratios (10%, 25%, and 50% of total solution used for the irrigation) in contrast to tap water as control and measurements were taken at an hourly basis on an 8-h irrigation cycle, that simulates most of the irrigation cycles occurring in a broad range of crops.

The native digestate liquid fraction used before the hydrocyclone filtering had a higher pH than the tap water used (pH = 8.23 vs. 7.84, respectively) and such pH slightly reduced after the hydrocyclone filtering. Information on the effect that the hydrocyclone filtering has on the pH of digestate liquid fraction and its total solids are scarce, nonetheless, centrifuge filtering was shown to have relatively high efficiency [25]. Thus, the pH reduction after the hydrocyclone treatment may have been due to the fractionation of the calcium carbonates or other high-weight solids in the digestate liquid fraction, including cations. The digester diet of the material in the present study was mainly composed of corn and barley biomasses (residual and dedicated) and to a lesser extent of cow slurry. This kind of digestate has high contents in potassium, chloride, carbonates, and proteins [20]. It is thus likely that the high pH of the HF-DLF under study was due to a high content of basic, high molecular weight proteins, which can be removed by hydrocyclone filtering. Such a hypothesis is corroborated by the further reduction of pH of the DLF found eight days after filtering, before the irrigation experiment started, which may have been due to oxidation of the organic material in the HF-DLF that in such time-lapse was resting. Hydrocyclone filtering, however, did not result in a reduction of the turbidity and such results could be due to the high total solid concentration in the DLF following incomplete filtering, as pointed by Guilayn et al. [25]. Indeed, in our study, the HF-DLF showed a dry matter concentration of 16.0‰ (on a weight basis) and a pH = 8.15 soon after the filtration. These traits suggested a low quality fraction for drip irrigators according to the early classification by Nakayama and Bucks [3]. This likely was a main cause of the differences among the amount of HF-DLF released by the emitters at increasing time and varying the HF-DLF ratios within the irrigation system.

The manufacturer declared the used emitters as and releasing 2 L h^{-1} at 100 kPa. When subjected to the 200 kPa pressure of the present study, we found that the amount of water released in the control ranged from 3.07 L in the first hour and such an amount increased on average by the 1.9% h^{-1}. Such variation were higher than those found by Bodole et al. [26]. The variation of the amount of HF-DLF each emitter released per hour also increased with time in the three dilution treatments (0.5–0.9% h^{-1}), but to a lesser extent compared to the control. Such variations likely depended on the usury of the system. In particular, the dilution at 10% and 25% released 1.9% and 3.5%, respectively, more HF-DLF per cycle than the water released by the control, whereas the dilution at 50% released 4.9% less HF-DLF per cycle than the control. Besides, the differences between each HF-DLF dilution and control in the amount of water released per unit time declined with time. The temperature of the tap water or the HF-DLFs was similar among the treatments and increased linearly during each irrigation cycle, starting at 9.42 °C ± 0.27 °C at the 9:00 a.m. (moment of the beginning of each experiment) and increased at a rate of 0.94 °C h^{-1} ± 0.05 °C h^{-1} (data not shown). This implies that differences by time can only partly be explained by a heating of the emitters and thus the expansion of their pore size. We hypothesize that the emitters used in this experiment rapidly lost their ability to compensate for the irrigation rate at the pressure we used. In addition, the 10% and 25% diluted HF-DLF did not likely consist in a strong occlusion of the emitters. This is consistent with the constant rate of the dry matter content of 10% and 25% HF-DLF and increasing rate of the dry matter content of the 50% HF-DLF, which progressively increased at a rate of 27.98 × 10^{-3} ± 0.95 × 10^{-3} pH units h^{-1}. When using the 50% dilution, 6.63% less HF-DLF was released, on average, if compared to the water released in the control or the HF-DLF in the 10% and 25% dilutions. Despite such difference, the potential effect on the pH (expressed as [OH^{-}] amount of a putative medium receiving the HF-DLF at the 50% dilution) strongly increased over time, whereas it did not vary in the in the control or the HF-DLF in the 10% and 25% dilutions

1.68 ± 0.04 Mol OH$^-$ (t HF-DLF)$^{-1}$ h^{-1}. Results from other experiments were variable and depended on the pressure, kind of emitter and kind of wastewater. In contrast to our study, Gamri et al. [27] found a strong reduction of the emitter performance with time and difference between the present and the one by Gamri et al. [27] experiment can be due to the higher pressure we used, which is two-fold compared to their study, and this occurred despite our HF-DLF had a solid concentration 2.3–9.8 fold higher than the synthetic wastewater composition used by Gamri et al. [27]. Nonetheless, these differences can be due to the high-frequency flushing in our experiment (one every 8 h). And indeed, Puig-Bargués et al. [14] showed that flushing every 540 h was sufficient to almost completely avoid the emitter clogging. Puig-Bargués et al. [14] also found, as in the present study, that dripline flow increased 8% and 25% over time when using a pressure compensating and a non-pressure compensating emitter, respectively, when used with a tertiary effluent from a wastewater treatment plant filtered with a 0.130 mm filtration level. Similarly, we found that coefficient of variation computed at an hourly basis was 1.653% (CI$_{95\%}$ 1.240–2.065%) in the control and 3.587% (CI$_{95\%}$ 2.713–6.444%) in the HF-DLF, with scarce differences among dilution, suggesting that accumulation of deposited material in the emitters affected the dripline flow performance. Such coefficient of variation was lower than those found in other similar studies [28,29] and can be marginally acceptable as indicated by Bodole et al. [26], according to which a test duration of more than 60 min is enough to minimize the uncertainty due to the initial fluctuation of the data. The lower variation of the HF-DLF is likely due to an anti-clogging shape of the present emitters if compared to other emitters [30].

5. Conclusions

In conclusions, hydrocyclone filtration scarcely affected the traits of the digestate liquid fraction used for the irrigation. Irrigation with hydrocyclone-filtered digestate liquid fraction (HF-DLF) injected in the system at 10% and 25% dilution did not affect the performance of the system nor the traits of the liquid fraction released by the emitters, whereas using 50% dilution of the HF-DLF consisted in a lower amount of liquid released at increasing pH. In particular, HF-DLF dilution at 10%, 25%, and 50% consisted in +1.9%, +3.5, and −4.9% amount of liquid released compared to the control. In 10% HF-DLF % and 25% HF-DLF, a constant pH difference of + 0.321 ± 0.014 pH units compared to the control was found, in the 50% HF-DLF pH increased by around a half point and such difference increased with time.

This implies that that highly concentrated digestate liquid fractions, i.e., low dilutions, can pose problems for the functioning of the system and may have potentially harmful effects on soils with high pH. Nonetheless, the use of digestate liquid fractions for irrigation purposes may be a valuable option in those areas with a high amount of biogas plants and digestate production, such as various nations in Europe, America and Asia including USA, China, Germany, United Kingdom, Italy, and France [31,32]. Results from the present study have beneficial implication on the on water conservation since digestate production by feeding the digester with barley and corn provide wastewater from the late spring to the early fall, when water requirements are high. At the one time, the ability to use such wastewaters with minimal impact on the irrigation system, and thus with reduced negative impacts due to the system maintenance and disposal.

The digestate liquid fraction used in the present study was previously subjected to an additional hydrocyclone filtering, that likely discarded the high-molecular weight fraction. However, since few differences were found between filtered and non-filtered liquid digestate fraction, it is likely that a dilution at least up to 25%, according to the present study, can allow for a direct use of the digestate liquid fraction in microirrigation system with a minimal harming of the system performances. However, since the present is a short-term experiment, these results would require additional experiments with unfiltered liquid fractions.

Supplementary Materials: The following are available online at http://www.mdpi.com/2073-4395/10/8/1150/s1, Table S1: LSmeans estimates and relative standard errors of the traits under study [Solution released; output of the turbidity sensor; dry matter concentration; [OH$^-$] concentration; and pH difference than control]. LSmeans

Agronomy **2020**, *10*, 1150

with a letter in common can't be considered different according to a conservative Tukey-grouping applied to the p-differences of the LSmeans.; Table S2: Linear models of the variation in time of the temperature in the whole experiment and of dry matter concentration, [OH$^-$] concentration and pH, and in the 50% diluted hydrocyclone filtered digestate liqud fraction; Table S3: Raw data of the amount of water or diluted hydrocyclone filtered digestate liqud fraction, the output of the turbidity sensor (mV), dry matter content (%), pH of the solution, and the amount of [OH$^-$] concentration.

Author Contributions: Conceptualization, S.B., M.B., E.R., S.S., L.P.; methodology, S.B., M.B., E.R., S.S.; software, S.B., E.R., S.S.; validation, S.B., S.S.; formal analysis, S.S.; investigation, S.B., M.B., E.R., S.S., P.C.; resources, M.C., P.T., C.B., L.P.; data curation, S.B., S.S., P.C.; writing—original draft, S.B., S.S.; writing—review and editing, S.B., S.S., P.C.; supervision, L.P.; project administration, L.P.; funding acquisition, L.P. All authors have read and agreed to the published version of the manuscript.

Funding: This research was carried out within the AGROENER project (D.D. n. 26329, 1 April 2016) funded by the Italian Ministry of Agriculture (MiPAAF).

Acknowledgments: The authors wish to thank Ivan Carminati, Gianluigi Rozzoni, Alex Filisetti and Elia Premoli for their assistance in performing the tests and for their professionalism and availability.

Conflicts of Interest: The authors have no conflicts of interest to disclose. The authors declare that they have no financial interests or personal relationships with the brands cited, nor endorse any of the brands cited or their products. Furthermore, the funders had no role in the design of the study; in the collection, analyses, or interpretation of data; in the writing of the manuscript, or in the decision to publish the results.

References

1. Elliott, J.; Deryng, D.; Müller, C.; Frieler, K.; Konzmann, M.; Gerten, D.; Glotter, M.; Flörke, M.; Wada, Y.; Best, N.; et al. Constraints and potentials of future irrigation water availability on agricultural production under climate change. *Proc. Natl. Acad. Sci. USA* **2013**, *111*, 3239–3244. [CrossRef] [PubMed]

2. Döll, P. Impact of Climate Change and Variability on Irrigation Requirements: A Global Perspective. *Clim. Chang.* **2002**, *54*, 269–293. [CrossRef]

3. Nakayama, F.; Bucks, D. Water quality in drip/trickle irrigation: A review. *Irrig. Sci.* **1991**, *12*, 12. [CrossRef]

4. Pereira, L.S.; Oweis, T.; Zairi, A. Irrigation management under water scarcity. *Agric. Water Manag.* **2002**, *57*, 175–206. [CrossRef]

5. Saia, S.; Fragasso, M.; De Vita, P.; Beleggia, R. Metabolomics Provides Valuable Insight for the Study of Durum Wheat: A Review. *J. Agric. Food Chem.* **2019**, *67*, 3069–3085. [CrossRef]

6. Rusan, M.J.M.; Hinnawi, S.; Rousan, L. Long term effect of wastewater irrigation of forage crops on soil and plant quality parameters. *Desalination* **2007**, *215*, 143–152. [CrossRef]

7. Tarchitzky, J.; Lerner, O.; Shani, U.; Arye, G.; Brener, A.; Chen, Y.; Lowengart-Aycicegi, A. Water distribution pattern in treated wastewater irrigated soils: Hydrophobicity effect. *Eur. J. Soil Sci.* **2007**, *58*, 573–588. [CrossRef]

8. Adin, A.; Sacks, M. Dripper-Clogging Factors in Wastewater Irrigation. *J. Irrig. Drain. Eng.* **1991**, *117*, 813–826. [CrossRef]

9. Elgallal, M.; Fletcher, L.; Evans, B. Assessment of potential risks associated with chemicals in wastewater used for irrigation in arid and semiarid zones: A review. *Agric. Water Manag.* **2016**, *177*, 419–431. [CrossRef]

10. Drechsel, P.; Scott, C.A.; Raschid-Sally, L.; Redwood, M.; Bahri, A. *Wastewater Irrigation and Health*; Bahri, A., Drechsel, P., Raschid-Sally, L., Redwood, M., Eds.; Routledge: London, UK; Sterling, VA, USA, 2009; 432p, ISBN 9781849774666.

11. Capra, A.; Scicolone, B. Emitter and filter tests for wastewater reuse by drip irrigation. *Agric. Water Manag.* **2004**, *68*, 135–149. [CrossRef]

12. Ahmed, B.A.O.; Yamamoto, T.; Fujiyama, H.; Miyamoto, K. Assessment of emitter discharge in microirrigation system as affected by polluted water. *Irrig. Drain. Syst.* **2007**, *21*, 97–107. [CrossRef]

13. Goyal, M.R. *Wastewater Management for Irrigation*; Goyal, M.R., Triphati, V.K., Eds.; Apple Academic Press & CRC Press: Boca Raton, FL, USA, 2016; ISBN 9780429152498.

14. Puig-Bargués, J.; Arbat, G.; Elbana, M.; Duran-Ros, M.; Barragan, J.; De Cartagena, F.R.; Lamm, F. Effect of flushing frequency on emitter clogging in microirrigation with effluents. *Agric. Water Manag.* **2010**, *97*, 883–891. [CrossRef]

15. Hills, D.J.; Brenes, M.J. Microirrigation of Wastewater Effluent Using Drip Tape. *Appl. Eng. Agric.* **2001**, *17*, 17. [CrossRef]

16. Cararo, D.C.; Botrel, T.A.; Hills, D.J.; Leverenz, H.L. Analysis of Clogging in Drip Emitters during Wastewater Irrigation. *Appl. Eng. Agric.* **2006**, *22*, 251–257. [CrossRef]

17. Makádi, M.; Tomcsik, A.; Orosz, V. Digestate: A New Nutrient Source—Review. *Biogas* **2012**, 1–17. [CrossRef]

18. Green, O.; Katz, S.; Tarchitzky, J.; Chen, Y. Formation and prevention of biofilm and mineral precipitate clogging in drip irrigation systems applying treated wastewater. *Irrig. Sci.* **2018**, *36*, 257–270. [CrossRef]

19. Barzee, T.J.; Edalati, A.; El-Mashad, H.; Wang, D.; Scow, K.; Zhang, R. Digestate Biofertilizers Support Similar or Higher Tomato Yields and Quality Than Mineral Fertilizer in a Subsurface Drip Fertigation System. *Front. Sustain. Food Syst.* **2019**, *3*, 3. [CrossRef]

20. Akhiar, A.; Battimelli, A.; Torrijos, M.; Carrere, H. Comprehensive characterization of the liquid fraction of digestates from full-scale anaerobic co-digestion. *Waste Manag.* **2017**, *59*, 118–128. [CrossRef]

21. Tahir, M.U.; Ahsan, S.M.; Arif, S.M.; Abdullah, M. GSM Based Advanced Water Quality Monitoring System Powered by Solar Photovoltaic System. In Proceedings of the 2018 Australasian Universities Power Engineering Conference (AUPEC), Auckland, New Zealand, 27–30 November 2018; pp. 1–5.

22. Standard Methods for the Examination of Water and Wastewater. In Proceedings of the AWRA's 1999 Annual Water Resources Conference: Watershed Management to Protect Declining Species, Seattle, WA, USA, 5–9 December 1999; APHA: Washington, DC, USA, 1999.

23. Saia, S.; Aissa, E.; Luziatelli, F.; Ruzzi, M.; Colla, G.; Ficca, A.G.; Cardarelli, M.; Rouphael, Y. Growth-promoting bacteria and arbuscular mycorrhizal fungi differentially benefit tomato and corn depending upon the supplied form of phosphorus. *Mycorrhiza* **2019**, *30*, 133–147. [CrossRef]

24. Giovino, A.; Militello, M.; Gugliuzza, G.; Saia, S. Adaptation of the tropical hybrid Euphorbia×lomi Rauh to the exposure to the Mediterranean temperature extremes. *Urban For. Urban Green.* **2014**, *13*, 793–799. [CrossRef]

25. Guilayn, F.; Jimenez, J.; Rouez, M.; Crest, M.; Patureau, D. Digestate mechanical separation: Efficiency profiles based on anaerobic digestion feedstock and equipment choice. *Bioresour. Technol.* **2019**, *274*, 180–189. [CrossRef] [PubMed]

26. Bodole, C.; Koech, R.; Pezzaniti, D. Laboratory evaluation of dripper performance. *Flow Meas. Instrum.* **2016**, *50*, 261–268. [CrossRef]

27. Gamri, S.; Soric, A.; Tomas, S.; Molle, B.; Roche, N. Biofilm development in micro-irrigation emitters for wastewater reuse. *Irrig. Sci.* **2013**, *32*, 77–85. [CrossRef]

28. Pinto, M.F.; Molle, B.; Alves, D.G.; Ait-Mouheb, N.; De Camargo, A.P.; Frizzone, J.A. Flow rate dynamics of pressure-compensating drippers under clogging effect. *Rev. Bras. Eng. Agríc. Ambient.* **2017**, *21*, 304–309. [CrossRef]

29. Dalri, A.B.; Santos, G.O.; Dantas, G.D.F.; De Faria, R.T.; Zanini, J.R.; Palaretti, L.F. Performance of drippers in two filtering systems using sewage treatment effluent. *Rev. Bras. Eng. Agríc. Ambient.* **2017**, *21*, 363–368. [CrossRef]

30. Zhang, J.; Zhao, W.; Tang, Y.; Lu, B. Anti-clogging performance evaluation and parameterized design of emitters with labyrinth channels. *Comput. Electron. Agric.* **2010**, *74*, 59–65. [CrossRef]

31. Deng, Y.; Xu, J.; Liu, Y.; Mancl, K. Biogas as a sustainable energy source in China: Regional development strategy application and decision making. *Renew. Sustain. Energy Rev.* **2014**, *35*, 294–303. [CrossRef]

32. Akhiar, A. Characterization of Liquid Fraction of Digestates after Solid-Liquid Separation from Anaerobic Co-Digestion Plants. Ph.D. Thesis, Université Montpellier, Montpellier, France, 2017. Submitted 15/01/2018 (NNT: 2017MONTS004).

Article

Estimation of Sensible and Latent Heat Fluxes Using Surface Renewal Method: Case Study of a Tea Plantation

Jizhang Wang [1], Noman Ali Buttar [1,2], Yongguang Hu [1,*], Imran Ali Lakhiar [1], Qaiser Javed [1] and Abdul Shabbir [1,2]

[1] School of Agricultural Engineering, Jiangsu University, Zhenjiang 212013, China; whxh@ujs.edu.cn (J.W.); noman_buttar@yahoo.com (N.A.B.); 5103160321@stmail.ujs.edu.cn (I.A.L.); qjaved_uaf@yahoo.com (Q.J.); abdulshabbir@uaf.edu.pk (A.S.)

[2] Department of Irrigation and Drainage, University of Agriculture, Faisalabad 3800, Pakistan

* Correspondence: deerhu@ujs.edu.cn; Tel.: +86-511-8879-7338

Abstract: An experiment of sensible and latent heat flux measurement was conducted in a tea plantation near the Yangtze River within Danyang of Jiangsu Province, China. High-frequency (~10 Hz) air temperature measurement with fine-wire thermocouples ($\oslash = 50$ μm) was used for the estimation of sensible heat flux (H), and latent heat flux (LE) was extracted as a residual of the energy balance equation using additional measurements of net radiation (R_n) and soil heat flux (G). Results were compared against the eddy covariance (EC) system under unstable conditions only, and days with high precipitation were excluded from further analysis. Half-hourly datasets of the sensible heat flux estimated using the surface renewal method (SR) (H_{SR}) and measured by the EC system (H_{EC}) were analyzed. Results showed good agreement with $R^2 = 0.80$, root mean square error ($RMSE$) = 27.87 W m^{-2}, relative error (RE) = 9.02%, and a regression slope of 0.68—this slope was used for the calibration of the uncalibrated H_{SR} estimated by SR. On the other hand, the half-hourly dataset of LE_{SR} was regressed against EC, and it showed good agreement with relatively high $R^2 = 0.93$, $RMSE = 32.99$ W·m^{-2}, and $RE = 5.67\%$. Hence, the SR method may estimate the surface fluxes at a relatively low cost, ultimately improving calculations of evapotranspiration. Thus, the SR method could provide an economical tool for improving crop water management of tea plantations.

Keywords: sensible and latent heat fluxes; surface renewal method; tea plantation; evapotranspiration; eddy covariance

Citation: Wang, J.; Buttar, N.A.; Hu, Y.; Lakhiar, I.A.; Javed, Q.; Shabbir, A. Estimation of Sensible and Latent Heat Fluxes Using Surface Renewal Method: Case Study of a Tea Plantation. *Agronomy* **2021**, *11*, 179. https://doi.org/10.3390/agronomy11010179

Received: 26 November 2020
Accepted: 14 January 2021
Published: 18 January 2021

Publisher's Note: MDPI stays neutral with regard to jurisdictional claims in published maps and institutional affiliations.

1. Introduction

Latent heat flux (LE) measurement is very critical in the field of micrometeorology for the efficient management of available water resources. The precise estimation of LE, also termed as evapotranspiration (ET), is vital due to its great influence on precipitation, plant growth, and amount of irrigation water runoff. There is constant pressure on the available water resources for different crops in arable agricultural areas. Therefore, there is always a need for accurate estimation of crop evaporation to maximize the available water resources and crop water use efficiency. Traditionally, crop water evaporation was estimated from climatic data and reference evapotranspiration methods and by using the crop coefficient approach [1]. In recent decades, the instrumentation, techniques, and different approaches for the estimation of ET have been improved a lot. In order to spread scientifically approved techniques into commercial practice, simpler techniques are preferred. Various micrometeorological methods have been used for estimating LE (i.e., scintillometer, eddy covariance (EC), Bowen ratio (BR) [2], lysimeter, surface renewal (SR), and flux variance (FV)) [3]. Use of a lysimeter is a reliable method for ET measurements, but it is seldom used outside of experimental stations [4,5]. The Bowen ratio (BR) method requires extensive fetch and responsiveness to the biases of the instrument used for estimating the air temperature and water vapor pressure at two levels [2]. The scintillometer method is a high-cost method based on the Monin–Obukhov similarity theory (MOST) and high skills are required for

141

correct operation. In addition, its estimations are disrupted by optical interception of rainfall, insects, frost, and vertical air temperature to differentiate between the ascending and descending directions of sensible heat flux (H) [6]. The EC system allows the direct measurement of surface fluxes (sensible and latent heat fluxes), and a lysimeter can also provide estimations of these fluxes with better accuracy. In spite of their many advantages, these techniques are not easily available for daily usage by farmers due to high cost and operational complexity; hence, simple and low-cost methods for ET measurement are required [7]. There are few studies of ET estimation over different homogeneous surfaces and crops using these methods. Most of these methods are costly and their instrumentation is sensitive to damage, requiring site homogeneity and relatively wide fetch. The SR method has been proposed as a reliable alternative to methods such as the EC system, first proposed by Van Atta in 1977 [8]. The theoretical basics of the SR method have been reviewed and applied over different crop canopies and surfaces with better results [9–16]. The SR method uses high-frequency (~10 Hz) air temperature measurements obtained from fine-wire thermocouples for the estimation of the sensible heat flux (H) above the plant canopy. The SR method has been deployed over natural surfaces with mixed results and is considered as an attractive replacement for other available meteorological methods due to its low-cost and relatively simple instrumentation, which makes it easier to apply close to the canopy surface and with limited fetch [17–19].

This study tests the performance of the surface renewal method for estimating the surface fluxes, including sensible and latent heat fluxes, over a tea plantation in Danyang. Tea plants grow well all year round in temperate and humid climates. During the last decade, sprinkling irrigation has been gradually applied to tea plantations and drip irrigation has been recently adopted with the extension of fertigation technology. Usually, growers start irrigation based on only soil moisture content, without knowing the exact crop water requirement [20]. This study was performed with the aim to provide a relatively low-cost technique for estimation of the crop water requirement of tea plants. The performance of the SR method was evaluated for the estimation of surface fluxes including sensible and latent heat fluxes and the results were compared against the measurements of the eddy covariance system.

2. Materials and Methods

2.1. Study Site and Climate

The experiment was conducted at a tea plantation located in Danyang, Jiangsu, P.R China (32.026177° N, 119.674201° E), at an elevation of 18.5 m above the sea. The study site is mainly dominated by homogeneous crops with trees on the boundaries (Figure 1). The tea plants were 2 years old at the time of experiment and plant height ranged from 0.7 to 0.8 m. The mean air temperature during the experimental seasons ranged from 10 to 15 °C in winter and 25 to 35 °C in summer, with mean annual precipitation of 2.73 mm.day^{-1}. An EC system was installed at the study station, comprising a 3D sonic anemometer (CSAT3, Sonic anemometer, Campbell Scientific, Logan, UT, USA) and an open-path infrared gas analyzer (IRGA, EC150, Campbell Scientific, Logan, UT, USA) for the surface flux measurements. Both the sonic anemometer and infrared gas analyzer were placed at height of 2.3 m above the ground in the wind dominant direction. The EC system can measure surface fluxes including H and LE directly. For the additional measurements of relative humidity and air temperature, a sensor (HC2S3-L, Campbell Scientific, Logan, UT, USA) was placed above the ground on a tower along with the EC system.

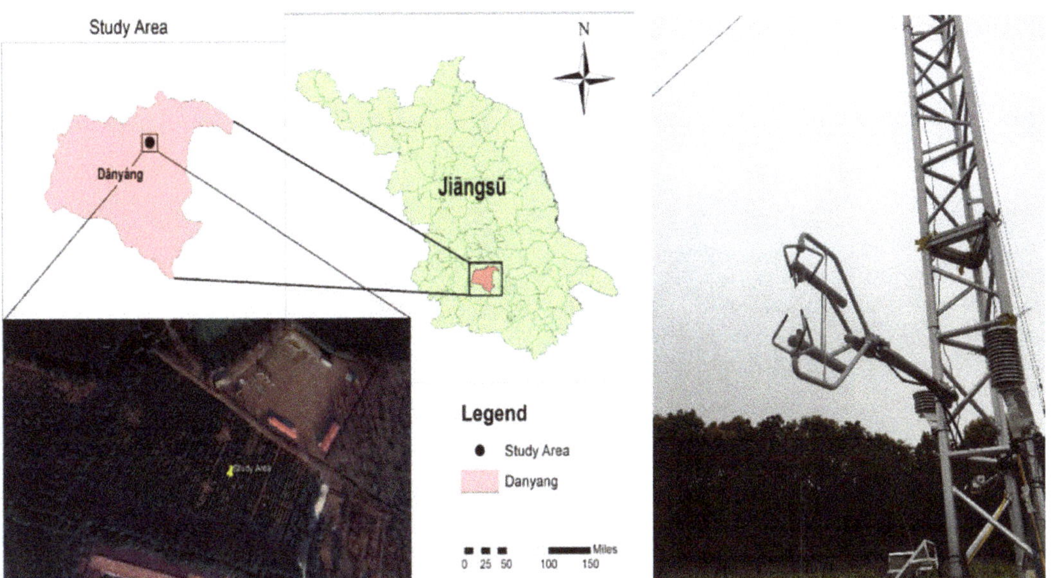

Figure 1. A map representing the location of the study area and the eddy covariance (EC) system installed at study site.

All pieces of equipment were supported with batteries connected with solar panels. A Net radiometer (CNR4-L, KIPP and ZENON) was placed at 2.3 m above the ground on the same pole with the EC system in the south direction to avoid the shading effect. For soil heat flux (G) measurement, two soil heat flux plates (HFT-3.1, TEBS, Seattle, WA, USA) were placed at a depth of 0.08 m. One plate was installed in wet soil between the plants and the other one was installed along the pathway. To calculate the change in heat storage (ΔS), two thermocouples were installed in the soil layer above each plate at a depth of 0.02 and 0.06 m, respectively. The installation and raw data calculation from these plates were performed following the instructions of Campbell Scientific, Inc. [21]. The calculation of soil heat flux was done using the soil properties, including the soil water content, soil heat capacity, and bulk density of the soil from the soil samples collected during the field visits, following the procedure recommended by Tanny, Haijun and Cohen (2006) [22]. For estimation by the SR method, two fine-wire thermocouples (type T), with a diameter of 50 μm (COCO-002, Omega Eng., Irlam, Manchester, UK), were placed at a height of 1.8 m above the ground in predominant wind direction (northwest). The raw data signals from both systems, eddy covariance and surface renewal, were sampled at a high frequency of 10 Hz because the fine-wire thermocouples cannot handle a higher sampling frequency, e.g., 20 Hz. Raw signals were stored on a datalogger (CR3000 from Campbell Scientific). All the recorded data were later analyzed to calculate the turbulence fluxes produced by the eddy covariance system, e.g., frictional velocity. Raw data of latent heat flux measured from the EC system were corrected for coordinate system rotation [23], sensor separation by applying the frequency response correction [24], and path averaging.

On the other hand, the sensible heat flux measured by the EC system was corrected for path averaging and coordinate rotation system. The sonic temperature was also converted to the thermodynamic temperature using high-frequency readings of water vapor concentration obtained through the open-path gas analyzer. Regular maintenance of the instruments was performed during the whole experimental duration. The net radiometer was cleared of dust that accumulates on its domes and its position was maintained horizontally. The sonic anemometer was checked regularly and kept safe from spider webs.

Finally, the thermocouples were checked regularly as they are very vulnerable; broken thermocouples were replaced with new ones, fortunately only once during the experiment.

2.2. Footprint Analysis

Footprint analysis was conducted for estimating the relative contribution of the upwind surface to the fluxes measured by the EC method [13]. In many agricultural practices, surfaces are limited in their area or surrounded by some trees or buildings. Therefore, estimation of the footprint for turbulent fluxes is crucial for proper and reliable execution of EC measurements. The following were input variables for the footprint analysis: measurement height (z_a), displacement height (d), mean wind speed ($m \cdot s^{-1}$), Obukhov length (L), standard deviation of horizontal wind speed ($m \cdot s^{-1}$), friction velocity (u*), and wind direction (o) [25–28]. The footprint model used for the estimation of the distance from which 90% of the measured flux originated, or the ratio of this distance to measurement height, is expressed as Equation (1) [29]:

$$\frac{x}{z_a} = \frac{2 \times 9.491}{z_a} x_{peak} \tag{1}$$

where x is the horizontal distance along the fetch from the EC system, z_a is the measurement height, and x_{peak} is the peak location of the footprint distribution function, expressed as Equation (2):

$$x_{peak} = \frac{Dz_u^P |L|^{1-P}}{2k^2} \tag{2}$$

Here, D and P are similarity parameters, and z_u is calculated as Equation (3):

$$z_u = (z_a - d) \cdot \frac{z_a - d}{z_a - (d + z_a)} \left[\ln \frac{z_a - d}{z_0} - 1 + \frac{z_0}{z_a - d} \right] \tag{3}$$

where z_0 is surface roughness length.

2.3. Surface Renewal (SR) Method

The SR method is based on the turbulent exchange of the sensible heat between the plant canopy and the atmosphere, caused by the instantaneous replacement of an air parcel interacting with the surface (Figure 2). The air parcel exchanges energy between air and canopy elements; then, the parcel is detached from the surface, and a new air parcel swings in to renew the removed air. Thus, understanding the features of this turbulence mechanism is vital for correct operation and analysis of this method. The signals of air temperature display well-managed coherent structures which resemble ramp events [30–32]. The SR method is constructed on the investigative energy budget of the coherent structures that exist within the crop canopy [31,32]. The exchange of air parcels between the surface and atmosphere is established as ramp-like shapes in the turbulence temperature; H_{SR} can be estimated as Equation (4):

$$H_{SR} = \alpha H'_{SR} = \alpha z \left(\rho c_p \right) \frac{a}{\tau} \tag{4}$$

where α is the calibration coefficient obtained through the slope of regression between H_{SR} (calibrated sensible heat flux) and the H'_{SR} (uncalibrated sensible heat flux); z is the measurement height (m), ρ is the specific air density ($kg \cdot m^{-3}$), c_p is the specific heat capacity ($J \cdot kg^{-1} \cdot K^{-1}$), a is the ramp amplitude (K) and τ is the ramp period (s). The SR estimations require calibration against any independent method (i.e., EC and BR).

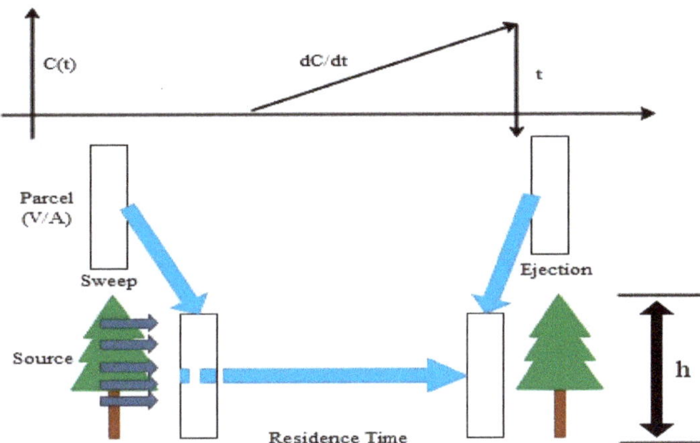

Figure 2. The concept of the surface renewal theory.

The calibration factor (α) is obtained as the slope of the linear regression forced through the origin, between the estimations of H_{SR} and the H_{EC}; the value of the calibration coefficient depends on the measurement height, canopy height and architecture, atmospheric stability, turbulence characteristics, and sensor dynamic response characteristics [9]. The structure–function analysis is calculated by Equation (5) [8]:

$$\overline{S^n(r)} = \frac{1}{m-j} \sum_{k=1}^{m-j} \left[\left(T_k - T_{k-j} \right)^n \right] \tag{5}$$

where S denotes the structure function, n is the order of the structure function (2nd, 3rd, or 5th in this case), r is the order of function, m is the total number of data points, j is the sample lag, and T_k is the kth element in the calculated temperature data. The structure–function values are used to determine the coefficient in the following cubic polynomial expressed as Equation (6) [8]:

$$a^3 + pa + q = 0 \tag{6}$$

where p is obtained as Equation (7) [8]:

$$p = \left[10S^2_{(r)} - \frac{S^5_{(r)}}{S^3_{(r)}} \right] \tag{7}$$

Here, q is obtained as Equation (8) [8]:

$$q = 10S^3_{(r)} \tag{8}$$

These equations can be solved analytically to obtain the ramp amplitude. The ramp period is calculated from the ramp amplitude, time lag, and third-order structure function using Equation (9) [8]:

$$\tau = -\frac{a^3_{(r)}}{S^3_{(r)}} \tag{9}$$

The shortened energy closure was used for the estimation of LE above the plant canopy using H_{SR} and the remaining parameters, including R_n and G, using Equation (10) [8]:

$$LE_{SR} = Rn - G - H_{SR} \tag{10}$$

The performance of the SR method was analyzed by linear regression analysis against the EC system with statistical errors including root mean square error (RMSE), the slope of regression, intercept, coefficient of determination, and relative error (RE), which indicate the performance of the SR method for estimating H and LE. The SR method overcomes the issues related to the fetch, leveling, orientation, and instrument placement, which can overcome the potential uncertainties in the EC system [33–37].

3. Results and Discussion

3.1. Energy Balance Closure

The energy balance closure is a typical method for evaluating the reliability of the EC system's measurements. A half-hourly linear regression analysis was performed between the available energy fluxes ($R_n - G$) and the turbulent fluxes ($LE + H$) measured from the EC system (Figure 3).

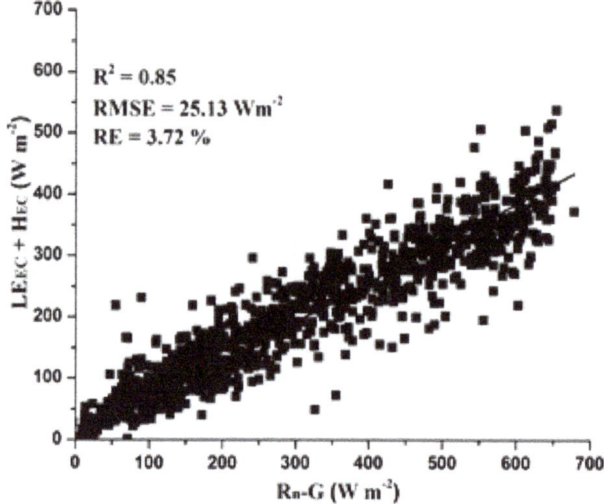

Figure 3. Half-hourly regression analysis between ($R_n - G$) and ($LE_{EC} + H_{EC}$).

In this study, measurements of the EC system were used to evaluate the performance of the SR method. Although the shortage of energy balance was within the accepted range, the estimation of LE using the SR method was based on the energy balance Equation (11):

$$LE_{EC} + H_{EC} = 0.64 \, (R_n - G) \tag{11}$$

The overall results showed that both fluxes were in good agreement, with relatively high $R^2 = 0.85$, a slope of regression of 0.64, and statistical errors $RMSE$ and RE of 25.13 W·m^{-2} and 3.72%, respectively. The energy balance slope was within the acceptable range related to the EC system application in the literature [38,39]. Hence, measurements of the EC system including the sensible heat flux and latent heat flux were used to calibrate the performance of the surface renewal method in this study.

The coherent movements of the fluxes cannot be justified by either the surface renewal or the eddy covariance method, and their role in the post-field data processing involving both of these methods remains unknown. However, this issue can play a critical role for elucidating the surface energy balance closure when the SR method is used.

3.2. The Footprint of EC Flux Measurements

A footprint model was applied to analyze the relative contribution of the windward distance to surface fluxes measured by the EC system. The footprint model was estimated by Equations (1)–(3), which provide the ratio between 90% flux footprint and measurement height; Equation (1). The analysis was performed mostly for the daytime (unstable conditions) [27]. Two days with different climatic conditions were selected from the experimental duration: One partly cloudy (27 November 2018) and the other one (18 October 2018) a sunny day. Diurnal variations of footprint/height ratios for these two days are shown in Figure 4:

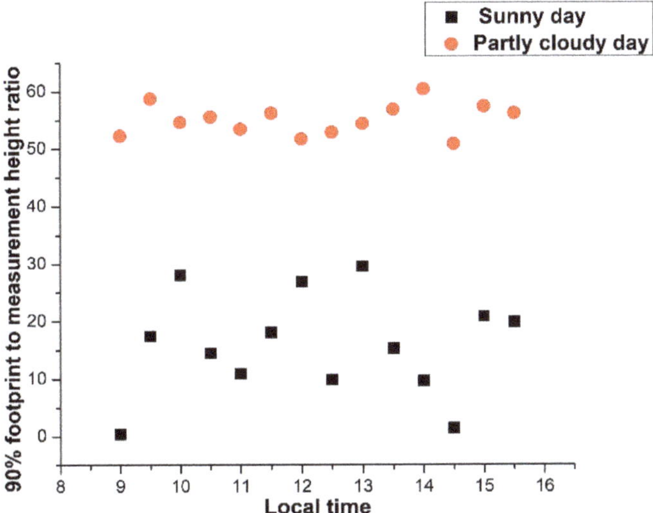

Figure 4. Variation of half-hourly 90% footprint measurement height ratio of two different days [23].

The results in Figure 4 show that the ratio during the sunny day was in the range 0–30, significantly lower than the ratio determined during the partly cloudy day which ranged between 50 and 60. This difference is presumably because during the sunny day, the surface was warmer, and the boundary layer was more unstable than during the partly cloudy day. This larger instability resulted in a shorter 90% flux footprint during the sunny than the partly cloudy day. Besides, Figure 4 shows that on both days, the ratio was smaller than the common 100:1 fetch/height ratio. This indicates that under the conditions of this experiment, most of the flux measured by the EC system originated from within the tea field under study. Hence, the EC data are reliable and can be used for the calibration of the SR method [13].

3.3. Sensible Heat Flux

The sensible heat flux was estimated using Equation (4), using high-frequency air temperature measurements of fine-wire thermocouples [34]. A linear regression analysis was performed between the half-hourly datasets of H_{SR} and H_{EC} under unstable conditions (Figure 5). Overall estimations of the SR method were in better agreement with the EC system with a coefficient of regression $R^2 = 0.80$. This performance was evaluated for the time-lag of 0.5 s, keeping in view the results of previous studies using the same time-lag and frequency for the estimation of surface fluxes using the SR method and sampling frequency (~10 Hz) [10,18]. The uncalibrated H'_{SR} was corrected using the calibration coefficient of 0.68 obtained from the slope of regression between the uncalibrated and calibrated sensible heat fluxes obtained from the SR and EC systems, respectively, as shown in Table 1 for the same duration under unstable conditions. The comparison was performed to assess the performance of the SR method in tea plants.

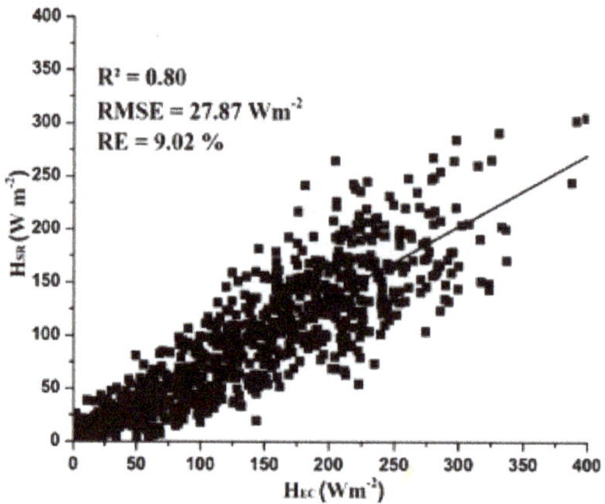

Figure 5. Half-hourly regression analysis between H_{SR} and H_{EC}.

Table 1. The regression statistics between the half-hourly dataset.

Fluxes	z (m)	Time-Lag (s)	Slope (α)	R^2	$RMSE$ (W·m^{-2})	RE (%)
H_{SR} vs. H_{EC}	1.8	0.5	0.68	0.80	27.87	9.02
LE_{SR} vs. LE_{EC}	1.8	0.5	1.21	0.93	32.99	5.67
$(R_n - G)$ vs. $(LE + H)$	2.3	0.5	0.64	0.85	25.13	3.72

The overall results showed good agreement between the estimations by the SR method and the EC system, with R^2 = 0.80, $RMSE$ = 27.87 W·m^{-2}, and RE = 5.67% (Table 1). Diurnal variation of the half-hourly estimations of H_{SR}, H_{EC}, and R_n were observed for two randomly selected days from the experiment duration, one being a clear day (day of year 125, 2019) (Figure 6a) and one being a day with variable clouds (day of year 101, 2019) (Figure 6b).

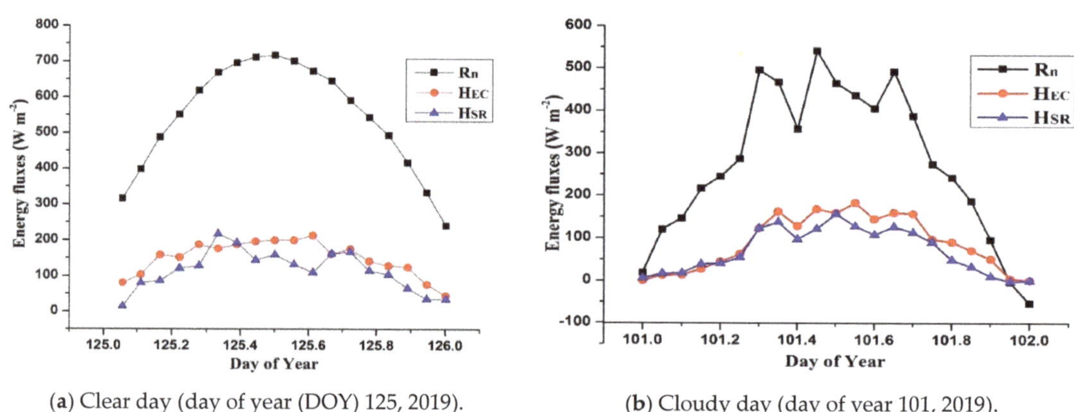

(**a**) Clear day (day of year (DOY) 125, 2019).

(**b**) Cloudy day (day of year 101, 2019).

Figure 6. Diurnal variation of the half-hourly H_{SR}, H_{EC}, and R_n (**a**,**b**).

On the clear day, the estimations were better correlated with the measurements of the EC system in the morning and in the later part of the day (evening). The overall estimation was relatively good throughout the day. On the other hand, the estimation of H_{SR} was changing all day, corresponding to the values of R_n. In the morning, H_{SR} was close to zero and it increased steadily with the increase in R_n; the estimation of H_{SR} was varied with the measurement of R_n, and same trend was observed throughout the day (Figure 6a).

This showed that the air temperature fluctuations present a ramp pattern. The ramps are different for the different climatic conditions and have a direct impact on the measurements of the net radiation.

3.4. Latent Heat Flux

Latent heat flux was extracted as a residual of the energy balance closure using the estimated H_{SR}, including the R_n and G measured separately. A linear regression analysis was performed between the half-hourly estimations of LE_{SR} and LE_{EC} under unstable conditions, and only positive estimations of LE were compared during the analysis because only positive LE corresponds to evapotranspiration, which mainly arises under unstable conditions [40].

The H_{SR} estimated by the SR method yielded a good estimate of the LE_{SR} using a time-lag of 0.5 s for every half-hourly dataset. The results were in good agreement with relatively high $R^2 = 0.93$ as shown in Figure 7. The performance of the SR method for estimating LE_{SR} was evaluated using statistical tools including $RMSE$ and RE. The best result of LE_{SR} was obtained at a height of 1.8 m above the ground, with slope of regression of 1.21. The statistical errors including $RMSE$ and RE were obtained as 32.99 W·m^{-2} and 5.67%, respectively. These results represent the performance of the SR method for the estimation of LE_{SR}. The $RMSE$ values were greater between the linear regressions of LE_{SR} vs. LE_{EC} as compared to those of H_{SR} vs. H_{EC}. This was observed due to the fact that the errors were related to the measurements of R_n and G in both the SR method and the EC system (Table 2).

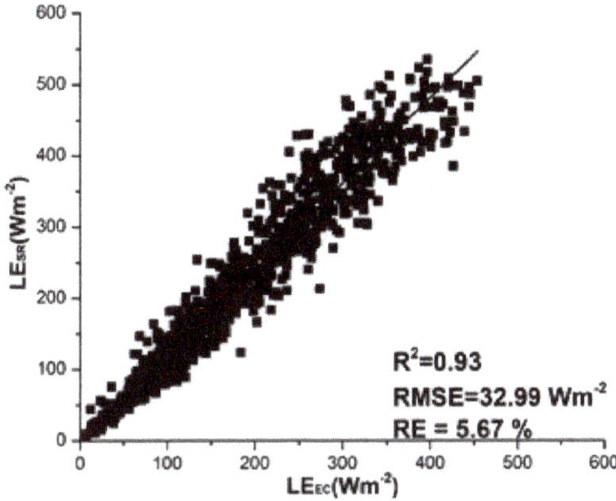

Figure 7. Half-hourly estimated LE_{SR} vs. LE_{EC} under unstable conditions.

A diurnal comparison was made of LE with the SR estimations and the measurements by the EC system at time-lag of 0.5 s and a measurement height of 1.8 m above the ground under unstable conditions. For the comparison, two different days were selected: one with a clear sky (day of year 125, 2019) and the other with variable clouds (day of year 101, 2019). The diurnal variation of LE_{SR} and LE_{EC} was observed with respect to the net radiation throughout the day, mainly in the daytime, usually from 8:00 to 16:00. The diurnal variation of LE_{EC} and LE_{SR} was observed throughout the day with respect to R_n, for on a clear day,

R_n was relatively high, with a maximum value of more than 700 W·m^{-2}, and the variation of LE_{SR} and LE_{EC} was in relatively strong agreement, mainly in the mid-day (Figure 8a).

Table 2. Experiment instruments and installation.

Observations	Notation	Unit	Installation Height (m)	Instruments
3D wind velocity, sonic temperature	u, v, w, T_s	m·s^{-1} °C	2.3	CSAT3, Sonic anemometer, Camp-bell Scientific, Logan, Utah, USA
H$_2$O and CO$_2$ concentrations	-	μmol·m^{-3}	2.3	EC150, Campbell Scientific, Logan, Utah, USA
Soil temperature	T_{soil}	°C	0.02–0.06 (Depth)	TCAV-L, Campbell Scientific, Logan, Utah, USA
Air temperature for SR analysis	T_a	°C	1.8	Fine-wire thermocouple, COCO-002, Omega, Eng., UK
Relative humidity	RH	%	2.1	HC2S3-L, Campbell Scientific, Logan, Utah, USA
Soil heat flux	G	W·m^{-2}	0.08	HFP01, Hukse flux plate sensor
Net radiation	R_n	W·m^{-2}	2.3	CNR4-L, KIPP and ZENON
Liquid precipitation	-	mm	2.1	TE525MM, Campbell Scientific Inc., Logan, Utah, USA
Soil water content	θ_v	m^3·m^{-3}	0.04 (Depth)	CS655, Campbell Scientific Inc., Logan, Utah, USA
Datalogger	CR3000			Campbell Scientific Inc., Logan, Utah, USA.

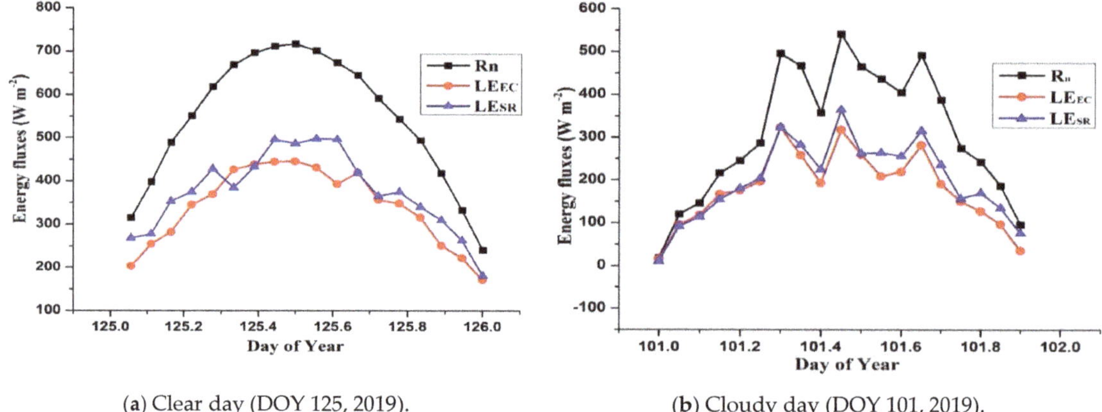

(**a**) Clear day (DOY 125, 2019).

(**b**) Cloudy day (DOY 101, 2019).

Figure 8. Variation of the R_n, LE_{EC}, and LE_{SR}.

On the other hand, the diurnal variation was in good agreement in the day with variable clouds and the estimations were directly influenced by the amount of R_n (Figure 8b). Overall, good correlation was observed throughout the day between the measurements of the EC system and the SR estimations.

4. Conclusions

The performance of the classical surface renewal method was examined in a tea plantation located in Danyang, P.R China. The conventional SR method was applied for the estimation of sensible heat flux using high-frequency air temperature measurement by fine-wire thermocouples, under unstable conditions only, and the results were compared against the measurements of the EC system. Analysis of both these methods showed that the estimated H_{SR} corresponded well with H_{EC} with $R^2 = 0.80$, $RMSE = 27.87$ W·m^{-2}, and $RE = 9.02\%$, the slope of regression forced through the origin was ($\alpha = 0.68$), and this slope was used for calibrating the uncalibrated sensible heat flux estimated through the SR method. The estimated LE_{SR} was in strong agreement with the latent heat flux measured by the eddy covariance system, with a relatively high coefficient of regression. Based on the results, the surface renewal method can provide simple and relatively inexpensive estimations of H and LE above tea plantations and, hence, evapotranspiration, which can help in improving the per capita production of tea plants with better irrigation application. In the future, this study can help in adopting this method for obtaining low-cost information about crop water requirements; furthermore, the SR method can be used independently in case the eddy covariance system is not available—for instance, in fields where the fetch requirement is very limited, and application of the EC system is not easy at the corner of the field. On the other hand, the SR method can be installed in a more appropriate way to obtain complete information of wind direction and other climatic factors, which can help growers to manage available irrigation resources at a relatively low cost. The results of this study were in good agreement with some previous studies performed for different crops and in different climatic conditions [5,11,17,18].

Author Contributions: Investigation, J.W. and N.A.B.; funding acquisition and supervision, Y.H.; writing—original draft, J.W. and N.A.B.; writing—review and editing, I.A.L., Q.J. and A.S. All authors have read and agreed to the published version of the manuscript.

Funding: This research was funded by the Key R&D program of Zhenjiang (NY2018007), China, and the Jiangsu Postdoctoral Science Foundations (2016M600376 and 1601032C) and Priority Academic Program Development of Jiangsu Higher Education Institutions (PAPD-2018-87).

Acknowledgments: We thank Luo Enyou for technical assistance during the experimental activities in the tea fields. We also thank the referees and editors who helped to improve the manuscript quality.

Conflicts of Interest: The authors declare no conflict of interest.

References

1. Allen, R.; Pereira, L.; Raes, D.; Smith, M. *Crop Evapotranspiration-Guidelines for Computing Crop Water Requirements-FAO Irrigation and Drainage Paper 56*; Food and Agriculture Organization of the United Nations: Rome, Italy, 1998; Volume 300.
2. Buttar, N.A.; Yongguang, H.; Shabbir, A.; Lakhiar, I.A.; Ullah, I.; Ali, A.; Aleem, M.; Yasin, M.A. Estimation of evapotranspiration using Bowen ratio method. *IFAC-PapersOnLine* **2018**, *51*, 807–810. [CrossRef]
3. Drexler, J.Z.; Anderson, F.E.; Snyder, R.L. Evapotranspiration rates and crop coefficients for a restored marsh in the Sacramento–San Joaquin Delta, California, USA. *Hydrol. Process.* **2008**, *22*, 725–735. [CrossRef]
4. Niaghi, A.R.; Jia, X.; Scherer, T.; Steele, D. Measurement of unirrigated turfgrass evapotranspiration rate in the red river valley. *Vadose Zone J.* **2019**, *18*, 1–11. [CrossRef]
5. Castellvi, F. Combining surface renewal analysis and similarity theory: A new approach for estimating sensible heat flux. *Water Resour. Res.* **2004**, *40*, W052011–W0520120. [CrossRef]
6. Savage, M.J. Estimation of evaporation using a dual-beam surface layer scintillometer and component energy balance measurements. *Agric. For. Meteorol.* **2009**, *149*, 501–517. [CrossRef]
7. Drexler, J.Z.; Snyder, R.L.; Spano, D.; Paw, K.T. A review of models and micrometeorological methods used to estimate wetland evapotranspiration. *Hydrol. Process.* **2004**, *18*, 2071–2101. [CrossRef]
8. Atta, V. Effect of coherent structures on structure functions of temperature in the atmospheric boundary layer. *Arch. Mech. Stosow.* **1977**, *29*, 161–171.
9. Snyder, R.L.; Duce, P.; Paw, K.T.U. Surface renewal analysis for sensible heat flux density using structure functions. *Agric. For. Meteorol.* **1997**, *86*, 259–271.
10. Hu, Y.; Buttar, N.A.; Tanny, J.; Snyder, R.L.; Savage, M.J.; Lakhiar, I.A. Surface Renewal Application for Estimating Evapotranspiration: A Review. *Adv. Meteorol.* **2018**, *2018*, 1–11. [CrossRef]
11. Mengistu, M.G.; Savage, M.J. Surface renewal method for estimating sensible heat flux. *Water SA* **2010**, *36*, 9–18. [CrossRef]
12. Paw, K.T.; Snyder, R.L.; Spano, D.; Su, H.B. Surface Renewal Estimates of Scalar Exchange. *CA Water Plan Update* **2009**, *4*, 1–65.
13. Buttar, N.A.; Hu, Y.; Tanny, J.; Akram, M.W.; Shabbir, A. Fetch Effect on Flux-Variance Estimations of Sensible and Latent Heat Fluxes of *Camellia sinensis*. *Atmosphere* **2019**, *10*, 299. [CrossRef]
14. Ullah, I.; Buttar, N.A.; Hu, Y.; Aleem, M. Height effect of air temperature measurement on sensible heat flux estimation using flux variance method. *Pak. J. Agric. Sci.* **2019**, *56*, 793–800.
15. Castellví, F. A method for estimating the sensible heat flux in the inertial sub-layer from high-frequency air temperature and averaged gradient measurements. *Agric. For. Meteorol.* **2013**, *180*, 68–75. [CrossRef]

16. Shabbir, A.; Mao, H.; Ullah, I.; Buttar, N.A.; Ajmal, M.; Lakhiar, I.A. Effects of drip irrigation emitter density with various irrigation levels on physiological parameters, root, yield, and quality of cherry tomato. *Agronomy* **2020**, *10*, 1685. [CrossRef]

17. Mekhmandarov, Y.; Pirkner, M.; Achiman, O.; Tanny, J. Application of the surface renewal technique in two types of screenhouses: Sensible heat flux estimates and turbulence characteristics. *Agric. For. Meteorol.* **2015**, *203*, 229–242. [CrossRef]

18. Poblete-Echeverría, C.; Sepúlveda-Reyes, D.; Ortega-Farías, S. Effect of height and time lag on the estimation of sensible heat flux over a drip-irrigated vineyard using the surface renewal (SR) method across distinct phenological stages. *Agric. Water Manag.* **2014**, *141*, 74–83. [CrossRef]

19. Wyngaard, J.C.; Coté, O.R. Cospectral similarity in the atmospheric surface layer. *Q. J. R. Meteorol. Soc.* **1972**, *98*, 590–603. [CrossRef]

20. Yongguang, H.; Chen, Z.; Pengfei, L.; Amoah, A.E.; Pingping, L. Sprinkler irrigation system for tea frost protection and the application effect. *Int. J. Agric. Biol. Eng.* **2016**, *9*, 17–23.

21. Campbell Scientific Inc. *Eddy Covariance System Doperator's Manual, CA27 and KH20*; Campbell Scientific Inc.: Logan, UT, USA, 1998.

22. Tanny, J.; Haijun, L.; Cohen, S. Airflow characteristics, energy balance and eddy covariance measurements in a banana screenhouse. *Agric. For. Meteorol.* **2006**, *139*, 105–118. [CrossRef]

23. Webb, E.K.; Pearman, G.I.; Leuning, R. Correction of flux measurements for density effects due to heat and water vapour transfer. *Q. J. R. Meteorol. Soc.* **1980**, *106*, 85–100. [CrossRef]

24. Moore, C.J. Frequency response corrections for eddy correlation systems. *Bound. Layer Meteorol.* **1986**, *37*, 17–35. [CrossRef]

25. Kljun, N.P.; Calanca, M.W.; Rotach, S.; Schmid, H.P. A simple two-dimensional parameterisation for Flux Footprint Prediction (FFP). *Geosci. Model. Dev.* **2015**, *8*, 3695–3713. [CrossRef]

26. Gash, J.H.C. A note on estimating the effect of a limited fetch on micrometeorological evaporation measurements. *Bound. Layer Meteorol.* **1986**, *35*, 409–413. [CrossRef]

27. Savage, M.J.; Everson, C.S.; Metelerkamp, B.R. *Evaporation Measurement Above Vegetated Surfaces Using Micrometeorological Techniques*; Water Research Commission: Pretoria, South Africa, 1997.

28. Hsieh, C.I.; Lai, M.C.; Hsia, Y.J.; Chang, T.J. Estimation of sensible heat, water vapor, and CO_2 fluxes using the flux-variance method. *Int. J. Biometeorol.* **2008**, *52*, 521–533. [CrossRef] [PubMed]

29. Savage, M.J.; Everson, C.S.; Odhiambo, G.O.; Mengistu, M.G.; Jarmain, C. *Theory and Practice of Evaporation Measurement, with Spatial Focus on SLS as an Operational Tool for the Estimation of Spatially-Averaged Evaporation*; Report No. 1335/1/04; Water Research Commission: Pretoria, South Africa, 2004; p. 204.

30. Deardorff, J.W. Observed characteristics of the outer layer. In *Short Course on the Planetary Boundary Layer*; American Meteorological Society: Boulder, CO, USA, 1978; p. 101.

31. Zhao, X.; Liu, Y.; Tanaka, H.; Hiyama, T. A Comparison of Flux Variance and Surface Renewal Methods with Eddy Covariance. *IEEE J. Sel. Top. Appl. Earth Obs. Remote Sens.* **2010**, *3*, 345–350. [CrossRef]

32. Albertson, J.D.; Parlange, M.B.; Katul, G.G.; Chu, C.R.; Stricker, H.; Tyler, S. Sensible Heat Flux from Arid Regions: A Simple Flux-Variance Method. *Water Resour. Res.* **1995**, *31*, 969–973. [CrossRef]

33. Snyder, R.L.; Spano, D.; Pawu, K.T. Surface renewal analysis for sensible and latent heat flux density. *Bound. Layer Meteorol.* **1996**, *77*, 249–266. [CrossRef]

34. Tha, P.U.K.; Jie, Q.; Hong-Bing, S.; Tomonori, W.; Yves, B. Surface renewal analysis: A new method to obtain scalar fluxes. *Agric. For. Meteorol.* **1995**, *74*, 119–137.

35. Spano, D.; Snyder, R.L.; Duce, P.; Paw, K.T. Estimating sensible and latent heat flux densities from grapevine canopies using surface renewal. *Agric. For. Meteorol.* **2000**, *104*, 171–183. [CrossRef]

36. Anandakumar, K. Sensible heat flux over a wheat canopy: Optical scintillometer measurements and surface renewal analysis estimations. *Agric. For. Meteorol.* **1999**, *96*, 145–156. [CrossRef]

37. Castellví, F.; Snyder, R.L.; Baldocchi, D.D. Surface energy-balance closure over rangeland grass using the eddy covariance method and surface renewal analysis. *Agric. For. Meteorol.* **2008**, *148*, 1147–1160. [CrossRef]

38. Aubinet, M.; Vesala, T.; Papale, D.; Kb, D.P.D.F.; William, J.M.; Loescher, H.W.; Luo, H.; Rebmann, C.; Kolle, O.; Heinesch, B.; et al. *Eddy Covariance: A Practical Guide to Measurement and Data Analysis*; Springer Science & Business Media: Dordrecht, The Netherlands, 2012.

39. Wilson, K.; Goldstein, A.; Falge, E.; Aubinet, M.; Baldocchi, D.; Berbigier, P.; Bernhofer, C.; Ceulemans, R.; Dolman, H.; Field, C.; et al. Energy balance closure at FLUXNET sites. *Agric. For. Meteorol.* **2002**, *113*, 223–243. [CrossRef]

40. Allen, R.G. Using the FAO-56 dual crop coefficient method over an irrigated region as part of an evapotranspiration intercomparison study. *J. Hydrol.* **2000**, *229*, 27–41. [CrossRef]

Article

Hybrid Bermudagrass and Tall Fescue Turfgrass Irrigation in Central California: I. Assessment of Visual Quality, Soil Moisture and Performance of an ET-Based Smart Controller

Amir Haghverdi [1,*], Maggie Reiter [2,3], Anish Sapkota [1] and Amninder Singh [1]

[1] Environmental Sciences Department, University of California Riverside, Riverside, CA 92521, USA; asapk001@ucr.edu (A.S.); asing075@ucr.edu (A.S.)

[2] University of California Division of Agriculture and Natural Resources, Cooperative Extension, Fresno, CA 93710, USA; reit0215@umn.edu

[3] Department of Horticultural Science, University of Minnesota, MN 55108, USA

[*] Correspondence: amirh@ucr.edu

Citation: Haghverdi, A.; Reiter, M.; Sapkota, A.; Singh, A. Hybrid Bermudagrass and Tall Fescue Turfgrass Irrigation in Central California: I. Assessment of Visual Quality, Soil Moisture and Performance of an ET-Based Smart Controller. *Agronomy* **2021**, *11*, 1666. https://doi.org/10.3390/agronomy11081666

Academic Editors: Aliasghar Montazar and Paola A. Deligios

Received: 10 August 2021
Accepted: 19 August 2021
Published: 21 August 2021

Publisher's Note: MDPI stays neutral with regard to jurisdictional claims in published maps and institutional affiliations.

Abstract: Research-based information regarding the accuracy and reliability of smart irrigation controllers for autonomous landscape irrigation water conservation is limited in central California. A two-year irrigation research trial (2018–2019) was conducted in Parlier, California, to study the response of hybrid bermudagrass and tall fescue to varying irrigation scenarios (irrigation levels and irrigation frequency) autonomously applied using a Weathermatic ET-based smart controller. The response of turfgrass species to the irrigation treatments was visually assessed and rated. In addition, turfgrass water response functions (TWRFs) were developed to estimate the impact of irrigation scenarios on the turfgrass species based on long-term mean reference evapotranspiration (ET_0) data. The Weathermatic controller overestimated ET_0 between 5% and 7% in 2018 and between 5% and 8% in 2019 compared with California Irrigation Management Information System values. The controller closely followed programmed watering-days restrictions across treatments in 2018 and 2019 and adjusted the watering-days based on ET_0 demand when no restriction was applied. The low half distribution uniformity and precipitation rate of the irrigation system were 0.78 and 28 mm h^{-1}, respectively. The catch-cans method substantially underestimated the precipitation rate of the irrigation system and caused over-irrigation by the smart controller. No water-saving and turfgrass quality improvement was observed owing to restricting irrigation frequency (watering days). For the hybrid bermudagrass, the visual rating (VR) for 101% ET_0 treatment stayed above the minimum acceptable value of six during the trial. For tall fescue, the 108% ET_0 level with 3 d wk^{-1} frequency kept the VR values in the acceptable range in 2018 except for a short period in mid-trial. The TWRF provided a good fit to experimental data with r values of 0.79 and 0.75 for tall fescue and hybrid bermudagrass, respectively. The estimated VR values by TWRF suggested 70–80% ET_0 as the minimum irrigation application to maintain the acceptable hybrid bermudagrass quality in central California during the high water demand months (i.e., May to August) based on long-term mean ET_0 data. The TWRF estimations suggest that 100% ET_0 would be sufficient to maintain the tall fescue quality for only 55 days. This might be an overestimation impacted by the relatively small tall fescue VR data in 2019 owing to minimal fertilizer applications and should be further investigated in the future.

Keywords: autonomous landscape irrigation; Hargreaves and Samani evapotranspiration model; water conservation

1. Introduction

The state of California, and in general the U.S. west, has some of the largest cities across the nation, making urban water demand a vital component of any integrated water resources management plan. Statistics indicate California's population rose to nearly

40 million in 2016 and projections by the California Department of Finance show an increase to more than 45 million by 2060 [1]. It is expected that, in the future, competition for water and land resources among urban, environmental, and agricultural uses will intensify as a result of increased population, coupled with changes in land use and climate [2]. In addition, climate change is altering precipitation and temperature patterns, making drought severity likely to increase in the American Southwest [3].

Irrigation demand is a significant component of total water use in the urban sector in California [4]; therefore, improving irrigation water use efficiency is a crucial water conservation strategy. Considerable water savings have been reported as a result of implementing emerging technologies for landscape irrigation management [5–7]. Soil moisture sensor-based and evapotranspiration (ET)-based smart irrigation controllers can help increase irrigation water use efficiency by maintaining the root zone soil moisture status within a programmed desired range and scheduling irrigation based on crop coefficient and reference ET (ET_o) data, respectively.

Water agencies in California often incentivize the adoption of smart irrigation controllers in residential areas. According to Singh et al. [8], in 2018–2019, approximately half of the major water agencies across the state provided rebates, ranging from $45 to $300, for installing ET-based smart irrigation controllers. However, the scientific research on the efficacy of smart irrigation controllers for autonomous irrigation scheduling and water conservation is limited in the region. Most scientific work on smart irrigation controllers has focused on avoiding over-irrigation in humid areas with abundant precipitation [9]. Davis et al. [6] compared ET-based smart controllers with a time-based treatment in Florida and reported, on average, a 43% reduction in applied water. Compared with timer-based fixed irrigation, the water-saving potential of ET-based controllers was also reported in other case studies in Florida and Nevada [10,11]. Recently, we [7] used an ET-based smart irrigation controller for autonomous irrigation scheduling of 'Tifgreen' hybrid bermudagrass in southern California and obtained promising results. In another study conducted in Southern California, Bijoor et al. [12] investigated the water budgets of lawns under three management scenarios, including the use of a smart soil moisture sensor-based controller. They concluded that the implementation of smart sensors was a more significant option than the choice of turfgrass species in irrigation efficiency.

An urban feature for potential water conservation is the turfgrass landscape, as it is a large component of urbanized land area. Turf has become an important crop based on the acreage planted, including residential, commercial, and institutional lawns, parks, golf courses, and athletic fields. Beyond recreational use, turf provides valuable ecosystem services that are in high demand by society, like capturing runoff, contributing to the abatement of the heat island, reducing dust and noise, and fostering biodiversity [13]. However, the turfgrass irrigation water requirement could be more significant than some alternative landscape species [14]. Consequently, it is crucial to precisely identify the minimum water requirement of commonly planted turfgrass species and study the water use of alternative warm-season turfgrass species that are more resistant to heat, drought, and salinity.

In a recently published study [7], we introduced the turfgrass water response function (TWRF) as an empirical statistical model to estimate the response of turfgrass species (based on the aesthetic values) to varying irrigation scenarios and ET_o demand. We used NDVI as the response variable owing to its overall high correlation to turfgrass health and growth. We estimated the response of hybrid bermudagrass to varying irrigation and water conservation scenarios in inland Southern California using long-term ET_o data.

This study was carried out to (i) determine the response of hybrid bermudagrass and tall fescue to varying irrigation scenarios (level and frequency) in central California, (ii) evaluate the use of an ET-based smart irrigation controller for autonomous irrigation scheduling, (iii) monitor and assess the dynamics of near-surface soil moisture over time under the imposed irrigation scenarios, and (iv) develop regression-based TWRFs and use them to estimate the response of turfgrass species to irrigation scenarios based on long term mean ET_o demand in the study region. Part two of this study focuses on applying ground-based remote sensing for turfgrass irrigation management [15].

2. Materials and Methods

2.1. Study Area

A two-year turfgrass irrigation research project (2018–2019) was conducted at the University of California Agricultural and Natural Resources Kearney Research and Extension Center (36°36′02.2″ N 119°30′38.8″ W) in Parlier, California. Figure 1 depicts the long-term mean (1983–2021) ET_o and precipitation data for the study site obtained from a nearby California Irrigation Management Information System (CIMIS) weather station (station number 39). The long-term mean annual ET_o and precipitation for the region were 1367 and 273 mm yr^{-1}, respectively. ET_o demand exceeds the natural precipitation, indicating the need to irrigate urban landscape species, particularly over the summer. The long-term data show that peak ET_o of 7 mm per day occurred in early July (Figure 1). The annual precipitation was 192 mm and ET_o was 1452 mm in 2018. The annual precipitation was 268 mm and ET_o was 1462 mm in 2019.

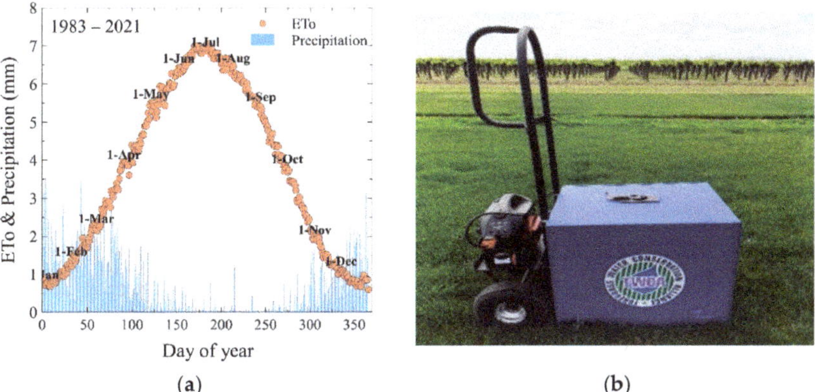

Figure 1. The long-term mean reference evapotranspiration (ET_o) and precipitation trends at the study site obtained from a nearby CIMIS weather station (**a**). The lightbox used in this study to collect digital images for the visual rating (**b**).

The soil at the research site is classified as Hanford fine sandy loam (websoilsurvey.sc. egov.usda.gov; accessed on 18 August 2021). Figure 2 shows the laboratory-determined soil water retention and hydraulic conductivity curves for the experimental site. Four undisturbed soil samples were taken from approximately the top 20 cm layer at the beginning of the trial. The samples were analyzed to determine the soil water retention and hydraulic conductivity curves using HYPROP and WP4C laboratory instruments [16]. The soil water retention (based on the composite data) at soil tensions of 10, 33, and 1500 kPa was 0.25 m³ m⁻³, 0.17 m³ m⁻³, and 0.06 m³ m⁻³, respectively. The saturated hydraulic conductivity was 11 mm day⁻¹.

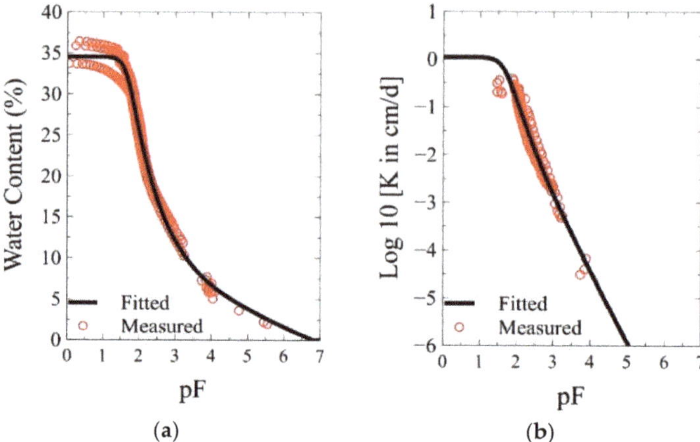

Figure 2. The soil water retention (**a**) and hydraulic conductivity (**b**) curves of the surface soil (0–20 cm) at the study site.

2.2. Irrigation Trials

The responses of hybrid bermudagrass ['Latitude 36' *Cynodon dactylon* (L.) Pers. × *C. transvaalensis* Burtt-Davy] and tall fescue [a blend of 'PennRK4', 'Rebel XLR', + 'Firecracker SLS' *Schedonorus arundinaceus* (Schreb.) Dumort.] to six irrigation treatments (3 irrigation levels × 2 irrigation frequency restrictions) were studied. A total of 36 plots (3.7 m × 3.7 m) formed two adjacent trials organized in a factorial randomized complete block design with repeated measures over time. Table 1 summarizes the irrigation treatments. The irrigation system was installed and the plots were prepared in early 2017. The turfgrass plots were established with sod in late July 2017. For several months afterward, the plots were under non-limiting irrigation for root development and grass establishment. A uniformity test was conducted using catch-can devices on 5 August 2017 with wind speed less than 0.9 m s^{-1} on eight randomly selected plots. The low half distribution uniformity (DU$_{lh}$) was 0.78 and the estimated precipitation rate was 23 mm h^{-1}.

Table 1. Irrigation treatments imposed throughout the 2-year tall fescue and hybrid bermudagrass irrigation research experiments conducted at the University of California Kearney Research and Extension Center.

2018 Trial, Start: 4 May 2018; End: 11 September 2018
Target Irrigation Levels (% ET$_o$): Tall Fescue: 50%, 65%, 80%; Hybrid Bermudagrass: 40%, 50%, 60%   Irrigation Efficiency: 78%   Watering Days: 2 days per week, 3 days per week
2019 Trial, Start: 22 June 2019; End: 26 August 2019
Target Irrigation Levels (% ET$_o$): Tall Fescue: 50%, 65%, 80%; Hybrid Bermudagrass: 40%, 50%, 60%   Irrigation Efficiency: 78%   Watering Days: 3 days per week, 7 days per week (no restriction)

The controller used the user-defined "plant type" information to convert ET$_o$ to irrigation application (irrigation application = plant type × ET$_o$). For each treatment, the plant type was calculated as the irrigation levels (% ET$_o$) divided by the irrigation efficiency of the system.

The target ET$_o$ levels varied from 40 to 80%. Considering DU$_{lh}$ equal to 0.78 as the irrigation efficiency, the programmed irrigation levels (i.e., target ET$_o$ levels divided by irrigation efficiency) ranged from 51% to 103%. The watering days were restricted to two (Sunday, Thursday) and three (Monday, Wednesday, and Saturday) days per week in 2018. In 2019, the watering days were changed to 3 (Sunday, Wednesday, and Friday) and 7 days per week. The 7-day treatment represented a no watering restriction scenario when the smart controller could irrigate as needed based on ET_o demand. To avoid light irrigation applications, the controller was programmed to use the default deficit threshold of approximately 4 mm as the lower deficit limit before any irrigation occurs.

The standard cultural practices were followed to maintain the plots throughout the study, including mowing the plots once a week (mowing heights for tall fescue and hybrid bermudagrass were set to 76 mm and 44 mm, respectively), applying fertilizer two times per year in spring and fall for tall fescue and once a year in the early summer for bermudagrass at a typical rate of 49 kg ha^{-1} nitrogen each application, and spraying the borders as needed with herbicides to control the weeds. The study was started on 4 May and data collection ended on 11 September in 2018. All plots were switched back to the uniform non-limiting irrigation for recovery before starting the second year of the experiment on 22 June 2019. On 26 August 2019, the main irrigation pipe broke and flooded the field, forcing the research team to terminate the trial.

A Weathermatic Smartline (SL) 4800 controller (Telsco Industries, Inc., Garland, TX, USA) was used to autonomously schedule irrigation. A Weathermatic SLFSI-T10 flow sensor (Telsco Industries, Inc., Garland, TX, USA) was installed and connected to the controller in 2019 to monitor the flow across treatments and detect leaks automatically. The controller used an onsite temperature sensor and latitude-based solar radiation information to calculate ET_o using the Hargreaves and Samani equation [17]. The controller used the user-defined "plant type" values and irrigation precipitation rate to calculate irrigation application for each treatment as fractions of ET_o and convert it to equivalent run times. The plant type for each treatment was calculated as the target ET_o level divided by the efficiency of the irrigation system.

Irrigation was done overnight and early morning to avoid evaporative loss and minimize wind drift. The smart controller performed a run/soak schedule to eliminate runoff and provide enough soak time. All three replications for each treatment were wired to the same zone on the controller to receive irrigation at the same time. Each plot was equipped with a TORO 252 Series solenoid valve (Toro Co., Bloomington, MN, USA). The solenoid valve supplied water to four Toro O-T-12-QP corner-pop-up 6″ sprinkler heads (152 mm tall) with an operating pressure range and flow rate of 276–517 kPa and 0.02–9.08 l min^{-1}, respectively. The sprinklers had factory-installed, pressure-compensating discs to ensure steady water application (Toro Co., Bloomington, MN, USA).

2.3. Data Collection and Statistical Analysis

The National Turfgrass Evaluation Program (NTEP) standard [18] was used as the guideline to visually assess and rate the turfgrass plots. The NTEP standard ranges from 1 to 9 representing dead and ideal turfgrass, respectively. The visual rating (VR) of six in this study was considered the minimum acceptable quality for residential areas. The visual assessment was done continuously approximately once a week during the trial for a total of 19 times in 2018 and 9 times in 2019. Visual ratings were taken from digital images of each plot. Figure 1 shows the enclosed lightbox used to collect digital images for visual rating. There were borders between the adjacent plots approximately 60 cm wide to avoid interference between them. The data were collected from the center of each plot to eliminate the plot edge effect. A total of 15 soil moisture readings were collected per treatment (5 readings per plot × 3 replications) from the top 12 cm soil layer using a handheld FieldScout TDR 300 Meter (Spectrum Technologies, Inc., Aurora, IL, USA).

The VR and soil moisture data were statistically analyzed using PROC GLIMMIX in SAS 9.4 software package [19]. Each year and species were independently analyzed for the treatment effects as the frequency restrictions and duration of the experiment differed across the years and species. For all the response variables, the fixed effects were the irrigation levels, irrigation frequencies, and the date of data collection. The random effects were block and its interaction with irrigation levels and irrigation frequencies. The treatment effects were considered significant at p-values ≤ 0.05. All graphs were created using the plotting software package Veusz 3.3.1 [20].

The daily ET_0 data were collected from CIMIS station #39, located approximately 170 m away from the experimental site in an adjacent field. The CIMIS ET_0 data were compared to the estimated ET_0 data by the smart controller. In addition, the Hargreaves and Samani equation [17] was used to calculate long-term daily ET_0 for the study site (via the PyETo software package: https://pyeto.readthedocs.io/en/latest/license.html; accessed on 18 August 2021) and compared against CIMIS ET_0.

$$ET_o = 0.0023R_a(T + 17.8)\sqrt{TR} \qquad (1)$$

where Ra is the extraterrestrial radiation (mm day^{-1}), TR is the difference between the daily maximum and minimum air temperatures (°C), and T is the mean air temperature (°C).

2.4. Turfgrass Water Response Function (TWRF)

A multiple linear regression model (with interactions and quadratic terms included) was used to develop TWRFs for hybrid bermudagrass and tall fescue species. The data for both years were combined. The primary input variables were the applied irrigation levels (%ET_0), irrigation frequency restrictions, and cumulative ET_0 (since the beginning of the experiment for each particular year). The mean VR values for treatments were used as the output variable. The SAS 9.4 software (SAS Institute Inc., Cary, NC, USA) was used to develop and rank all possible regression equations based on correlation coefficients (with 0.7 as the minimum acceptable value). Multiple regression diagnostics, including the Shapiro–Wilk W statistic (to check the normality of the residuals), the condition index (to monitor the collinearity between the variables), and the first and second moment specification test (to check the equal residual variance) were used to finalize the list of input variables of the top model. The long-term mean daily ET_0 values were obtained from the CIMIS station #39 and used to estimate the response of tall fescue and hybrid bermudagrass to varying ET-based irrigation scenarios (60–100% ET_0). The simulation was done for four months, from May to August, using the TWRFs.

The root mean square error (*RMSE*), mean absolute error (*MAE*), mean bias error (*MBE*), and correlation coefficient (*r*) were calculated to evaluate the TWRFs.

$$RMSE = \sqrt{\frac{1}{n}\sum_{i=1}^{n}(E_i - M_i)^2} \qquad (2)$$

$$MAE = \frac{\sum_{i=1}^{n}|E_i - M_i|}{n} \qquad (3)$$

$$MBE = \frac{\sum_{i=1}^{n}(E_i - M_i)}{n} \qquad (4)$$

$$r = \frac{\sum_{i=1}^{n}(E_i - \overline{E})(M_i - \overline{M})}{\sqrt{\sum_{i=1}^{n}(E_i - \overline{E})^2 \sum_{i=1}^{n}(M_i - \overline{M})^2}} \qquad (5)$$

where E and M are estimated and measured visual rating values, respectively. $\overline{M}$ and $\overline{E}$ are the mean-measured and the mean-estimated visual rating values, respectively, and n is the total number of measured data points for the entire experiment (n = 162 for each species).

3. Results

3.1. Performance of the Smart ET-Based Controller

The Weathermatic controller overestimated ET_0 by 5–7% in 2018 and by 5–8% in 2019 compared with CIMIS ET_0 values. On average, across all treatments, MAE was 2.8 mm day^{-1} (4% ET_0) and 2.9 mm day^{-1} (5% ET_0) in 2018 and 2019, respectively. We ran a flow test at the end of the trial, which revealed that the actual precipitation rate of the irrigation system was 28 mm h^{-1}, 21% higher than the 23 mm h^{-1} precipitation rate initially estimated using the catch-cans. Consequently, the applied irrigation was recalculated using the irrigation run time data recorded by the controller for the duration of the experiment, as listed in Table 2. The adjusted irrigation levels varied between 83% and 129% of ET_o for the tall fescue plots and between 65% and 101% of ET_o for the hybrid bermudagrass treatments.

Table 2. Target irrigation treatments (T1–T3) versus programmed and applied irrigation levels for tall fescue and hybrid bermudagrass irrigation research experiments.

	Tall Fescue			Hybrid Bermudagrass		
Irrigation	**T1**	**T2**	**T3**	**T1**	**T2**	**T3**
Treatment	50%	65%	80%	40%	50%	60%
Programmed	64%	83%	103%	51%	64%	77%
Applied	83%	108%	129%	65%	84%	101%

Programmed irrigation levels are equal to target treatment levels divided by the irrigation efficiency of 0.78 (i.e., the low half distribution uniformity of the irrigation system). Applied irrigation levels were recalculated based on the irrigation run time data retrieved from the controller and precipitation rate of 28 mm day^{-1} measured for the system at the end of the trial.

The controller closely followed programmed watering days restrictions across treatments in 2018 and 2019. For the 7 d wk^{-1} treatment (no frequency restriction scenario) in 2019, the controller adjusted the actual irrigation days based on the evaporative demand and minimum allowed water deficit. For example, on average, plots were irrigated five days per week for 65% ET_o irrigation treatment with no frequency restriction.

Figure 3 depicts the performance of the Hargreaves and Samani equation [17] against CIMIS ET_o based on the long-term data (1983–2019) obtained from the CIMIS station #39. Overall, there was a strong agreement between the ET_o values obtained using the Hargreaves and Samani [17] and CIMIS methods, as depicted by high correlation ($r = 0.96$) and well-scattered data points around the identity line (1:1). The MAE between Hargreaves and Samani and CIMIS ET_o varied between 0 and 3.98 mm day^{-1} with an average value of 0.55 mm day^{-1}. The mean annual MAE values fluctuated over the years between 0.4 mm day^{-1} and 0.8 mm day^{-1}. The MAE values were relatively higher in peak ET_o months (i.e., May to September) than during the rest of the year.

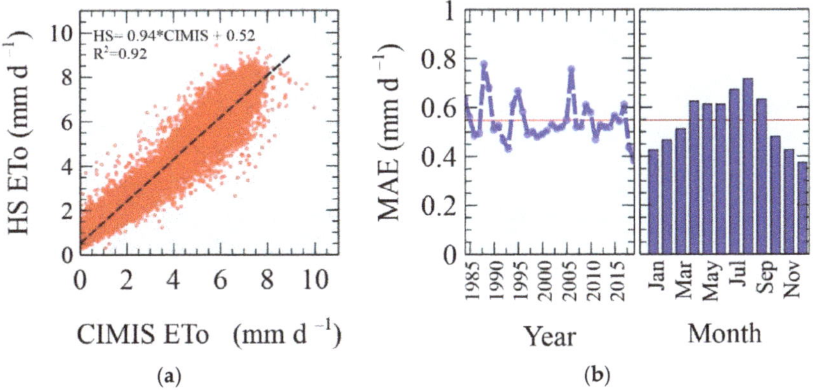

Figure 3. Scatter plot of ET_o estimated using the Hargreaves and Samani (HS) [17] versus CIMIS values (**a**) and variation in long-term annual and seasonal mean absolute error (MAE) values (**b**). MAE shows the difference between the HS and CIMIS models and the red horizontal line in figure (**b**) represents the mean MAE.

3.2. Impact of Irrigation Levels and Frequency on Turfgrass Visual Ratings

Table 3 summarizes the statistical analysis of the VR data for both species in 2018 and 2019. Figure 4 illustrated the dynamics of VR values over time for tall fescue and hybrid bermudagrass species across the imposed irrigation treatments. The VR values greater than six were considered acceptable for residential areas and are highlighted in green.

Table 3. Statistical analysis of the hybrid bermudagrass and tall fescue response (visual rating) to irrigation treatments imposed in 2018 and 2019 (each year and species were analyzed separately).

	Tall Fescue			Hybrid Bermudagrass	
Treatment	**2018**	**2019**	**Treatment**	**2018**	**2019**
Level			**Level**		
129% ET$_o$	7.4 a	5.6 a	101% ET$_o$	7.4 a	6.8 a
108% ET$_o$	6.9 b	5.6 ab	84% ET$_o$	7.1 ab	6.8 a
83% ET$_o$	6.3 c	4.9 b	65% ET$_o$	6.7 b	6.3 a
Frequency			**Frequency**		
2 d wk^{-1}	6.7 b		2 d wk^{-1}	6.9 a	
3 d wk^{-1}	7.0 a	5.2 a	3 d wk^{-1}	7.1 a	6.4 a
7 d wk^{-1}		5.5 a	7 d wk^{-1}		6.8 a
Model effect	**2018**	**2019**	**Model effect**	**2018**	**2019**
I	***	NS	I	*	NS
F	*	NS	F	NS	NS
I × F	NS	NS	I × F	NS	NS
T	***	***	T	***	***
I × T	***	*	I × T	***	NS
F × T	NS	NS	F × T	NS	NS
I × F × T	NS	NS	I × F × T	NS	NS

NS, *** and * are non-significant or significant at $p \leq 0.001$ and 0.05, respectively. Means sharing a similar letter are not significantly different, based on Turkey's test at $\alpha = 0.05$. I, F, and T in the table refer to irrigation levels, irrigation frequency restrictions, and time (i.e., repeated measures of visual rating each year over time), respectively.

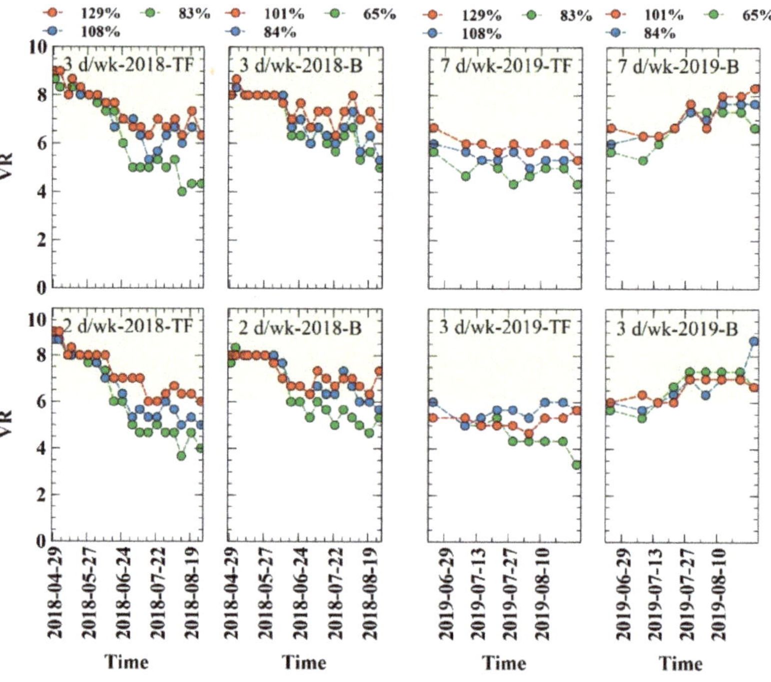

Figure 4. The dynamics of visual rating (VR) values over time showing the response of hybrid bermudagrass (B) and tall fescue (TF) turfgrass to varying irrigation treatments (percentages of ET$_o$) imposed in 2018 and 2019. d/wk: days per week.

For hybrid bermudagrass, in 2018 and 2019, the VR values ranged from 5 to 9. The effect of irrigation levels was only significant in 2018 ($p < 0.05$). Irrigation frequency restriction showed no significant impact on either of the years. The interaction of irrigation levels and frequency restrictions was not statistically significant in either of the years. In 2018 and for the 3 d wk^{-1} frequency restriction, the VR values started to fluctuate noticeably beginning in June, yet stayed above or very close to the minimum threshold for most of the trial (Figure 4). Toward the end of the experiment, VR values fell below 6 for the 65% ET$_o$ and 84% ET$_o$ treatments. The VRs for 101% ET$_o$ treatment stayed in the acceptable range during the trial. The trend was very similar for the 2 d wk^{-1} frequency restriction treatment. In 2019, the VRs were similar for both 7 and 3 d wk^{-1} frequency restrictions. It showed a gradual and constant increase from VR values close to 6 up to 7–8 at the end of the trial. The dynamics of VR values over time were very similar across the irrigation levels in 2019.

For tall fescue in 2018, the VR values ranged from 4 to 9. In 2019, the VR values ranged from 3 to 8. The irrigation level ($p < 0.001$) and frequency restriction ($p < 0.05$) significantly impacted the VR values in 2018, but not in 2019. The interaction of irrigation levels and frequency restrictions was not statistically significant in either of the years. In 2018, the tall fescue VR values were higher than the minimum threshold (VR = 6) for the 129% ET$_o$ treatment and 108% ET$_o$, except for the early July period. For the 83% ET$_o$ treatment, the VR values decreased as the trial progressed and eventually fell below the minimum accepted values of 6. The trends were similar for the 2 d wk^{-1} and 3 d wk^{-1} irrigation frequency restrictions. In 2019, the VR fluctuated less for the 7 d wk^{-1} treatments (no frequency restriction), yet only stayed above the threshold for the highest irrigation treatments of 129% ET$_o$ for most of the experiment. The lowest quality ratings were observed for the irrigation level of 83%. For the 3 d wk^{-1} restriction, all treatments had VR values below the acceptable threshold, and the 109% treatment had slightly higher VR values than the 129% treatment.

3.3. Impact of Irrigation Treatments on Near-Surface Soil Moisture Dynamics

The near-surface soil volumetric water content dynamics across the irrigation treatments for both turfgrass species are shown in Figure 5. For tall fescue in 2018, the minimum and maximum moisture values were 17.8% and 43.8%, respectively. In 2019, the minimum and maximum moisture values were 21.1% and 36.0%, respectively. In 2018, the soil moisture values showed some reduction at the beginning of the trial across treatments. The soil moisture for the 83% ET$_o$ was noticeably lower than the other treatments. This difference was more pronounced for the 3 d wk^{-1} treatments compared with the 2 d wk^{-1} treatments. However, the soil moisture trends for the 129% ET$_o$ and 108% ET$_o$ treatments were similar; they increased (more pronounced for the 3 d wk^{-1} frequency restriction) early in the trial and then stabilized with some decline toward the end of the experiment. In 2019, the greatest and smallest soil moisture values belonged to the highest and lowest irrigation treatments, respectively, for both frequency restrictions scenarios, as expected. The two lesser irrigation levels caused a gradual decline in soil moisture over time, which was more pronounced for the 7 d wk^{-1} treatment (no frequency restriction).

For hybrid bermudagrass in 2018, the minimum and maximum moisture values were 15.3% and 35.7%, respectively. In 2019, the minimum and maximum moisture values were 11.5% and 33.4%, respectively. In 2018, soil moisture showed no substantial fluctuations over time, except for an initial decrease in all treatments at the beginning of the trial (more pronounced for the 3 d wk^{-1} irrigation restriction treatment). After that, the soil moisture showed minor fluctuations across treatments. In 2019, the soil moisture showed a gradual decrease over time for both watering restriction scenarios. The greatest and smallest irrigation levels caused the highest and lowest near-surface soil moisture, respectively, as expected.

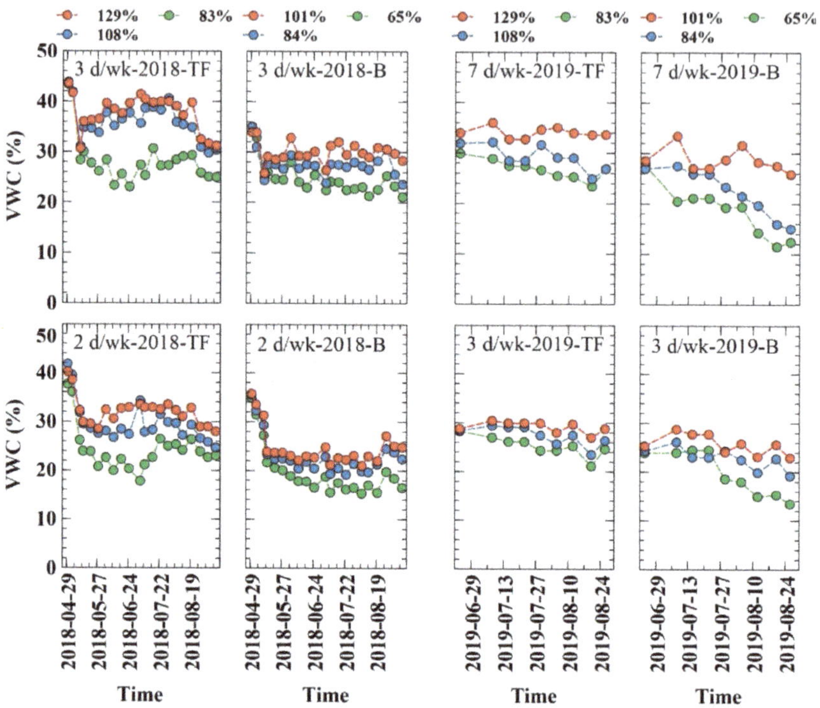

Figure 5. The near-surface (12 cm) volumetric soil water content (VWC) dynamics in 2018 and 2019 across the irrigation treatments (percentages of ET_o). TF: tall fescue; B: hybrid bermudagrass; d/wk: days per week.

3.4. Turfgrass Water Response Function (TWRF)

The TWRFs developed using the combined two years of data for tall fescue (Equation (6)) and hybrid bermudagrass (Equation (7)) species are as follows:

$$VR = 9.81 - 0.02(CET_o) - 0.56(F) + 7.08 \times 10^{-6}\left(CET_o^2\right) + 0.01(I \times CET_o) + 1.06 \times 10^{-3}(I \times F) \tag{6}$$

$$VR = 9.11 - 0.01(CET_o) - 0.46(F) + 2.23 \times 10^{-6}\left(CET_o^2\right) + 0.01(I \times CET_o) + 1.57 \times 10^{-3}(I \times F) \tag{7}$$

where VR is the visual rating, CET_o is the cumulative ET_o over time (mm), F is the irrigation frequency restriction (days wk^{-1}), and I is the irrigation level (ET_o percentages).

Figure 6 shows the performance of the fitted TWRFs developed for tall fescue and hybrid bermudagrass species. Table 4 summarizes the performance statistics for the fitted TWRFs.

Table 4. Performance statistics for the turfgrass water response functions developed for tall fescue and hybrid bermudagrass species.

	r [1]	RMSE	MBE	MAE
Tall Fescue	0.79	0.64	0.0005	0.59
Hybrid Bermudagrass	0.75	0.37	−0.0032	0.47

[1] r = correlation coefficient; RMSE: root means square error; MBE: mean bias error; MAE: mean absolute error.

The TWRFs estimated VR with acceptable accuracy for both species, as illustrated by well-scattered data points around the 1:1 line (Figure 6). The RMSE, MAE, and r values were equal to 0.64, 0.59, and 0.79, respectively, for tall fescue and 0.37, 0.47, and 0.75, respectively, for hybrid bermudagrass species. No systematic bias was observed for any of the TWRFs, given the negligible MBE values of 0.0005 and −0.0032 for tall fescue and hybrid bermudagrass species, respectively.

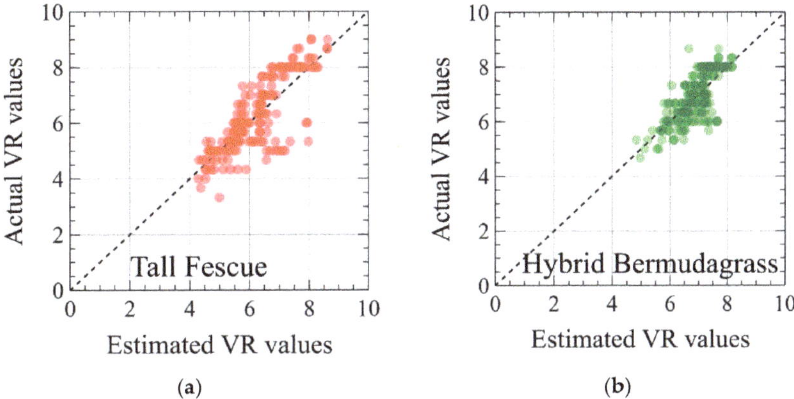

Figure 6. Performance of turfgrass water response functions developed for tall fescue (**a**) and hybrid bermudagrass (**b**) species using a combination of 2018 and 2019 experimental data. VR: visual rating values.

Figure 7 illustrates the estimated impact of multiple irrigation scenarios ranging from 50% to 100% ET_o on turfgrass (VR) using TWRF. The long-term mean ET_o data were obtained from the CIMIS station #39. The estimated period is from May to August. Hybrid bermudagrass maintained its quality above minimum acceptable value for irrigation levels more than 70% ET_o. Tall fescue held its rate above the threshold for approximately 40 and 55 days for the 60% and 100% ET_o irrigation scenarios, respectively.

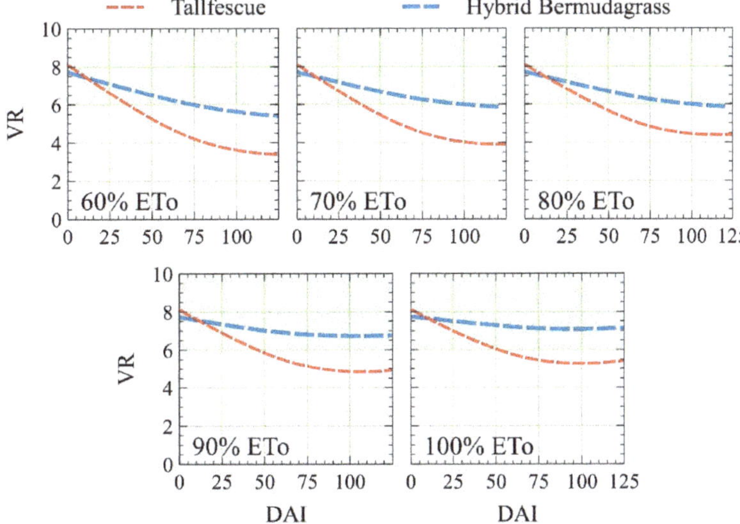

Figure 7. Response of tall fescue and hybrid bermudagrass to multiple irrigation levels based on long-term mean ET_o data obtained from a weather station located nearby the experimental site. VR: visual rating values, DAI: days after initiation.

4. Discussion

4.1. Performance of the Smart ET-Based Controller

It is crucial to evaluate the manufacturer programmed and user-defined settings of the available ET-based commercial controllers and investigate the reliability of their algorithms for efficient autonomous irrigation scheduling [11,21]. The actual water applied by the irrigation controller in this study in both years was substantially higher than the initial target irrigation levels across the treatments. The main reason for overirrigation was the initial underestimation of the irrigation precipitation rate using the catch-cans method. We [7] also reported the same issue about underestimating the irrigation precipitation rate using the catch-cans test. The inadequacy of the catch-cans method might be attributed to spatial variation in water applied within each plot and limited sampling areas represented by the catch devices. Moreover, a substantial fraction of applied water may not get into catch devices because of water hitting the cans at an angle and thus splashing out. Accurate estimation of irrigation precipitation rate and proper selection of minimum plant factors are the most critical factors for reliable autonomous irrigation scheduling using a Weathermatic ET-based controller [7].

The second reason for the overirrigation was the overestimation of ET_0 by the controller compared with CIMIS ET_0. We [7] obtained on average 5.7% overestimation compared with CIMIS ET_0 for the same Weathermatic controller equipped with an on-site temperature sensor and latitude-based solar radiation estimations in inland Southern California. This error range is very close to the 4–5% ET_0 differences observed in this study, indicating an acceptable estimation of ET_0 by the controller. Hargreaves and Samani [17] performed even better when long-term weather data were used. Further studies are needed to determine the performance of temperature-based ET_0 models in different climate regions across the state. Both this study and our recently published study in Southern California [7] focused on the summer months with the highest irrigation demand when no rainfall is typically received, and therefore, water conservation is essential in Central and Southern California. The higher reported overirrigation values of 10% in North Carolina [22] and 32% in Florida [23] suggest that considerable precipitation may negatively impact the efficiency of ET-based controllers as the incorporation of rain into irrigation scheduling by these controllers is often very simplistic.

4.2. Turfgrass Irrigation Management

In their review paper, Colmer and Barton [24] gathered 29 bermudagrass crop coefficient values ranging from 0.40 to 1.27 in well-watered conditions and from 0.52 to 0.94 under deficit irrigation. Variation in water requirements reported for warm-season grass species shows that local crop coefficient and irrigation recommendations information should be developed, and utilizing a nominal value for all cultivars and locations is not optimal. The TWRF estimated VR values based on long-term mean ET_0 demand suggested 70–80% ET_0 as the minimum irrigation application to maintain the acceptable hybrid bermudagrass quality in central California during the high water demand months (i.e., May to August). Reducing the irrigation level to 60% ET_0 only maintains the turf quality for approximately 75 days before it falls below the minimum acceptable value of 6. We [7] conducted a similar analysis using TWRF developed based on three years of experimental data in inland southern California. In that study, we reported 75% ET_0 as the minimum requirement to maintain the 'Tifgreen' hybrid bermudagrass quality in peak summer months, which agrees with the findings of this study. Wherley et al. [25] also reported that a commonly used crop coefficient of 0.6 for warm-season grass underestimated water requirement for 'Tifway' hybrid bermudagrass in Florida.

Richie et al. [26] conducted a two-year tall fescue field irrigation trial in Riverside, California. They reported visual qualities between 4 and 6 for most of the experimental periods (~mid-June to mid-November) for irrigation application of 79–85% ET_0. Brown et al. [27] conducted a research project to evaluate the response of 'Monarch' tall fescue to different irrigation regimes in Las Vegas, NV, USA. They reported 80% ET_0 as the optimum irrigation level to maintain color and coverage of tall fescue turfgrass at an acceptable condition. Ervin and Koski [28] conducted a study near Fort Collins, CO, USA and reported 75% ET_0 as the safe irrigation level to maintain an acceptable quality for tall fescue turfgrass (a blend of 'Rebel Jr.', 'Crewcut', and 'Monarch'). In our study, the 83% ET_0 irrigation level only maintained the visual rating above the minimum threshold for two months in 2018. The 108% ET_0 level with 3 d wk^{-1} frequency kept the VR values in the acceptable range in 2018, except for a short period in mid-trial. Both Brown et al. [27] and Ervin and Koski [28] estimated the irrigation volumes applied by measuring the volume of water collected in catch-can devices. If we use the precipitation rate estimated by the catch-cans test, the 108% ET_0 treatment (which maintained the turf quality for much of the trial in 2018) will be reduced to 89% ET_0. However, the catch-cans-based precipitation rate estimations were proved to be inaccurate in our study. The same issue regarding the inadequacy of the catch-cans method to estimate the precipitation rate of landscape irrigation systems was also reported in our recently published study conducted in Southern California [7]. Therefore, we recommend the water savings and crop coefficients reported in the literature based on the catch-cans-estimated precipitation rates be re-evaluated and used cautiously.

In 2019, the VR values of tall fescue plots were relatively low across all irrigation treatments. This might be related to the minimal fertilizer application rates, diminishing growth and greenness of tall fescue in 2019. The positive impact of higher nitrogen fertilization treatment on color ratings was also reported by [27,29]. The TWRF estimations suggest that 60% ET_0 would be sufficient to maintain the tall fescue quality within the acceptable range for approximately 40 days. Hong et al. [30] conducted a two-year dry down study in Kansas and reported 50% ET_0 irrigation held 'Seed Research 8650' tall fescue quality above minimum acceptable rating (VR = 6) for 45 and 82 days in two years of their study. The TWRF estimations suggest that 100% ET_0 would be only sufficient to maintain the tall fescue quality for approximately 55 days. This might be an underestimation impacted by the low VR data in 2019 and, therefore, should be further investigated in the future.

Limiting irrigation watering days to specific days per week is a common strategy that is particularly popular and imposed by water agencies and municipalities in California during droughts to help conserve water in urban areas. We, however, observed no substantial water saving associated with restricting irrigation frequency (watering days). In fact, for both species and in both years, less restrictive watering days improved the VR values. We [7] also found no turfgrass quality improvement or water conservation associated with restricting the watering days in Southern California. This finding is attributed partly to sandy soils with low water holding capacity at both sites and the fact that most of the turfgrass roots are expected to be in the topsoil layer. Su et al. [31] reported that 86% of all root length of 'Dynasty' tall fescue in the field was in the upper 30 cm under well-watered condition and silt loam soil in Kansas, USA. Sinclair et al. [32] reported low amounts of root mass in the deepest soil layer during sod establishment for 'Tifway 419' hybrid bermudagrass grown in tubes filled with loamy sand soil in Florida, USA. The Weathermatic smart controller can dynamically adjust the irrigation frequency based on the actual ETo demand, so restricting irrigation frequency is unnecessary. When irrigation frequency is not restricted, programming a minimum deficit threshold is crucial to avoid light irrigation applications and prevent excessive evaporative loss.

The irrigation scenarios directly impacted the dynamics of near-surface soil moisture. However, the near-surface soil moisture fluctuation over time was not always adequate to explain the variation in turfgrass quality. For example, in 2019, the visual quality of tall fescue for the 83% ET_0 treatment decreased over time, but the soil moisture showed minimum fluctuations. Moreover, in 2019, the continuous late green-up of hybrid bermudagrass through June and a fertilizer application in mid-July caused the quality ratings to improve as the trial progressed despite a constant decrease in the near-surface soil moisture of the hybrid bermudagrass plots. Consequently, the near-surface soil moisture data should be interpreted carefully and in conjunction with other parameters such as turfgrass physiology and fertilizer applications. The in situ soil moisture values were higher than the laboratory-measured water retention data (Figure 2). This could be related to differences between laboratory and field sensors and the small size of the laboratory samples. Note that the reported results in this study are for a pop-up sprinkler system with autonomous irrigation scheduling using the Weathermatic ET-based smart irrigation controller. Further studies are needed to determine whether lower irrigation amounts might be sufficient when irrigation scheduling is done based on actual root zone soil water holding capacity information and for irrigation systems with higher potential efficiency, such as underground drip irrigation systems.

5. Conclusions

Our results suggest that applying typically recommended values of 60% ET_0 for hybrid bermudagrass and 80% ET_0 for tall fescue is insufficient to maintain the acceptable quality over high ET_0 demand months in Central California. The TWRFs fitted to the experimental data suggested 80% ET_0 as the minimum requirement to maintain the quality of hybrid bermudagrass above the minimum acceptable VR value of 6 for four months (May to August). The TWRFs estimations suggested that applying 100% ET_0 was sufficient to maintain the tall fescue quality above the threshold for only 50 days. This finding might be an overestimation as tall fescue ratings were negatively affected by minimal fertilizer application in 2019. The Weathermatic controller showed promising results by providing acceptable ET_0 estimations (5–8% higher than CIMIS ET_0) only using onsite temperature measurements. The controller also closely followed programmed watering days restrictions and adjusted the watering days based on ET_0 demand when no watering restriction was applied. The efficient irrigation scheduling by the smart controller also depends on the accurate calculation of the precipitation rate of the irrigation system, which was substantially underestimated in our study using the widely used catch-cans method. Therefore, researchers should be cautious when using this approach to estimate the precipitation of the irrigation systems.

Author Contributions: Conceptualization, A.H.; methodology, A.H.; software, A.H., A.S. (Anish Sapkota), and A.S. (Amninder Singh); validation, A.H., A.S. (Anish Sapkota), and A.S. (Amninder Singh); formal analysis, A.H., M.R., A.S. (Anish Sapkota), and A.S. (Amninder Singh); investigation, A.H., M.R., A.S. (Anish Sapkota), and A.S. (Amninder Singh); resources, A.H. and M.R; data curation, A.H., M.R., A.S. (Anish Sapkota), and A.S. (Amninder Singh); writing—original draft preparation, A.H.; writing—review and editing, A.H., M.R., A.S. (Anish Sapkota), and A.S. (Amninder Singh); visualization, A.H.; supervision, A.H.; project administration, A.H. and M.R.; funding acquisition, A.H. and M.R. All authors have read and agreed to the published version of the manuscript.

Funding: This study was supported by the University of California Division of Agriculture and Natural Resources competitive grant (ID#: 17-5021) and the United States Geological Survey (ID#: 2017CA371B).

Institutional Review Board Statement: Not applicable.

Informed Consent Statement: Not applicable.

Conflicts of Interest: The authors declare no conflict of interest.

References

1. CSDO. *Population Projections Methodology (2019 Baseline)*; 2020. Available online: https://www.dof.ca.gov/forecasting/demographics/projections (accessed on 20 August 2021).
2. UCANR. STRATEGIC VISION 2025: University of California Division of Agriculture and Natural Resources. 2019. Available online: https://ucanr.edu/files/906.pdf (accessed on 20 August 2021).
3. Cook, B.I.; Ault, T.R.; Smerdon, J.E. Unprecedented 21st century drought risk in the American Southwest and Central Plains. *Sci. Adv.* **2015**, *1*, e1400082. [CrossRef]
4. Cooley, H. *Urban and Agricultural Water Use in California, 1960–2015*; Pacific Institute: Oakland, CA, USA, 2020.
5. Cardenas, B.; Dukes, M.D. Soil moisture sensor irrigation controllers and reclaimed water; Part I: Field-plot study. *Appl. Eng. Agric.* **2016**, *32*, 217–224.

6. Davis, S.; Dukes, M.D.; Miller, G. Landscape irrigation by evapotranspiration-based irrigation controllers under dry conditions in Southwest Florida. *Agric. Water Manag.* **2009**, *96*, 1828–1836. [CrossRef]

7. Haghverdi, A.; Singh, A.; Sapkota, A.; Reiter, M.; Ghodsi, S. Developing irrigation water conservation strategies for hybrid bermudagrass using an evapotranspiration-based smart irrigation controller in inland southern California. *Agric. Water Manag.* **2021**, *245*, 106586. [CrossRef]

8. Singh, A.; Haghverdi, A.; Nemati, M.; Hartin, J. *Efficient Urban Water Management: Smart Weather-Based Irrigation Controllers*; UCANR Publication, 2020; p. 8674. Available online: https://anrcatalog.ucanr.edu/pdf/8674.pdf (accessed on 20 August 2021).

9. Dukes, M. Water conservation potential of landscape irrigation smart controllers. *Trans. ASABE* **2012**, *55*, 563–569. [CrossRef]

10. Devitt, D.; Carstensen, K.; Morris, R. Residential water savings associated with satellite-based ET irrigation controllers. *J. Irrig. Drain. Eng.* **2008**, *134*, 74–82. [CrossRef]

11. Davis, S.L.; Dukes, M.D. Importance of ET controller program settings on water conservation potential. *Appl. Eng. Agric.* **2016**, *32*, 251–262.

12. Bijoor, N.S.; Pataki, D.E.; Haver, D.; Famiglietti, J.S. A comparative study of the water budgets of lawns under three management scenarios. *Urban Ecosyst.* **2014**, *17*, 1095–1117. [CrossRef]

13. Service, N.P. Benefits of Turf Grass. In *Turf Management*; 2018. Available online: https://www.nps.gov/subjects/turfmanagement/benefits.htm (accessed on 20 August 2021).

14. Hartin, J.; Oki, L.; Fujino, D.; Reid, K.; Ingels, C.; Haver, D.; Baker, W. UC ANR research and education influences landscape water conservation and public policy. *Calif. Agric.* **2019**, *73*, 25–32. [CrossRef]

15. Haghverdi, A.; Reiter, M.; Singh, A.; Sapkota, A. Hybrid Bermudagrass and Tall fescue Irrigation in Central California: II. Assessment of NDVI, CWSI and Canopy Temperature Dynamics. *Agronomy* **2021**. submit.

16. Haghverdi, A.; Najarchi, M.; Öztürk, H.S.; Durner, W. Studying Unimodal, Bimodal, PDI and Bimodal-PDI Variants of Multiple Soil Water Retention Models: I. Direct Model Fit Using the Extended Evaporation and Dewpoint Methods. *Water* **2020**, *12*, 900. [CrossRef]

17. Hargreaves, G.H.; Samani, Z.A. Reference crop evapotranspiration from temperature. *Appl. Eng. Agric.* **1985**, *1*, 96–99. [CrossRef]

18. Morris, K.N.; Shearman, R.C. NTEP turfgrass evaluation guidelines. In Proceedings of the NTEP Turfgrass Evaluation Workshop, Beltsville, MD, USA, 17 October 1998; pp. 1–5.

19. SAS Institute. *Base SAS 9.4 Procedures Guide*; SAS Institute, 2015. Available online: https://documentation.sas.com/doc/en/pgmsascdc/9.4_3.5/procstat/titlepage.htm (accessed on 20 August 2021).

20. Sanders, J. *Veusz-a Scientific Plotting Package*. 2008. Available online: https://veusz.github.io/ (accessed on 20 August 2021).

21. Davis, S.; Dukes, M. Irrigation scheduling performance by evapotranspiration-based controllers. *Agric. Water Manag.* **2010**, *98*, 19–28. [CrossRef]

22. Grabow, G.; Ghali, I.; Huffman, R.; Miller, G.; Bowman, D.; Vasanth, A. Water application efficiency and adequacy of ET-based and soil moisture–based irrigation controllers for turfgrass irrigation. *J. Irrig. Drain. Eng.* **2013**, *139*, 113–123. [CrossRef]

23. Davis, S.; Dukes, M.D.; Vyapari, S.; Miller, G.L. Evaluation and demonstration of evapotranspiration-based irrigation controllers. In Proceedings of the World Environmental and Water Resources Congress 2007: Restoring Our Natural Habitat, Tampa, FL, USA, 15–19 May 2007; pp. 1–18.

24. Colmer, T.D.; Barton, L. A review of warm-season turfgrass evapotranspiration, responses to deficit irrigation, and drought resistance. *Crop Sci.* **2017**, *57*, S-98–S-110. [CrossRef]

25. Wherley, B.; Dukes, M.; Cathey, S.; Miller, G.; Sinclair, T. Consumptive water use and crop coefficients for warm-season turfgrass species in the Southeastern United States. *Agric. Water Manag.* **2015**, *156*, 10–18. [CrossRef]

26. Richie, W.; Green, R.; Klein, G.; Hartin, J. Tall fescue performance influenced by irrigation scheduling, cultivar, and mowing height. *Crop Sci.* **2002**, *42*, 2011–2017. [CrossRef]

27. Brown, C.; Devitt, D.; Morris, R. Water use and physiological response of tall fescue turf to water deficit irrigation in an arid environment. *HortScience* **2004**, *39*, 388–393. [CrossRef]

28. Ervin, E.H.; Koski, A.J. Drought avoidance aspects and crop coefficients of Kentucky bluegrass and tall fescue turfs in the semiarid west. *Crop Sci.* **1998**, *38*, 788–795. [CrossRef]

29. Schiavon, M.; Pornaro, C.; Macolino, S. Tall Fescue (*Schedonorus arundinaceus* (Schreb.) Dumort.) Turfgrass Cultivars Performance under Reduced N Fertilization. *Agronomy* **2021**, *11*, 193. [CrossRef]

30. Hong, M.; Bremer, D.; Keeley, S. Minimum water requirements of cool-season turfgrasses for survival and recovery after prolonged drought. *Crop Sci.* **2020**, *5*. [CrossRef]

31. Su, K.; Bremer, D.J.; Keeley, S.J.; Fry, J.D. Rooting characteristics and canopy responses to drought of turfgrasses including hybrid bluegrasses. *Agron. J.* **2008**, *100*, 949–956. [CrossRef]

32. Sinclair, T.R.; Schreffler, A.; Wherley, B.; Dukes, M.D. Irrigation frequency and amount effect on root extension during sod establishment of warm-season grasses. *HortScience* **2011**, *46*, 1202–1205. [CrossRef]

US, landscape water use in the summer months can reach up to 90% of the total municipal water use [1]. Turfgrass is a large component of urban landscapes that provides valuable recreation areas and ecosystem services [2]. Across the nation, the largest sector of turfgrass is residential lawns. Therefore, developing recommendations for efficient irrigation management of turfgrass is crucial for maintaining urban green infrastructure.

In the last two decades, documented research on urban irrigation management has mainly focused on the implementation of irrigation technologies to enhance irrigation water use efficiency [3–8], the use of low-quality water for irrigation to alleviate freshwater demand [9–11], and the applications of remote sensing (RS) techniques to detect drought injury and manage irrigation [12–14]. The latter is particularly timely considering the rapidly emerging advancements in novel RS platforms. Some promising results have been reported on the application of multispectral and thermal RS techniques to predict the green leaf area index of turfgrass [12], estimate the crop water stress index (CWSI) [15], calculate the location and rate of urban irrigation [13], and monitor turfgrass water stress and use [14]. For instance, Taghvaeian et al. [14] used ground-based optical and thermal RS to study the quality response, water stress, and water consumption of multiple turfgrass species under different soil and irrigation treatments in northern Colorado, USA. RS helps scale the findings of plot-based research projects providing the decision-making information necessary for assessing the urban irrigation footprint for large metropolitan areas. For instance, Chen et al. [16] used RS techniques plus water use records to estimate that in 2005–2007, 7% of the postal carrier routes in Los Angeles, California were overwatered in dry years and 43% were overwatered in wet years.

Monitoring canopy temperature can help quantify plant water stress in a fast and non-destructive way, which could be used for efficient irrigation management [17]. In addition, measuring canopy temperature variations due to deficit irrigation is necessary to understand the tradeoffs between water conservation and the vital role of the irrigated urban landscape to mitigate the urban heat island phenomenon. The irrigated landscape, through the process of evapotranspiration, can reduce daytime heat storage and enhance nighttime cooling, thereby moderating the climate of urban areas and creating localized cool islands [18]. Bonfils and Lobell [19] showed the significant cooling effect that irrigation expansion has had on summertime average daily daytime temperatures in California. Broadbent et al. [20] studied the cooling benefits of irrigation in a suburb of Adelaide, Australia and found that the diurnal average air temperature was reduced by up to 2.3 °C, but that increasing irrigation had a non-linear effect on cooling. Wang et al. [21] reported 4.52 ± 0.77 mm day^{-1} °C^{-1} surface air cooling in urban areas over the contiguous United States due to irrigation.

CWSI is a dimensionless temperature-based index [22] that has shown success in quantifying turfgrass water stress [14]. The empirical approach of Idso et al. [22] requires establishing the lower and upper-temperature baselines for non-water-stressed and non-transpiring conditions, respectively. The reported CWSI baselines for turfgrass in the literature vary widely. Therefore, specific baselines for each climatic region should be developed.

In the first part of this study [23], we used a visual rating (VR) to assess turfgrass response to a wide range of irrigation scenarios in central California. VR is the traditional method of rating turfgrass quality ranging from 1 (worst) to 9 (best) that has been used by researchers and turfgrass managers worldwide for turfgrass evaluation [24]. However, it is subjective and can be inaccurate since different observers may rate the turfgrass differently, or even an identical evaluator may give different ratings to the same plots over time, and therefore, such quality ratings are nonreproducible [25,26]. RS is an alternative approach that can provide a more accurate, consistent, and reliable evaluation of overall turfgrass quality, growth, and health. In addition, the recent advancement in unmanned aerial vehicles makes RS a superior method for scouting and identifying drought injury on large irrigated turfgrass areas such as parks and golf courses. The normalized difference vegetation index (NDVI) has been used as a replacement to assess the response of turfgrass to irrigation scenarios and has been shown to be well correlated with the visual rating of different grass species, including 'Dynasty' tall fescue [27] and 'Tifgreen' hybrid bermudagrass [8]. In addition, Bremer et al. [28] reported high correlation values between NDVI and percentage green cover ($r = 0.91$) and shoot density ($r = 0.88$) using data obtained from multiple cool-season grass species.

This study was conducted to (i) monitor the changes in NDVI and canopy temperature of tall fescue and hybrid bermudagrass under varying irrigation treatments (amount and frequency) in central California, (ii) develop empirical CWSI and study its variability over time for both turfgrass species, and (iii) investigate the relationship between NDVI and visual rating values reported in the companion paper [23]. We focused on both cool-season and warm-season turfgrass species in this study because they are grown in different settings in urban areas in central California. Warm-season species are considered superior because of their relatively lower water requirement, but they could be less appealing to certain groups since they go dormant over the winter.

2. Materials and Methods

2.1. Experimental Site

This study was conducted at the University of California Agricultural and Natural Resources Kearney Research and Extension Center (36°36′02.2″ N 119°30′38.8″ W) in Parlier, California (Figure 1). The study site consisted of 36 plots. Each plot was roughly 14 m^2 with an approximately 60 cm border between the adjacent plots.

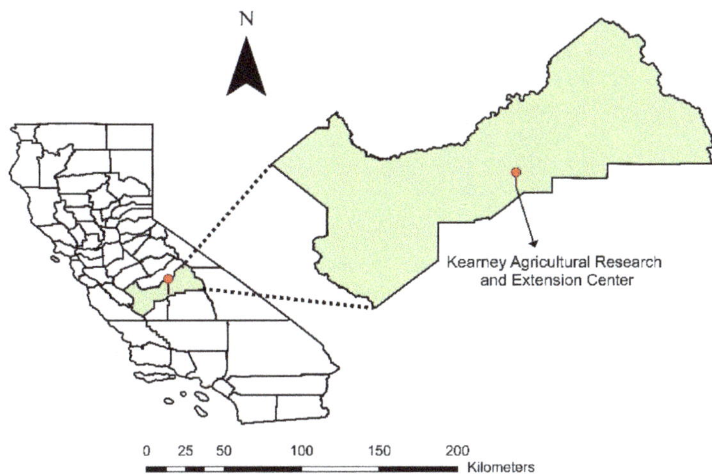

Figure 1. The location of the study area.

Figure 2a depicts the long-term mean cumulative reference evapotranspiration (ET$_o$) and precipitation data measured by the California Irrigation Management Information System (CIMIS) weather station #75 located close to the study site. Long-term annual weather data show that ET$_o$ is roughly five times greater than the precipitation received in this area. Irrigation is necessary to keep the urban landscape species alive, particularly during the summer months when evaporative demand is highest. The soil at the research site is classified as Hanford fine sandy loam (websoilsurvey.sc.egov.usda.gov, accessed on 27 August 2021). The laboratory analysis of undisturbed samples collected from the top 20 cm revealed that the available soil water-holding capacity is 0.19 m^3 m^{-3} for 10 and 1500 kPa that represent field capacity and permanent wilting point, respectively [29].

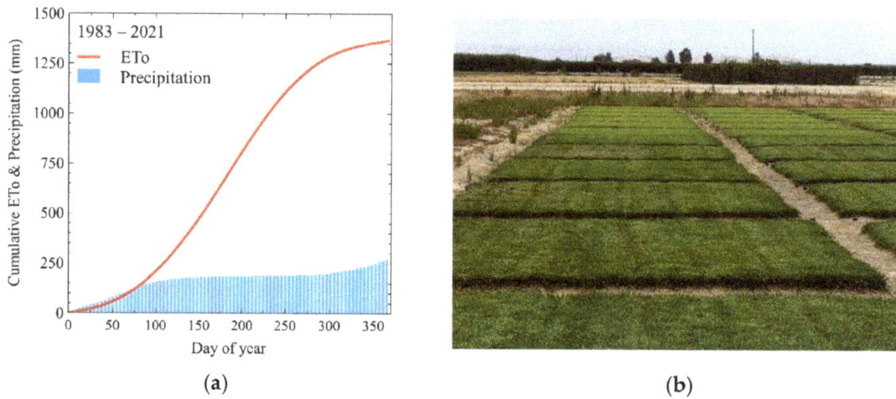

Figure 2. The long-term mean reference evapotranspiration (ET$_o$) and precipitation trends at the study site from a nearby weather station (**a**). A photo taken on 30 May 2018 showing the research plots (**b**).

2.2. Irrigation Trials

Two adjacent hybrid bermudagrass ('Latitude 36' *Cynodon dactylon* (L.) Pers. × *C. transvaalensis* Burtt-Davy) and tall fescue (A blend of 'PennRK4', 'Rebel XLR', + 'Firecracker SLS' *Schedonorus arundinaceus* (Schreb.) Dumort.) irrigation trials were arranged in a factorial randomized complete block design with repeated measures of canopy temperature and NDVI. Each trial consisted of 18 plots (3.7 m × 3.7 m) forming three blocks (replications) to impose six irrigation treatments. To minimize evaporative loss and wind drift, irrigation was done overnight and early morning. In addition, the smart controller performed an automatic run/soak schedule to eliminate runoff and provide enough soak time. All three replications for each treatment were irrigated at the same time by wiring to the same zone on the controller. Each plot was equipped with a TORO 252 Series solenoid valve (Toro Co., Bloomington, MN, USA) supplying water to four Toro O-T-12-QP corner-pop-up 6″ sprinkler heads (152 mm tall). The sprinkler nozzles had an operating pressure range and flow rate of 276–517 kPa and 0.02–9.08 L min^{-1}, respectively. To achieve steady water application, the sprinklers were equipped with factory-installed pressure-compensating discs (Toro Co., Bloomington, MN, USA).

The irrigation treatments consisted of three ET$_o$-based irrigation levels and two irrigation frequencies (Table 1). A Weathermatic Smartline (SL) 4800 controller (Telsco Industries, Inc., Garland, TX, USA) was used to control irrigation treatments and schedule irrigation throughout the study autonomously. The controller used an onsite temperature sensor and latitude-based solar radiation information to calculate ETo using the Hargreaves and Samani equation [30]. Irrigation efficiency (i.e., low half distribution uniformity) of 0.78 and irrigation precipitation rate of 23 mm h^{-1} was calculated using a catch-cans test performed before conducting the trial in year 1. Irrigation was non-limiting to ensure actively growing, non-stress turfgrass prior to initiating irrigation treatments. The experiment started on 4 May 2018, and data collection ended on 11 September 2018. All plots were switched back to the uniform non-limiting irrigation for recovery before starting the trial on 22 June 2019. On 26 August 2019, the main irrigation pipe broke and flooded the field, forcing the research team to terminate the trial. More information about the irrigation system characteristics and establishment of the plots is provided in the companion paper [23].

Table 1. Irrigation treatments throughout the 2-year tall fescue and hybrid bermudagrass irrigation research experiments at the University of California Kearney Research and Extension Center.

2018 Trial, Start: 4 May 2018 | End: 11 September 2018
Target Irrigation Levels (% ET$_o$): Tall Fescue: 50%, 65%, 80% | Hybrid Bermudagrass: 40%, 50%, 60% Irrigation Efficiency: 78% Watering Days: 2 days per week, 3 days per week
2019 Trial, Start: 22 June 2019 | End: 26 August 2019
Target Irrigation Levels (% ET$_o$): Tall Fescue: 50%, 65%, 80% | Hybrid Bermudagrass: 40%, 50%, 60% Irrigation Efficiency: 78% Watering Days: 3 days per week, 7 days per week (no restriction)

The controller used the user-defined "plant type" information to convert ET$_o$ to irrigation application (irrigation application = plant type × ET$_o$). For each treatment, the plant type was calculated as the irrigation levels (% ET$_o$) divided by the irrigation efficiency of the system.

Table 2 summarizes the irrigation application data. All treatments were over irrigated mainly due to the inaccurate estimation of the irrigation precipitation rate using the catch-cans method at the beginning of the trial. The actual applied irrigation rate was calculated at the end of the trial based on the revised irrigation precipitation rate of 18 mm h^{-1}. The performance of the smart controller is discussed in detail in the companion paper [23].

Table 2. Target irrigation treatments (T1–T3) versus programmed and applied irrigation levels throughout the 2-year tall fescue and hybrid bermudagrass irrigation research experiments conducted at the University of California Kearney Research and Extension Center.

	Tall Fescue			**Hybrid Bermudagrass**		
Irrigation	**T1**	**T2**	**T3**	**T1**	**T2**	**T3**
Treatment	50%	65%	80%	40%	50%	60%
Programmed	64%	83%	100%	51%	64%	77%
Applied	83%	108%	129%	65%	84%	101%

Programmed irrigation levels are equal to target treatment levels divided by the irrigation efficiency of 0.78 (i.e., the low half distribution uniformity of the irrigation system). Applied irrigation levels were recalculated based on the irrigation run time data retrieved from the controller and precipitation rate of 28 mm day^{-1} measured for the system at the end of the trial.

2.3. Data Collection and Statistical Analysis

Figure 3 illustrates an overview of the sensors and tools used in this study and the companion paper [8]. The active light source optical GreenSeeker handheld sensor (Trimble Inc., Sunnyvale, CA, USA) was used to collect NDVI data. The sensor has a measurement range of 0 to 0.99 and a roughly 51 cm wide oval field of view when held 122 cm above the ground. Canopy temperature was recorded using the Fluke 64 Max Infrared Thermometer (Fluke Corporation, Everett, WA, USA). According to the manufacturer, the thermometer has a measurement range of -30 to 500 °C and a spectral band of 8–14 microns with an accuracy of 1.5 °C or 1.5% of the reading. The resolution of the thermometer was 0.1 °C with an 87 mm field of view when held 150 cm above the ground. During the data collection, both sensors were held at approximately 1 m height and moved over the center of each plot ($\approx$3–4 m^2) while the trigger remained engaged to continuously scan and obtain an average representative value for each plot. The air temperature and relative humidity were recorded using the Fluke 971 Temperature Humidity Meter (Fluke Corporation, Everett, WA, USA) over each experimental plot during the data collection process. According to the manufacturer, the Fluke 971 handheld sensor has a temperature measurement accuracy of $\pm$0.5 °C in the 0 to 45 °C range and an RH measurement accuracy of $\pm$2.5% in the 10% to 90% RH range. The resolution of the temperature/humidity meter was 0.1 °C and 0.1% RH. The measurements were done within two hours of solar noon under cloud-free conditions. The handheld data were collected roughly once a week throughout the trial, 21 times in 2018 (from 30 April to 22 September), and 10 times in 2019 (from 24 June to 26 August).

Figure 3. An overview of the sensors and tools used in this study and the companion paper [23].

The NDVI and canopy temperature data were statistically analyzed using PROC GLIMMIX in SAS 9.4 software package (SAS Institute, 2014). The irrigation frequency and duration of the experiment differed in years 1 and 2. Therefore, each year was independently analyzed for the treatment effects to accommodate differences in experimental duration and irrigation regimes. The fixed effects were the irrigation levels, irrigation frequencies, and the date of data collection. The random effects were block and its interaction with irrigation levels and irrigation frequencies. The treatment effects were considered significant at p-values ≤ 0.05. The plotting software package Veusz 3.3.1 (https://veusz.github.io/, accessed on 27 August 2021) [31] was used to create all graphs. The NDVI data were compared against the turfgrass visual rating (VR) values presented in the companion study [23]. The rating was based on The National Turfgrass Evaluation Program (NTEP) standards [32], with the minimum and maximum scores of 1 and 9 assigned to dead and ideal turfgrass, respectively.

2.4. Crop Water Stress Index (CWSI)

The CWSI relies on the temperature difference between the canopy and air, dt °C (Tc–Ta), and it is defined as:

$$\text{CWSI} = \frac{dt_m - dt_{lb}}{dt_{ub} - dt_{lb}} \tag{1}$$

where m, lb, and ub indicate the measured, lower baseline (non-water-stressed), and upper baseline (non-transpiring) of dt, respectively.

The empirical CWSI is based on the linear relationship between the lower baseline temperature difference and vapor pressure deficit (VPD) [22,33]:

$$dt_{lb} = a(\text{VPD}) + b \tag{2}$$

The VPD is calculated as follows:

$$\text{VPD} = e_s - e_a \tag{3}$$

where e_s is the saturation vapor pressure (kPa) and e_a is the actual vapor pressure (kPa) calculated as:

$$e_s = 0.6108 * exp\left(\frac{17.27 \times T_a}{237.3 + T_a}\right) \tag{4}$$

$$e_a = (RH/100) * e_s \tag{5}$$

The CWSI was calculated for each species separately in this study. The mean canopy temperature data obtained from the highest irrigation levels were used to estimate the lower baselines over all the well-watered plots of each species. The mean air temperature and RH values collected over all the plots using the handheld Fluke 971 m were used to calculate VPD. The upper baseline was calculated as the mean temperature difference between air and severely stressed tall fescue grass [17]. The baseline was established using the canopy temperature data collected in both years from a plot of non-irrigated tall fescue turfgrass adjacent to the study field. The non-transpiring canopy temperature was assumed to be similar between tall fescue and hybrid bermudagrass species. This assumption is based on the data collected by the research team from side-by-side non-transpiring tall fescue and hybrid bermudagrass plots sprayed with glyphosate in southern California (data not published). Different baselines were established for each year to determine their stability over time.

3. Results

3.1. NDVI

Figure 4 shows the response of both species (dynamics of NDVI values) to the applied irrigation treatments in 2018 and 2019. Table 3 summarizes the results of the statistical analysis for both species in the years 2018 and 2019. For tall fescue, the NDVI values ranged between 0.30 and 0.80 in 2018 and between 0.23 and 0.69 in 2019. The irrigation level ($p < 0.001$) and frequency ($p < 0.05$) had significant effects on NDVI values in 2018. In 2019, only the irrigation level had a significant effect ($p < 0.01$) on turfgrass quality. The interaction between irrigation levels and frequency was not significant in neither of the years. In 2018, the dynamics of NDVI values over time for all treatments were somewhat similar, showing a slight decline as the trial progressed. However, for 83% ET_o treatment (2 d wk^{-1} frequency), a more noticeable reduction in NDVI values was observed toward the end of the experimental period. In 2019, the NDVI values for the lowest irrigation application of 83% ET_o started to decline around mid-July for both 3 and 7 d w^{-1} irrigation frequency. The NDVI values showed no substantial change for the other irrigation treatments.

For hybrid bermudagrass, the NDVI valued varied between 0.53 and 0.76 in 2018 and between 0.34 and 0.80 in 2019. The irrigation levels had no significant effect on NDVI values in 2018 and 2019 (Table 3). The impact of irrigation frequency was only significant ($p < 0.01$) in 2019. The interaction between irrigation levels and frequency was not significant in either of the years. In 2018, the NDVI for all treatments stayed fairly stable with no substantial fluctuations over time. In 2019, NDVI values increased over time with no noticeable differences between the irrigation treatments (Figure 4).

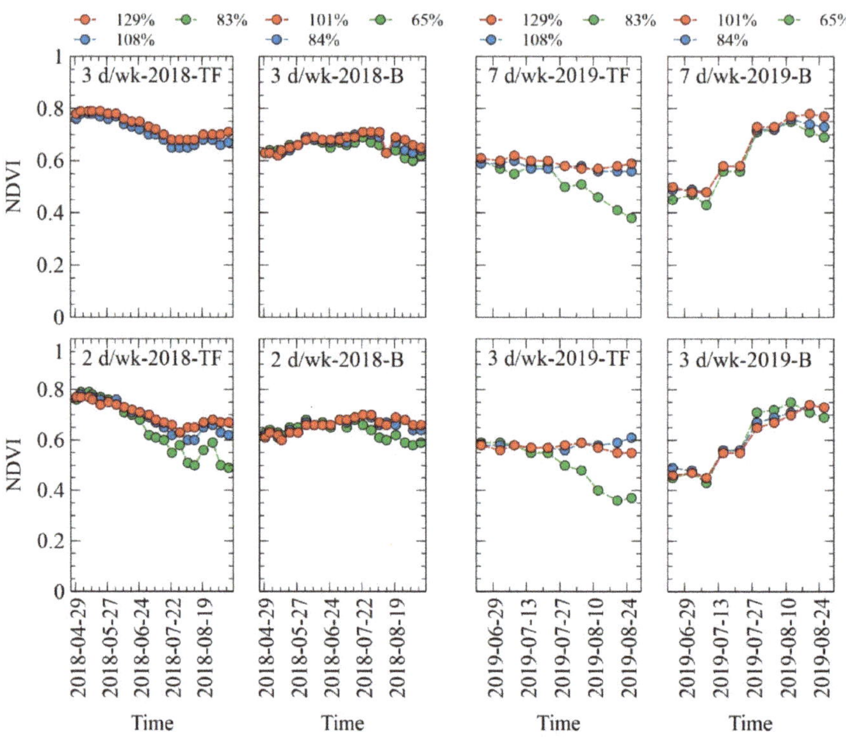

Figure 4. NDVI values showing the response of hybrid bermudagrass and tall fescue turfgrass to varying irrigation levels (ET$_o$%) and frequency (d/wk: days per week) imposed in 2018 and 2019. TF: tall fescue, B: hybrid bermudagrass.

Table 3. Statistical analysis of the hybrid bermudagrass and tall fescue response (NDVI and canopy temperature) to irrigation treatments imposed in years 2018 and 2019 (each year was analyzed separately).

	Tall Fescue					Hybrid Bermudagrass			
	NDVI		Canopy Temp.			NDVI		Canopy Temp.	
Treatment	2018	2019	2018	2019	Treatment	2018	2019	2018	2019
129% ET$_o$	0.72a	0.59a	31.8b	37.2b	101% ET$_o$	0.66a	0.62a	34.5a	39.4a
108% ET$_o$	0.70a	0.58a	31.9b	37.4b	84% ET$_o$	0.66a	0.62a	34.5a	39.4a
83% ET$_o$	0.65b	0.51b	33.0a	38.1a	65% ET$_o$	0.64a	0.59a	35.0a	39.5a
Frequency					**Frequency**				
2 d w^{-1}	0.68b		32.5a		2 d w^{-1}	0.65a		35.0a	
3 d w^{-1}	0.70a	0.55a	31.9b	37.8a	3 d w^{-1}	0.66a	0.59b	34.3b	39.7a
7 d w^{-1}		0.56a		37.4a	7 d w^{-1}		0.63a		39.2b
Model effect	**2018**	**2019**	**2018**	**2019**	**Model effect**	**2018**	**2019**	**2018**	**2019**
I	***	**	**	*	I	NS	NS	NS	NS
F	*	NS	*	NS	F	NS	**	*	*
I × F	NS	NS	NS	NS	I × F	NS	NS	NS	NS
T	***	***	***	***	T	***	***	***	***
I × T	***	***	NS	NS	I × T	***	NS	NS	NS
F × T	NS	NS	NS	NS	F × T	NS	NS	***	NS
I × F × T	NS	NS	NS	NS	I × F × T	NS	NS	NS	NS

NS, ***, **, and * are non-significant or significant at $p \leq 0.001$, 0.01, and 0.05, respectively. Means sharing a similar letter are not significantly different, based on Turkey's test at $\alpha = 0.05$. I, F, and T in the table refer to irrigation levels, frequency, and time (i.e., repeated measures of visual rating each year over time), respectively.

3.2. Canopy Temperature and CWSI

Figure 5 illustrates the tall fescue and hybrid bermudagrass canopy temperature fluctuations over time in response to the irrigation treatments imposed in 2018 and 2019. Table 3 summarizes the results of the statistical analysis for both species in the years 2018 and 2019. The canopy temperature readings were very similar in early and mid-trials in both years for both species across the treatments, while the fluctuations in canopy temperature values over time were more pronounced in 2018. The non-irrigated plot of turfgrass adjacent to the study field had on average 19 °C and 17 °C higher canopy temperature than the irrigated plots in 2018 and 2019, respectively.

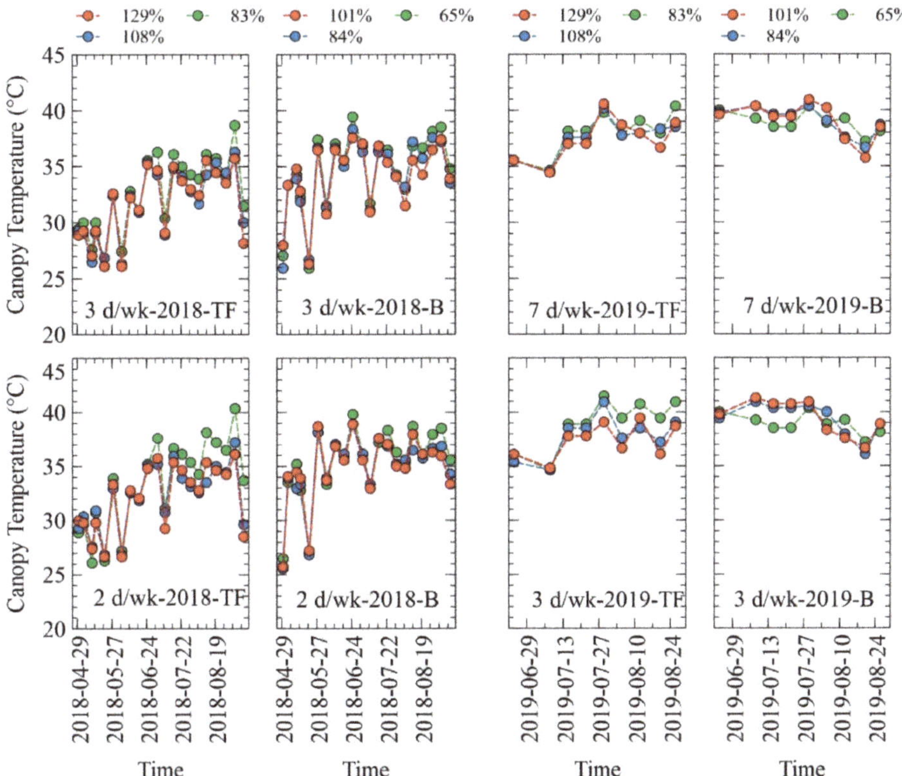

Figure 5. The canopy temperature dynamics of hybrid bermudagrass and tall fescue turfgrass plots under varying irrigation scenarios imposed in 2018 and 2019. d/wk: indicates irrigation frequency in days per week.

For the tall fescue plots, the minimum and maximum canopy temperature values were 24 °C and 49 °C in 2018 and 33 °C and 43 °C in 2019. Irrigation levels significantly impacted the canopy temperature in 2018 ($p < 0.01$) and in 2019 ($p < 0.05$) (Table 3). On average, there was a 1.2 °C in 2018 and 0.9 °C in 2019 temperature difference between the highest (129% ET_o) and lowest (83% ET_o) irrigation levels. The irrigation frequency had a significant effect in 2018 ($p < 0.05$) but not in 2019. However, in both years, the canopy temperature was slightly lower for the greater irrigation frequencies. The interaction of the irrigation level and irrigation frequency had no significant effect on canopy temperature values. In both years, 83% ET_o treatment started showing higher temperature values toward the end of the trial compared to the other irrigation treatments (Figure 5).

For the hybrid bermudagrass plots, the minimum and maximum canopy temperature values were 23 °C and 42 °C in 2018 and 36 °C and 43 °C in 2019. The irrigation levels had no significant effect on canopy temperature in neither of the years (Table 3). The irrigation frequency significantly impacted the canopy temperature ($p < 0.05$) in both years, such that more frequent irrigation reduced the canopy temperature. The mean canopy temperature was 0.5 °C in 2018 and 0.1 °C in 2019 lower in 101% ET_o than in 65% ET_o treatment. The interaction of the irrigation level and irrigation frequency had no significant effect on canopy temperature values. The 65% ET_o treatment in 2018 started showing higher canopy temperature values toward the end of the trial compared to other irrigation levels (Figure 5). However, in 2019, the 65% ET_o treatment showed lower canopy temperature early in the trial, but the temperature values were similar across the treatments toward the end of the experiment.

Figure 6 illustrates the lower and upper CWSI baselines established for tall fescue and hybrid bermudagrass in 2018 and 2019. Table 4 summarizes the coefficients for the lower baselines. For the tall fescue plots and 129% ET_o treatment, dt (i.e., canopy minus air temperature) varied between −7.1 and 5.2 °C in 2018 and between −0.6 and 7.9 °C in 2019. For the 83% ET_o treatment, dt varied between −3.8 and 6.4 °C in 2018 and −0.6 °C to 9.2 °C in 2019. There was a moderate correlation of 0.64 in 2018 and 0.88 in 2019 between dt and VPD for the lower baseline. The slope of the lower baseline was −2.52 in 2018 and −4.22 in 2019. The intercept was two times higher in 2019 compared to 2018.

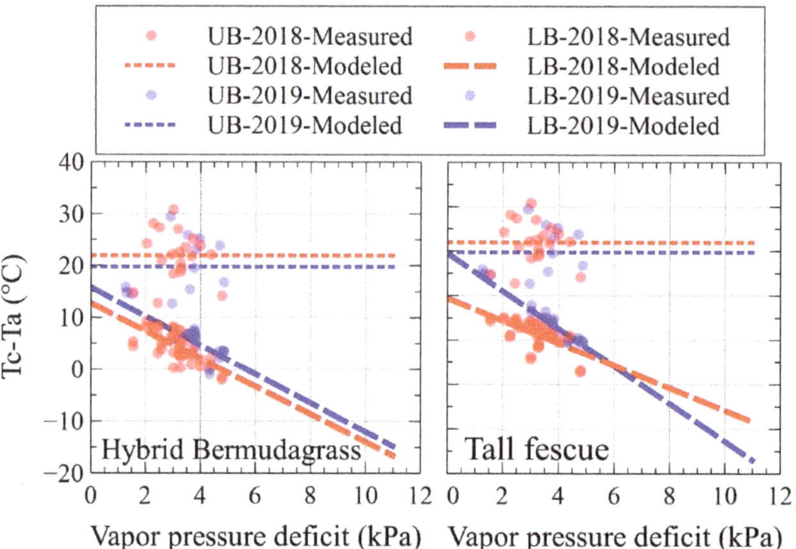

Figure 6. Graphical illustration of the lower (LB) and upper (UB) baselines of canopy temperature (Tc) minus air temperature (Ta) difference versus vapor pressure deficit for hybrid bermudagrass and tall fescue species in central California.

Table 4. Lower CWSI baselines for tall fescue and hybrid bermudagrass.

	Tall Fescue			Hybrid Bermudagrass		
	a	b	r	a	b	r
2018	−2.52	9.42	0.64	−2.67	12.78	0.69
2019	−4.22	19.5	0.88	−2.78	15.82	0.64

r: correlation coefficient; a: slope, b: intercept.

For hybrid bermudagrass, dt varied between −2.0 and 9.2 °C for the 101% ET_o level in 2018 and between −0.9 and 7.8 °C in 2019. For the 65% ET_o treatment, dt varied between 0.5 and 8.9 °C in 2018 and between 0.6 and 7.7 °C in 2019. The correlation between dt and

VPD for hybrid bermudagrass was 0.69 in 2018 and 0.64 in 2019. The lower CWSI baselines had a somewhat similar slope in both years, but the intercept was 3 °C higher in 2019.

The upper baseline (set to mean *dt* for the severely stressed non-irrigated tall fescue plot) was equal to 21.9 ± 4.7 °C (mean ± standard deviation, SD) in 2018 and 19.7 ± 5.3 °C in 2019. The combined upper baseline data for both years had a mean ± standard deviation *dt* of 21.1 ± 4.9 °C.

Figure 7 depicts the dynamics of tall fescue and hybrid bermudagrass CWSI over time in response to the irrigation treatments imposed in 2018 and 2019. The CWSI dynamics over time for both species are similar to the canopy temperature fluctuations (Figure 8), with minor differences mainly toward the end of 2019. For tall fescue, the CWSI values varied between −0.3 and 0.29 in 2018 and between −0.59 and 0.27 in 2019. For hybrid bermudagrass, the CWSI values ranged from −0.34 to 0.24 in 2018 and from −0.26 to 0.56 in 2019.

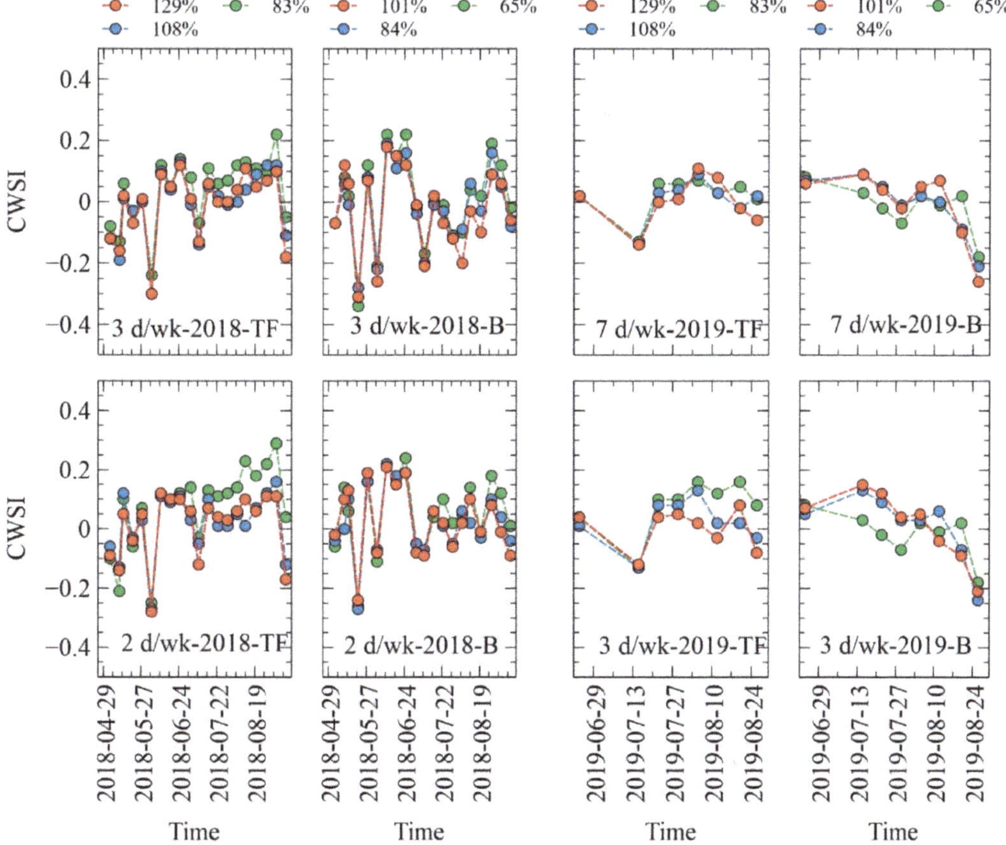

Figure 7. The crop water stress index (CWSI) dynamics for the hybrid bermudagrass (B) and tall fescue (TF) turfgrass plots under varying irrigation scenarios imposed in 2018 and 2019. d/wk: indicated irrigation frequency in days per week.

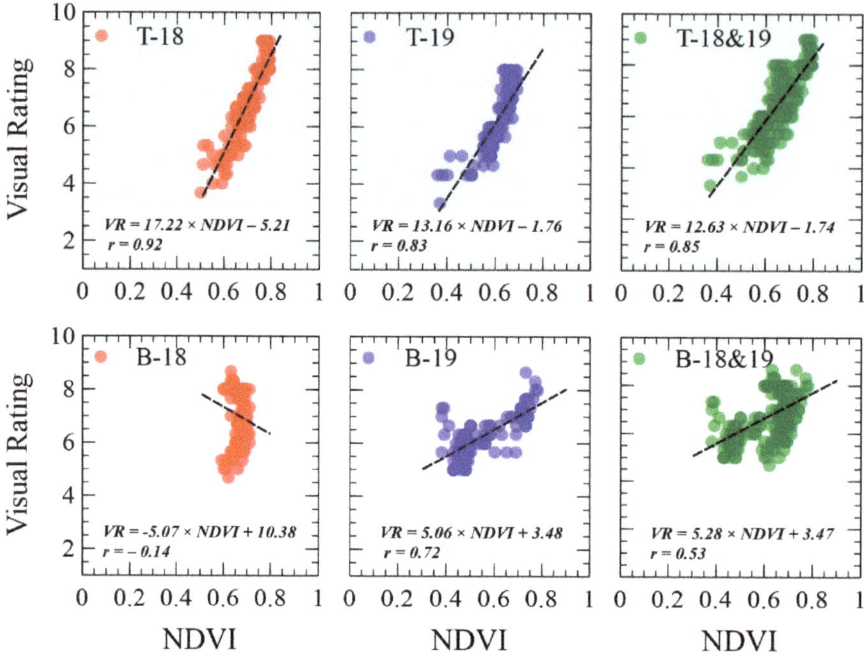

Figure 8. Relationship between visual rating (VR) and NDVI data collected in 2018 and 2019 from hybrid bermudagrass (B) and tall fescue (T) irrigation trials conducted at the University of California Kearney Research and Extension Center.

4. Discussion

4.1. NDVI and Visual Rating

For tall fescue, the NDVI values ranged between 0.30 and 0.80 in 2018 and between 0.23 and 0.69 in 2019. The lower NDVI values in 2019 agree with the lower VR values reported in the companion paper [23]. This is attributed to the minimal fertilizer application in 2019, diminishing growth and greenness of tall fescue. NDVI and VR were well correlated for tall fescue in 2018 ($r = 0.92$) and in 2019 ($r = 0.83$). Bremer et al. [27] conducted a 3-year study near Manhattan, Kansas and reported r-value of 0.75 between VR and NDVI values of 'Dynasty' tall fescue. The slope of the intercept of fitted regression lines in our study differed between 2018 and 2019. Bremer et al. [27] also obtained different models for each turfgrass in each year of their study and mentioned that as a potential practical limitation to estimate visual quality using NDVI values.

For hybrid bermudagrass, the NDVI values varied between 0.53 and 0.76 in 2018 and between 0.34 and 0.8 in 2019. In 2019, the NDVI values increased as the trial progressed in response to the late green-up of hybrid bermudagrass through June and a fertilizer application in mid-July. This trend agrees with an increase in VR values reported in the companion paper [23]. The correlation between NDVI and VR was moderate ($r = 0.72$) in 2019, which is on the lower end of the reported values in the literature. We [8] obtained $r = 0.84$ between NDVI and VR values for hybrid bermudagrass based on a 3-year composite dataset in Riverside, California. Bell et al. [34] conducted a two-year study and reported an annual r of 0.8 between NDVI and VR of hybrid bermudagrass (49 cultivars) in Stillwater, Oklahoma. Trenholm et al. [35] studied three hybrid bermudagrass cultivars in Griffin, Georgia and reported r-values ranging from 0.70 to 0.90 between turfgrass VR and NDVI.

In 2018, no meaningful correlation was observed between NDVI and VR values, which is attributed to the narrow NDVI range of variation (0.53–0.76). However, the range of

VR values was relatively wide (4–9) in 2018 [23]. We also noticed (data not presented in this study) the mean coefficient of variation between replications of the same irrigation treatment on average was 60% higher for VR values than NDVI values. The quality variation among replications of the same irrigation treatment is expected to be minimal. These results suggest that NDVI can be a more stable and repeatable parameter than VR to assess the overall response of the turfgrass to irrigation regimes.

Our results suggest the NDVI values of 0.6–0.65 for tall fescue and 0.5 for hybrid bermudagrass to maintain acceptable quality (VR = 6) in the central California region. We [8] obtained NDVI of 6 as the minimum threshold for hybrid bermudagrass in inland southern California. Further investigation is needed to verify the thresholds obtained in this study, particularly for hybrid bermudagrass, since the recommendation is only based on 2019 data.

4.2. Canopy Temperature and CWSI

The reported results in our companion paper [23] show that when water conservation is concerned, hybrid bermudagrass is the superior species since it can sustain its quality better than tall fescue when irrigation is limited, as expected. The tall fescue treatments received more water (equal to roughly 20% ET_o) than the hybrid bermudagrass and still could not continuously maintain the same VR values during the summer months in central California. Culpepper et al. [36] conducted two greenhouse trials in Texas, USA, to compare the response of bermudagrass, buffalograss (*Buchloe dactyloides* (Nutt.) Engelm.), and tall fescue to water deficit. They reported that tall fescue (a C_3 cool-season turfgrass) under heat and drought stress showed the most rapid decline in quality and photosynthetic rates compared to the C_4 warm-season grasses, demonstrating the benefits of the C_4 versus C_3 photosynthetic pathway.

On the other hand, the result of this study showed that the mean canopy temperature was higher for hybrid bermudagrass than tall fescue across the treatments in both years. On average, in both years, the hybrid bermudagrass plots were 1.6 °C warmer than tall fescue plots when they received the same irrigation treatments of 83–84% ET_o. Further comparative studies are needed to evaluate the potential water conservation and cooling benefits of irrigated warm-season and cool-season species versus alternative groundcover species in California.

For the same irrigation levels, increasing irrigation frequency (number of watering days) slightly (0.6 °C on average) but consistently decreased canopy temperature without compromising the turfgrass quality. We attribute this to a more pronounced evaporative cooling associated with higher irrigation frequencies while minimizing runoff and deep percolation. Consequently, as suggested in the companion paper [23], when ET-based smart controllers are used, we recommend no watering restrictions so the controller can adjust the watering days based on the actual weather conditions and evaporative demand. Then, a minimum deficit threshold should be programmed to avoid unnecessary evaporative loss due to light irrigation applications.

The correlation between dt and VPD for well-watered treatments (lower baselines of CWSI) was moderate in this study. We observed r-values of 0.64 and 0.88 for tall fescue and 0.69 and 0.64 for hybrid bermudagrass in two years of our research. Jalali-Farahani et al. [37] reported r = 0.87 between VPD and dt for well-watered hybrid bermudagrass in Tucson, Arizona. Payero et al. [38] observed r-values ranging from 0.92 to 0.95 between dt and VPD at different solar radiation levels for tall fescue grass at Kimberly, Idaho. Taghvaiean et al. [14] conducted a field study in Berthoud, Colorado, on multiple turfgrass species and mentioned that the effect of solar radiation is negligible when dt data are collected under the clear sky and close to solar noon. On the other hand, multiple studies reported a high correlation between CWSI and solar radiation [37,38]. We leave it to future studies to determine the potential improvement in hybrid bermudagrass and tall fescue CWSI lower baselines developed in this study when additional weather parameters such as solar radiation are considered.

The CWSI values are expected to vary from 0 to 1, representing no transpiration and maximum transpiration rates, respectively. In our study, the CWSI values ranged from -0.34 to 0.56 for hybrid bermudagrass and -0.59 to 0.29 for tall fescue. Jalali-Farahani [37] reported violations of the theoretical range one-fourth of the time. In addition, Al-Faraj et al. [39] studied the CWSI of 'Falcon' tall fescue in a controlled environment and concluded that there is no easy way to ensure that the empirical CWSI consistently stays within the theoretical range of zero (no stress) and one (severe water stress).

In our study, the mean $\pm$ SD CWSI values were 0 ± 0.13 and 0 ± 0.10 for well-watered tall fescue (129% ET_o) and hybrid bermudagrass (101% ET_o), respectively. Jalali-Farahani [37] reported a mean CWSI value of -0.02 and an SD of 0.28 for well-watered hybrid bermudagrass. In our study, for the 65% ET_o treatment, the mean $\pm$ SD hybrid bermudagrass CWSI was 0.1 ± 0.12. Somewhat similar to our results, Emekli et al. [15] reported 0.09 and 0.10 as seasonal mean CWSI of hybrid bermudagrass under 100% pan evaporation ($\approx$75% ET_o) and 75% pan evaporation ($\approx$56% ET_o) irrigation treatments in Antalya, Turkey. They reported a good relationship between VR and CWSI and suggested a CWSI of approximately 0.1 to maintain hybrid bermudagrass quality. However, we can not recommend a CWSI threshold to maintain turf quality in the acceptable range because of the high variability of CWSI values over time and their low correlation with VR values (not reported in this study). The canopy temperature and CWSI dynamics were very similar in this study, and CWSI did not provide much extra information regarding the dynamic impact of irrigation regimes on turfgrass quality.

5. Conclusions

It is vital to establish landscape irrigation water conservation strategies while determining its impact on the cooling effect of irrigated landscape in the US southwest. Our two-year field study focused on ground-based remote sensing of hybrid bermudagrass and tall fescue under varying irrigation scenarios autonomously imposed by an ET-based smart irrigation controller in central California. When the NDVI range of variation was high, it was well correlated to VR values for both species. Overall, the NDVI showed less variability between replications of the same treatments for both species when compared to VR. This finding suggests NDVI as a consistent and objective proxy of overall turfgrass quality in response to varying irrigation regimes. Hybrid bermudagrass was a superior species to tall fescue when water conservation was concerned. However, it showed 1.6 °C higher canopy temperature than tall fescue when it received the same amount of water. Our results suggested the NDVI values of 0.6–0.65 for tall fescue and 0.5 for hybrid bermudagrass to maintain acceptable quality (VR = 6) in the central California region. Further investigation is needed to verify the thresholds obtained in this study, particularly for hybrid bermudagrass, since the recommendation is only based on 2019 data. No CWSI minimum threshold could be identified to maintain the quality of the selected species due to its high variability and low correlation with VR. Given their ease of use for small plot data collection, we selected handheld sensors in this study to measure canopy temperature and NDVI. However, collecting data from larger irrigated areas in practice using handheld sensors might be time-consuming and challenging. Further studies are needed to explore the utility of unmanned aerial vehicles and advanced multispectral and thermal cameras and compare their readings with handheld sensors used in this study.

Author Contributions: Conceptualization, A.H.; methodology, A.H.; software, A.H., A.S. (Amninder Singh), and A.S. (Anish Sapkota); validation, A.H., A.S. (Amninder Singh), and A.S. (Anish Sapkota); formal analysis, A.H., M.R., A.S. (Amninder Singh) and A.S. (Anish Sapkota); investigation, A.H., M.R., A.S. (Amninder Singh) and A.S. (Anish Sapkota); resources, A.H. and M.R; data curation, A.H., M.R., A.S. (Amninder Singh) and A.S. (Anish Sapkota); writing—original draft preparation, A.H.; writing—review and editing, A.H., M.R., A.S. (Amninder Singh) and A.S. (Anish Sapkota); visualization, A.H.; supervision, A.H.; project administration, A.H. and M.R; funding acquisition, A.H. and M.R. All authors have read and agreed to the published version of the manuscript.

Funding: This study was supported by the University of California Division of Agriculture and Natural Resources competitive grant (ID#: 17-5021) and by the United States Geological Survey (ID#: 2017CA371B).

Data Availability Statement: Not applicable.

Conflicts of Interest: The authors declare no conflict of interest.

References

1. Cooley, H.; Gleick, P.H. *Urban Water-Use Efficiencies: Lessons from United States Cities*; Island Press: Washington, DC, USA, 2009.
2. Monteiro, J.A. Ecosystem services from turfgrass landscapes. *Urban For. Urban Green.* **2017**, *26*, 151–157. [CrossRef]
3. Blonquist, J., Jr.; Jones, S.B.; Robinson, D. Precise irrigation scheduling for turfgrass using a subsurface electromagnetic soil moisture sensor. *Agric. Water Manag.* **2006**, *84*, 153–165. [CrossRef]
4. Cardenas-Lailhacar, B.; Dukes, M.D.; Miller, G.L. Sensor-based automation of irrigation on bermudagrass, during wet weather conditions. *J. Irrig. Drain. Eng.* **2008**, *134*, 120–128. [CrossRef]
5. Davis, S.L.; Dukes, M.D. Importance of ET controller program settings on water conservation potential. *Appl. Eng. Agric.* **2016**, *32*, 251–262.
6. Grabow, G.L.; Ghali, I.E.; Huffman, R.L.; Miller, G.L.; Bowman, D.; Vasanth, A. Water application efficiency and adequacy of ET-based and soil moisture–based irrigation controllers for turfgrass irrigation. *J. Irrig. Drain. Eng.* **2013**, *139*, 113–123. [CrossRef]
7. Qualls, R.J.; Scott, J.M.; DeOreo, W.B. Soil moisture sensors for urban landscape irrigation: Effectiveness and reliability. *JAWRA J. Am. Water Resour. Assoc.* **2001**, *37*, 547–559. [CrossRef]
8. Haghverdi, A.; Singh, A.; Sapkota, A.; Reiter, M.; Ghodsi, S. Developing irrigation water conservation strategies for hybrid bermudagrass using an evapotranspiration-based smart irrigation controller in inland southern California. *Agric. Water Manag.* **2021**, *245*, 106586. [CrossRef]
9. Cardenas, B.; Dukes, M.D. Soil moisture sensor irrigation controllers and reclaimed water; Part. I: Field-plot study. *Appl. Eng. Agric.* **2016**, *32*, 217–224.
10. Cardenas, B.; Dukes, M.D. Soil moisture sensor irrigation controllers and reclaimed water; Part. II: Residential evaluation. *Appl. Eng. Agric.* **2016**, *32*, 225–234.
11. Chen, W.; Lu, S.; Pan, N.; Wang, Y.; Wu, L. Impact of reclaimed water irrigation on soil health in urban green areas. *Chemosphere* **2015**, *119*, 654–661. [CrossRef] [PubMed]
12. An, N.; Goldsby, A.L.; Price, K.P.; Bremer, D.J. Using hyperspectral radiometry to predict the green leaf area index of turfgrass. *Int. J. Remote Sens.* **2015**, *36*, 1470–1483. [CrossRef]
13. Johnson, T.D.; Belitz, K. A remote sensing approach for estimating the location and rate of urban irrigation in semi-arid climates. *J. Hydrol.* **2012**, *414*, 86–98. [CrossRef]
14. Taghvaeian, S.; Chávez, J.L.; Hattendorf, M.J.; Crookston, M.A. Optical and thermal remote sensing of turfgrass quality, water stress, and water use under different soil and irrigation treatments. *Remote Sens.* **2013**, *5*, 2327–2347. [CrossRef]
15. Emekli, Y.; Bastug, R.; Buyuktas, D.; Emekli, N.Y. Evaluation of a crop water stress index for irrigation scheduling of bermudagrass. *Agric. Water Manag.* **2007**, *90*, 205–212. [CrossRef]
16. Chen, Y.-J.; McFadden, J.P.; Clarke, K.C.; Roberts, D.A. Measuring spatio-temporal trends in residential landscape irrigation extent and rate in Los Angeles, California Using SPOT-5 satellite imagery. *Water Resour. Manag.* **2015**, *29*, 5749–5763. [CrossRef]
17. Irmak, S.; Haman, D.Z.; Bastug, R. Determination of crop water stress index for irrigation timing and yield estimation of corn. *Agron. J.* **2000**, *92*, 1221–1227. [CrossRef]
18. Shashua-Bar, L.; Pearlmutter, D.; Erell, E. The cooling efficiency of urban landscape strategies in a hot dry climate. *Landsc. Urban Plan.* **2009**, *92*, 179–186. [CrossRef]
19. Bonfils, C.; Lobell, D. Empirical evidence for a recent slowdown in irrigation-induced cooling. *Proc. Natl. Acad. Sci. USA* **2007**, *104*, 13582–13587. [CrossRef] [PubMed]
20. Broadbent, A.M.; Coutts, A.M.; Tapper, N.J.; Demuzere, M. The cooling effect of irrigation on urban microclimate during heatwave conditions. *Urban Clim.* **2017**, *23*, 309–329. [CrossRef]
21. Wang, C.; Wang, Z.-H.; Yang, J. Urban water capacity: Irrigation for heat mitigation. *Comput. Environ. Urban Syst.* **2019**, *78*, 101397. [CrossRef]
22. Idso, S.B.; Jackson, R.D.; Pinter, P.J., Jr.; Reginato, R.J.; Hatfield, J.L. Normalizing the stress-degree-day parameter for environmental variability. *Agric. Meteorol.* **1981**, *24*, 45–55. [CrossRef]
23. Haghverdi, A.; Reiter, M.; Sapkota, A.; Singh, A. Hybrid. Bermudagrass and Tall fescue Turfgrass Irrigation in Central California: I. Assessment of Visual Quality, Soil Moisture and Performance of an ET-based Smart Controller. *Agronomy* **2021**, *11*, 1666. [CrossRef]
24. Leinauer, B.; VanLeeuwen, D.M.; Serena, M.; Schiavon, M.; Sevostianova, E. Digital image analysis and spectral reflectance to determine turfgrass quality. *Agron. J.* **2014**, *106*, 1787–1794. [CrossRef]
25. Bell, G.E.; Martin, D.L.; Wiese, S.G.; Dobson, D.D.; Smith, M.W.; Stone, M.L.; Solie, J.B. Vehicle-mounted optical sensing: An objective means for evaluating turf quality. *Crop Sci.* **2002**, *42*, 197–201. [CrossRef]
26. Horst, G.; Engelke, M.; Meyers, W. Assessment of visual evaluation techniques [1]. *Agron. J.* **1984**, *76*, 619–622. [CrossRef]

27. Bremer, D.J.; Lee, H.; Su, K.; Keeley, S.J. Relationships between normalized difference vegetation index and visual quality in cool-season turfgrass: I. Variation among species and cultivars. *Crop Sci.* **2011**, *51*, 2212–2218. [CrossRef]

28. Bremer, D.J.; Lee, H.; Su, K.; Keeley, S.J. Relationships between normalized difference vegetation index and visual quality in cool-season turfgrass: II. Factors affecting NDVI and its component reflectances. *Crop Sci.* **2011**, *51*, 2219–2227. [CrossRef]

29. Haghverdi, A.; Najarchi, M.; Öztürk, H.S.; Durner, W. Studying Unimodal, Bimodal, PDI and Bimodal-PDI Variants of Multiple Soil Water Retention Models: I. Direct Model. Fit. Using the Extended Evaporation and Dewpoint Methods. *Water* **2020**, *12*, 900. [CrossRef]

30. Hargreaves, G.H.; Samani, Z.A. Reference crop evapotranspiration from temperature. *Appl. Eng. Agric.* **1985**, *1*, 96–99. [CrossRef]

31. Sanders, J. *Veusz-a Scientific Plotting Package*; 2008; Available online: https://veusz.github.io/ (accessed on 28 August 2021).

32. Morris, K.N.; Shearman, R.C. NTEP turfgrass evaluation guidelines. In Proceedings of the NTEP Turfgrass Evaluation Workshop, Beltsville, MD, USA, 17 October 1998; pp. 1–5.

33. Jackson, R.D.; Idso, S.B.; Reginato, R.J.; Pinter, P.J., Jr. Canopy temperature as a crop water stress indicator. *Water Resour. Res.* **1981**, *17*, 1133–1138. [CrossRef]

34. Bell, G.E.; Martin, D.L.; Koh, K.; Han, H.R. Comparison of turfgrass visual quality ratings with ratings determined using a handheld optical sensor. *HortTechnology* **2009**, *19*, 309–316. [CrossRef]

35. Trenholm, L.; Carrow, R.; Duncan, R. Relationship of multispectral radiometry data to qualitative data in turfgrass research. *Crop Sci.* **1999**, *39*, 763–769. [CrossRef]

36. Culpepper, T.; Young, J.; Montague, D.T.; Sullivan, D.; Wherley, B. Physiological responses in C3 and C4 turfgrasses under soil water deficit. *HortScience* **2019**, *54*, 2249–2256. [CrossRef]

37. Jalali-Farahani, H.; Slack, D.C.; Kopec, D.M.; Matthias, A.D. Crop water stress index models for Bermudagrass turf: A comparison. *Agron. J.* **1993**, *85*, 1210–1217. [CrossRef]

38. Payero, J.; Neale, C.; Wright, J. Non-water-stressed baselines for calculating crop water stress index (CWSI) for alfalfa and tall fescue grass. *Trans. ASAE* **2005**, *48*, 653–661. [CrossRef]

39. Al-Faraj, A.; Meyer, G.E.; Horst, G.L. A crop water stress index for tall fescue (Festuca arundinacea Schreb.) irrigation decision-making—A traditional method. *Comput. Electron. Agric.* **2001**, *31*, 107–124. [CrossRef]

Article

Water-Use Efficiency and Productivity Improvements in Surface Irrigation Systems

Carlos Chávez [1,*], **Isaías Limón-Jiménez** [2], **Baldemar Espinoza-Alcántara** [3], **Jacobo Alejandro López-Hernández** [3], **Emilio Bárcenas-Ferruzca** [3] and **Josué Trejo-Alonso** [1]

[1] Water Research Center, Department of Irrigation and Drainage Engineering, Autonomous University of Queretaro, Cerro de las Campanas SN, Col. Las Campanas, 76010 Queretaro, Mexico; josue.trejo@uaq.mx
[2] National Agrarian Registry (RAN), Delegation Queretaro, Cto. Moisés Solana 189, Balaustradas, 76079 Queretaro, Mexico; isaiaslj@yahoo.com.mx
[3] Water Users Association of Second Unit of Module Two, Irrigation District No. 023, Av. Panamericana No. 86, El Sauz Alto, Pedro Escobedo, 76700 Queretaro, Mexico; baldemar_e@hotmail.com (B.E.-A.); jackbuitre@gmail.com (J.A.L.-H.); eb_f@hotmail.com (E.B.-F.)
* Correspondence: chagcarlos@uaq.mx; Tel.: +52-442-192-1200 (ext. 6036)

Received: 15 October 2020; Accepted: 11 November 2020; Published: 12 November 2020

Abstract: In Mexico, agriculture has an allowance of 76% of the available water (surface and underground), although the average application efficiencies are below 50%. Despite the fact that in recent years modern pressurized irrigation systems have been the best option to increase the water-use efficiency (WUE), the gravity irrigation system continues to be the most used method to provide water to crops. This work was carried out during the 2014–2019 period in three crops, namely, barley, corn, and sorghum, in an irrigation district, showing the results of a methodology applied to gravity irrigation systems to increase the WUE. The results show that, with an efficient design, by means of irrigation tests, characterization of the plot, and the calculation of the optimal flow through an analytical formula, it was possible to reduce the irrigation times per hectare and the irrigation depth applied. Application efficiencies increased from 43% to 95%, while the WUE increased by 27, 38, and 47% for sorghum, barley, and corn, respectively. With this methodology, farmers are more attentive in irrigation because the optimal flow in each furrow or border is, in general, higher than that applied in the traditional way and they take less time to irrigate their plots. For farmers to adopt this methodology, the following actions are required: (a) be aware that the water that comes from dams is as valuable as the water from wells; (b) increase the irrigation quota; (c) seek government support to increase the WUE; and (d) show them that with less water they can have better yields.

Keywords: water-use efficiency; analytical formula; efficient design; application efficiency; gravity irrigation

---

1. Introduction

In Mexico there are 6.5 million hectares under irrigation. Of this, 3.3 million are distributed in 86 irrigation districts and 3.2 million are distributed in just over 40 thousand irrigation units located mainly in the center and north of the country. Gravity surface irrigation (border or furrows) is the most used method in these areas [1].

According to recent statistics and studies that have been registered by the National Water Commission (CONAGUA, according to its Spanish acronym), application efficiencies in Mexico are less than 50% [1–3], 5% below the global average [4,5], and consequently the water productivity (Kg/m^3) is very low compared to pressurized irrigation systems [1,6]. Despite the fact that support has been given to modernize the traditional sprinkler and drip irrigation systems, there are still various factors that cause farmers not to opt for these technologies; for example, a lack of investment, little government support, payment of trained personnel to operate the equipment, social factors, risks in the acquisition of equipment that they do not know, among others [4,6–8].

The low application efficiencies obtained by gravity irrigation systems are mainly due to water losses associated with deep infiltration, coleus, and flooding in some parts of the soil as a result of a poor flow design at the entrance of the furrow [2,3], as well as unevenness in the plots and little knowledge about the water depth that must be supplied to the crops so that they can develop fully. However, in a sector where the demand for water occupies a high percentage of the available water, the efficient use of irrigation water and the increase in productivity are crucial factors that must be addressed as soon as possible to sustainably manage the water and adaptation to climate change [5,9–11].

Faced with this problem, during the years 2015–2019 CONAGUA maintained a modernization program in gravity irrigation systems (RIGRAT for its acronym in Spanish) in an area of 200,000 ha in some irrigation districts of Mexico. The objective of the program was to increase the efficiency of water application in the crops through two main axes: a timely design for optimal flow at the entrance of the furrows and a reduction in irrigation time in each plot.

A correct design of the flow rate that must be applied to each border or furrow requires knowledge of the characteristics of the plots (length, moisture content, and apparent density), the established crop, phenological stage, and the irrigation depth to be applied, as well as the mean parameters of the infiltration equation being used: Richards or Green and Ampt [12].

Several simulation models can be found in the literature to model gravity irrigation, which are completely empirical [13–15] to those that use the complete Barré de Saint-Venant and Richards equations to model runoff and infiltration, respectively [16–18]. The use of these models (numerical or analytical) helps to better understand the behavior of water during the processes of advancing, storage, and recession [12,17,19]. However, the complexity with which they were developed or the limitations they have make their use impractical and only used for research purposes, leaving aside their practical application.

In recent studies, it has been reported that it is possible to have high application efficiencies in gravity irrigation systems, by applying the optimal flow rate at the entrance of the border or furrow and maintaining a high uniformity coefficient in the plot, called the Christiansen Uniformity Coefficient [2,3,12].

This methodology was applied by Chávez and Fuentes [2,3] for seven crops planted in 1010 ha: *Zea mays* L., *Sorghum vulgare* Pers., *Medicago sativa* L., *Phaseolus vulgaris* L., *Pachyrhizus erosus* L., *Hordeum vulgare* L., *Triticum aestivum* L. and *Allium cepa* L. They found that with the application of the optimal flow rate of the calculated irrigation, the irrigation depths decreased on average 19 cm. The irrigation times decreased on average 11.76 h ha^{-1} per irrigation event and, in addition, the average volume saved was 2000 m^3 ha^{-1} per irrigation event, which increased the average efficiency from 51 to 86%. However, to have an efficient design, knowledge of the plots and the water requirements of the crops established in the different phenological stages are necessary [12].

The objective of this work is to show that it is possible to increase the application efficiency and productivity of water in gravity irrigation systems in Irrigation District 023, San Juan del Río, Querétaro, Mexico. For the simulation, the kinematic wave model is used, and an analytical formula is used to calculate the optimal flow rate that takes into account the characteristics of the plot (length, moisture content, density, and texture), the parameters of the equation of infiltration (K_s and H_f), and the net irrigation depth to be applied to the crops in their different phenological stages.

2. Materials and Methods

2.1. The Kinematic Wave Model

The kinematic wave model considers that, in the Barré de Saint-Venant momentum equation, the inertial and pressure terms are negligible with respect to the friction and gravity terms [2]:

$$\frac{\partial A}{\partial t} + \frac{\partial Q}{\partial x} = -W \tag{1}$$

$$S_f = S_o \tag{2}$$

where $A = A(x,t)$ is the hydraulic area (L^2); $Q = Q(x,t)$ is the flow ($L^3 T^{-1}$); W is the infiltrated volume per unit of furrow length in unit time ($L^3 T^{-1}$); t is the time (T); S_o is the slope of the bottom of the furrow (LL^{-1}); and S_f is the friction slope (LL^{-1}).

2.2. Green and Ampt Equations

The Green and Ampt model [20] is established from the continuity equation and Darcy's law, with the following hypotheses: (a) the initial moisture profile in a soil column is uniform $\theta = \theta_o$; (b) water pressure at the soil surface is hydrostatic: $\psi \geq 0$, where h is the water depth; (c) there is a well-defined wetting front characterized by negative pressure: $\psi = -h_f < 0$, where h_f is the suction at the wetting front; and (d) the region between the soil surface and the wetting front is completely saturated (plug flow)): $\theta = \theta_s$ and $K = K_s$, where K_s is the hydraulic saturation conductivity, that is, the value of the hydraulic conductivity of Darcy's law corresponding to the volumetric saturation content of water. The resulting ordinary differential equation is as follows:

$$V_I = \frac{dI}{dt} = K_s\left(1 + \frac{h + h_f}{z_f}\right) I(t) = z_f \Delta\theta(t) \tag{3}$$

where $\Delta\theta = \theta_s - \theta_o$ is the storage capacity; I is the accumulated infiltrate volume per unit of soil surface or infiltrated depth; and z_f is the position of the wetting front.

2.3. Analytical Representation to Calculate Optimal Flow

According to Fuentes and Chávez [12], the analytical representation of the optimal irrigation expenditure is a function of the border length, the hydrodynamic properties, and the soil moisture constants, maintaining a maximum value of the uniformity coefficient. In this way,

$$q_o = \alpha_u K_s L \qquad \alpha_u = \frac{\ell_n}{\ell_n - \frac{S^2}{2K_s}\ln\left(1 + \frac{2K_s}{S^2}\ell_n\right)} \tag{4}$$

in which it should be noted that $K_s L = q_m$ represents the minimal unit flux necessary for the water to reach the final part of the border; S is the sorptivity of the medium expressed by $S^2 = 2 K_s h_f (\theta_s - \theta_o)$; and ℓ_n is the net irrigation depth. The optimal flow per row is calculated as $Q_o = b\, q_o$, where b is the width of the furrow. This analytical formula has been applied in field experiments with good results [2,3].

2.4. Case Study

Irrigation District 023 is located between the municipalities of San Juan del Río and Pedro Escobedo in the state of Querétaro, México (Figure 1), and has an area of 11,048 ha. The water for irrigation is obtained from the San Ildefonso, Constitución de 1917, La Llave and La Venta dams, and from 54 deep wells. It is legally constituted by three irrigation modules and the RIGRAT program was carried out in modules II and III on an area of 5021 ha. Its predominant climate is semiarid with summer rains, with an annual precipitation average of 599 mm and annual average temperature of 21 °C [1].

The water is conducted through open channels. The main channels are lined with concrete, but all the laterals channels that carry water to the plots are unlined. The separation of the plots in some cases are by trees, unlined channels, drains, or roads.

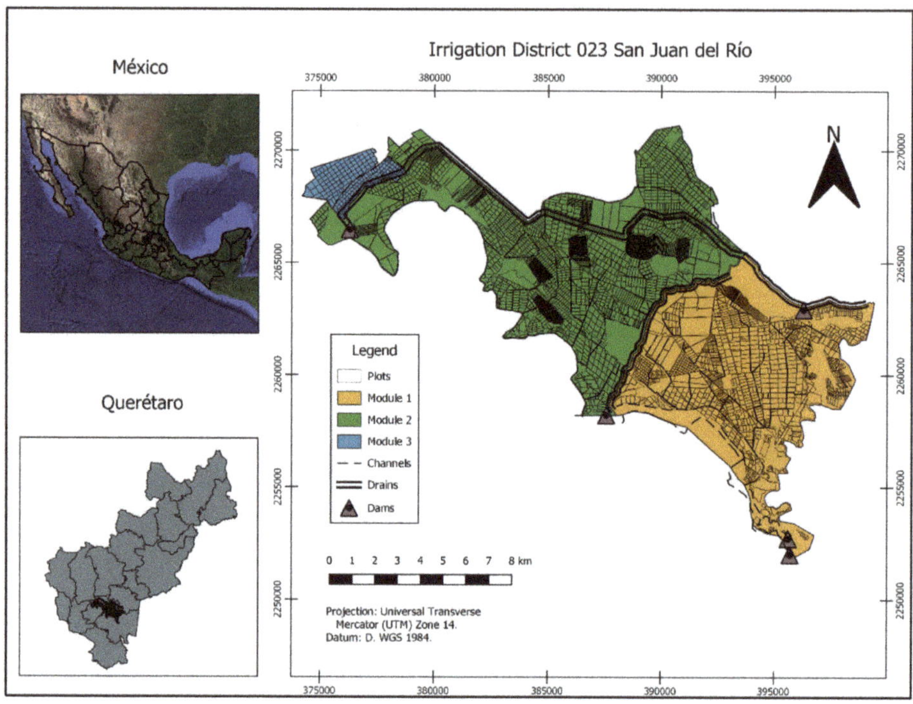

Figure 1. Location map of the Irrigation District 023 San Juan del Río Querétaro.

2.5. Variables Measurements

In the plots where the study was carried out, the length, longitudinal and transverse slope, texture, apparent density, initial moisture contents, and saturation were measured. The first three were measured using digital topography equipment, the texture was obtained in the laboratory through mesh analysis and the Bouyoucos hydrometer, the apparent density (ρ_a) was obtained using the known volume cylinder method, the initial moisture content by means of a calibrated sensor TDR 300® (Spectrum Technologies, Inc., Aurora, IL, USA), and the moisture saturation content was assimilated to the total porosity of the soil obtained through the relationship $\phi = 1 - \rho_a/\rho_s$, where ρ_s is the density of the quartz particles taken as 2.65 g/cm^3. For the laboratory measurements, six samples were collected from each plot at random at a depth of 0–30 cm. Later they were mixed, and a homogeneous sample was the one that was analyzed. The USDA triangle was used to classify the soil samples.

2.6. Irrigation Tests

The irrigation tests were carried out in the plots in order to observe the behavior of the water and the volumes of water used in each irrigation event. The process, in general, is as follows: (1) the initial moisture content in the plot was measured; (2) along the length of the border or furrow distances were marked at every 20 or 30 m, which depended on the length of each of the plots; (3) farmers do not use siphons to flood the plots, instead, they open the unlined channel for the water to pass through and lead it to the furrows or borders with the help of a shovel; (4) the time in which the water front reached each of the established marks (advance front) was counted while the water flow was counted at the entrance of the border or furrow using an ultrasonic Doppler effect sensor (FluxSense®) (SEHIDRO Inc., Qro., MEX); and (5) once the water reached the end of the border or furrow, the water inlet was cut off at the beginning and the time in which the recession wave arrived at all the marked points was taken. The irrigation tests were carried out on irrigation lines that varied from 10 to 60 furrows, depending on the expense at the entrance to the canal. The results of the advance and recession times correspond to the average of the furrows evaluated in each irrigation line.

With the data collected from the plots (length, longitudinal slope, apparent density, initial moisture contents and saturation, and flow at the entrance of furrow or border) and the measured times of the advance and recession phase, we proceeded to find the parameters of the Green and Ampt equation (Ks and hf) values that minimized the sum of the squares of the errors in this test. For this stage, we used the nonlinear Levenberg–Marquardt optimization algorithm [21].

2.7. Application Efficiency and Water-Use Efficiency

Application efficiency (η_A) is defined as [3]:

$$\eta_A = \frac{V_n}{V_b} = \frac{\ell_n}{\ell_b} \tag{5}$$

where V_n is the volume of water stored in the root zone and V_b is the total volume of water applied. The first is obtained with the expression $V_n = \ell_n A_r$, where ℓ_n is the net irrigation depth, defined according to the crop irrigation requirements, and A_r is the irrigated area considered. The second is obtained as $V_b = \ell_b A_r$, where ℓ_b is the gross irrigation depth. The net irrigation depth was calculated with the FAO-56 Penman–Monteith evapotranspiration method [22] and the crop phenology was estimated using the FAO methodology [23].

The water-use efficiency (WUE) is the relationship that exists between the biomass present in a crop per unit of water used by it, although recent studies refer to this term as the productivity of irrigation water [11,24]. This relationship is an indicator that allows us to calculate the economic value of the irrigation water in the area. High values of this index will indicate that we are managing to produce a greater amount of organic matter with less use of water. In this work we focus on the productive component of the harvest, which is why we will use the dry biomass yield (<15% of moisture) obtained in kg of the product per m^3 of water used:

$$WUE = \frac{\text{dry biomass yield (Kg)}}{\text{total volume of water applied (m}^3)} \tag{6}$$

3. Results and Discussion

3.1. Soil Texture

Soil texture is an indicator of the amount of water that soil can store and, consequently, the irrigation interval with which crops must be watered. Figure 2 shows the general classification of textures found in the study area. The predominant soils were loam (21.33%), silty clay loam (18.26%), silty loam (16.44%), and sandy loam (11.71%). This measure allowed us to detect the soils in which more water was used and explain the reasons why the application efficiencies were lower than other soils with similar or different textures.

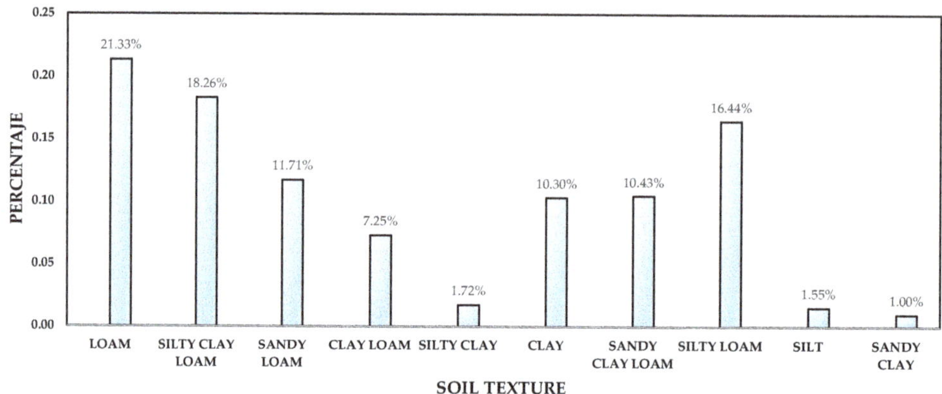

Figure 2. Soil textural classification in the studied areas.

3.2. Irrigations Test

During the study, 475 irrigation tests were carried out on a surface in different phenological stages of the crops: sowing, growth, flowering, and fruiting. The crops in which they were made were corn (*Zea mayz*), sorghum (*Sorghum vulgare*), and barley (*Hordeum vulgare*) in the spring–summer (S–S) and autumn–winter (A–W) cycles (2014–2019). The results show that, in 25% of the cases, the irrigation depth applied by irrigation is 30 cm, which represents an application efficiency of 43%, while 50% of the farmers apply an average irrigation depth of 23 cm per irrigation (η_A = 69%), and the remaining 25% applies an irrigation depth greater than 40 cm, which in some cases has reached up to 98 cm, which is equivalent to an application efficiency ranging from 16 to 40%.

The irrigation depth applied is directly related to the amount of water that reaches the plots, since the average value is close to 58 L/s, but there is a minimum value of 3.78 L/s that goes up to the maximum value of 160 L/s. This brings with it a very strong problem for the farmer, since he wants to apply the same furrow laying regardless of the flow that he has at the entrance of the plot, which leads to the irrigation time per hectare being 2–6 days and, consequently, there are low application efficiencies.

3.3. Irrigation Design

With the information of the established crop and the phenological stage in which it was found, the length of the borders or furrows, the parameters found in the irrigation tests (K_s and H_f), and the net irrigation depth to be applied in the next laying was calculated with the Equation (4). The entry flow in the plot was divided by the result obtained and the number of furrows per laying was found with which it should be watered for that specific crop and plot. The plots were split in half to apply traditional and designed irrigation.

As an example, in the case of Plot 778 (Table 1), the results of the evaluation and design were (1) for growing corn, 12 cm needed to be applied but the farmer applied 35.46 cm (η_A = 33.84%); (2) once the advance and recession phase had been calibrated, the optimal flow was calculated with Equation (4); (3) with these data, the farmer is told that with the flow he has at the entrance of the plot, he must only open 44 furrows instead of the 52 with which he had been irrigating in the traditional way; (4) the farmer had a reduction of 120 min in the irrigation time per set and consequently a reduction of 22.86 in the irrigation depth; and (5) finally, he had an increase in application efficiency, going from 33.84 to 95.23%.

Table 1. Results of the design of Plot 778 (η_A is the application efficiency).

Condition	No. Furrows Per Set	Irrigation Time (min)	Irrigation Depth Applied (cm)	η_A (%)
Traditional Irrigation	52	225	35.46	33.84
Designed Irrigation	44	105	12.50	95.23

With this methodology developed to obtain the optimal flow in each border or furrow of the plots, it was achieved that the irrigation time required to apply the expenses that have been designed decrease by more than half (Figure 3), where for example in soils with a silty loam texture, where previously the farmer took an average of 35 h per hectare, now it does so in less than 10 h. In the loam and silty clay loam texture plots, they exceed 25 h/ha and, in some cases, they have reached up to 89 h/ha. In general, in the plots where the study was carried out, it was found that the irrigation times decreased.

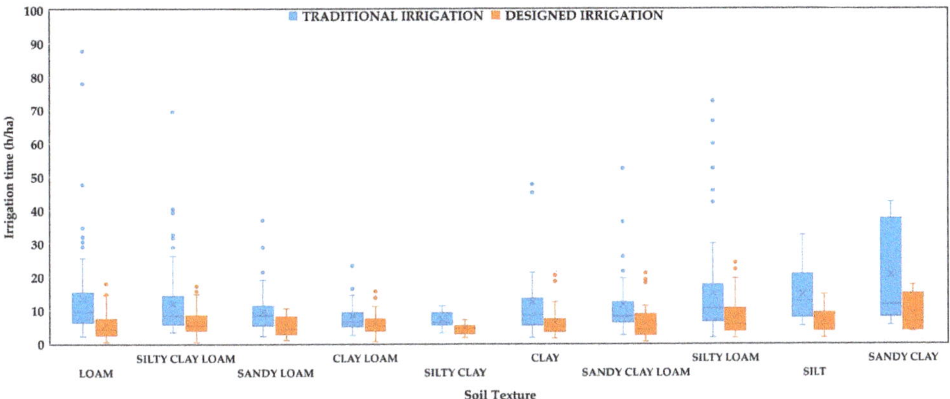

Figure 3. Irrigation time by textural class.

Irrigation times per hectare decreased significantly, but these times are a function of the expense that farmers receive at the entrance to the plot, so it is not the same to irrigate a hectare with 50 L/s to another that has an expense of 10 L/s. However, the same farmers stated that their lines are progressing evenly, but the only drawback they see is that now they must be more attentive. In some areas where the design was carried out, they had lengths of more than 300 m, and in these cases recommendations were given to modify the irrigation line, otherwise the irrigation will be deficient and the efficiencies will be questionable, despite the advice and design that under these conditions could be generated.

3.4. Reduction in Irrigation Depths

In the plots where the irrigation tests were carried out and the irrigation recipes calculated with Equation (4), it was possible to lower the irrigation depth by more than half, as shown in Figure 4. In general, it can be noted that the farmers have a tendency to apply more water in the plots with a loam texture (with an average of 40 cm/ha/irrigation and a maximum of 97 cm/ha/irrigation); however, after the recipe is shown and the information on the number of furrows per irrigation to be applied is given, this irrigation depth is reduced at 15 cm/ha/irrigation on average, which brings considerable savings to the irrigation district. On the other hand, in the plots with a silty clay texture, the irrigation depth applied is, on average, 19 cm/ha/irrigation.

The furrows per irrigation that must be opened depending on the flow at the entrance of the plot and the optimal flow are less than those that the farmers usually apply. This implies more work for him and sometimes he does not cooperate mainly because the payment that they receive is a function of the number of hectares they can irrigate. According to their experience, the slower they pour the water into the plots, the more plots they can serve. In this sense, in coordination with the irrigation district authorities, incentives were sought to convince the farmers to apply the irrigation recipe.

In addition, in some cases the irrigation lengths are greater than 300 m, and in these plots the irrigation depth that is applied per hectare exceed 40 cm. In Figure 4 appears atypical points, meaning that, in some plots, it has been detected that they apply almost 100 cm of depth per irrigation event. Here, they were given the recommendation to make two or three sections along their original irrigation lines.

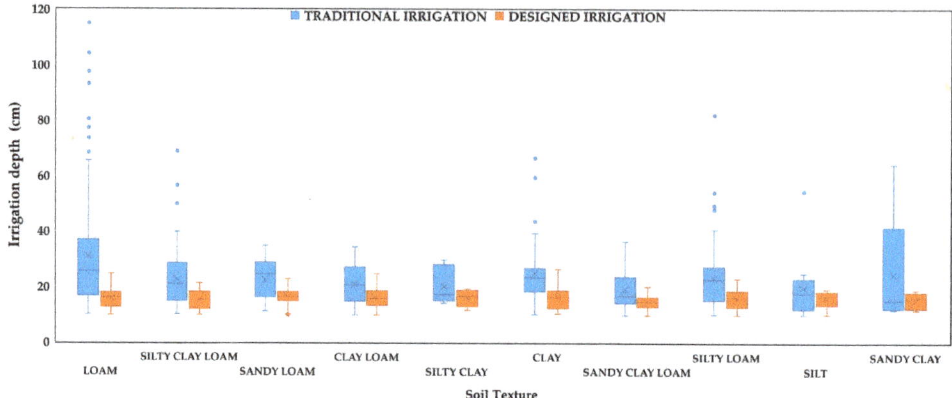

Figure 4. Irrigation depth applied in a traditional and design irrigation.

The efficient design that was carried out allowed us to lower the total amount of water used in each of the studied crops. In all cases it was observed that by applying a correct design the amount of water used per growing period was reduced. Figure 5 shows the average of the evaluations carried out during the years of this study, which in general shows a reduction of 35% for barley and 48% for corn and sorghum crops. The savings obtained for the corn and sorghum crops represented 93% of the amount of water required to plant the same crop in another hectare, while in the case of barley the savings obtained in 2 ha provided enough water to plant an additional 1 ha.

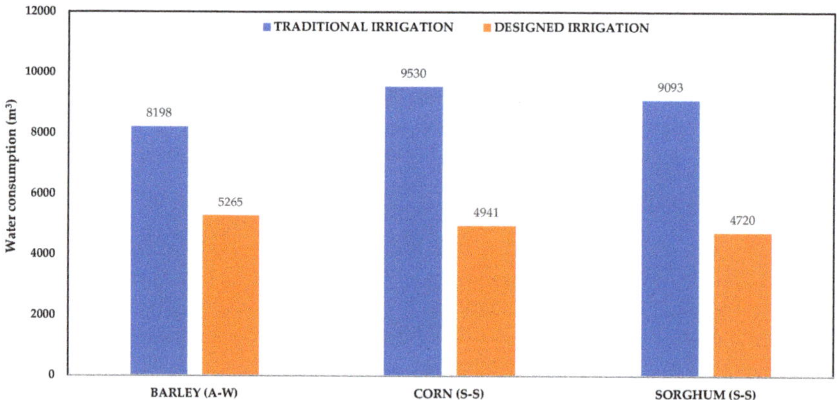

Figure 5. Water consumption in a traditional and design irrigation.

3.5. Water-Use Efficiency

The harvest of each crop was measured in entire plots and the WUE calculated with Equation (6) is shown in Figure 6 for the barley, corn, and sorghum crops. This value increased over the years, mainly due to the fact that the farmers and the supervisory personnel accepted the design of the optimal flow provided. Accepting the irrigation design represented a generational change in the way of watering for farmers, since most mentioned that the way of watering was as their father had taught them, and that in several cases there was no knowledge of irrigation water needs. However, in the last year of evaluation, increases in the WUE of 54.0%, 43.8%, and 23.0% were achieved for barley, corn, and sorghum, respectively.

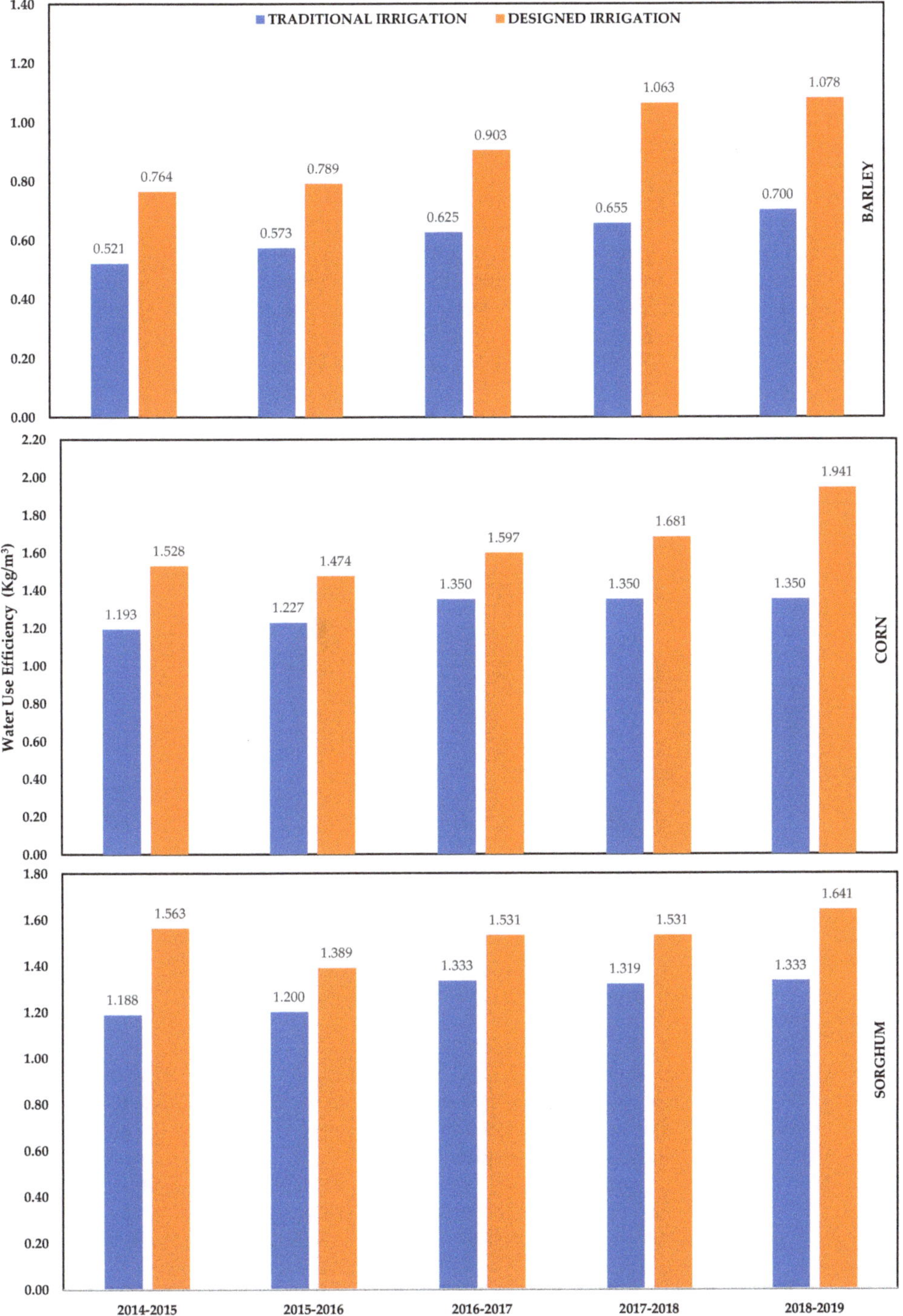

Figure 6. The water-use efficiency obtained by cycle.

The WUE in the evaluated gravity irrigation systems was also affected by various factors: environmental, social, management practices, payment for the service, land rent, among others. These results coincide with other studies where they show that farmers who do not use irrigation devices such as probes or tensiometers for estimating the soil water potential or volumetric water content tend to over-irrigate [9,25]. In this sense, in the plots with the most unfavorable slopes and those where they had waterlogging problems, it was recommended to perform land leveling (1200 ha). This practice helped to increase the efficiency in the use of water in the following agricultural cycle, as well as to obtain a better yield in the harvest.

The water savings obtained with an efficient design helped us to have increases in crop productivity, since without a prescription in the case of barley for the last year, 0.700 kg was harvested for each cubic meter of water used, and with the designed irrigation, for the same cubic meter of water, 1078 kg of grain were harvested; in the case of sorghum, this increase represented an additional 0.307 kg and in the case of corn it corresponded to 0.591 kg.

With the efficient design of optimal flow, we managed to reduce the amount of water used to produce 1 kg of biomass (Figure 7). In the case of barley, this reduction was 47.45%, going from 1.96 m^3/kg to 0.93 m^3/kg, while in the case of corn and sorghum crops, the reduction was 38% and 27%, respectively. These data are an important indicator in areas with essential water resources and allow calculating the economic value of the irrigation water that can be maximized, and therefore will be one of the main requirements in making decisions about the distribution and use policies of water in food production.

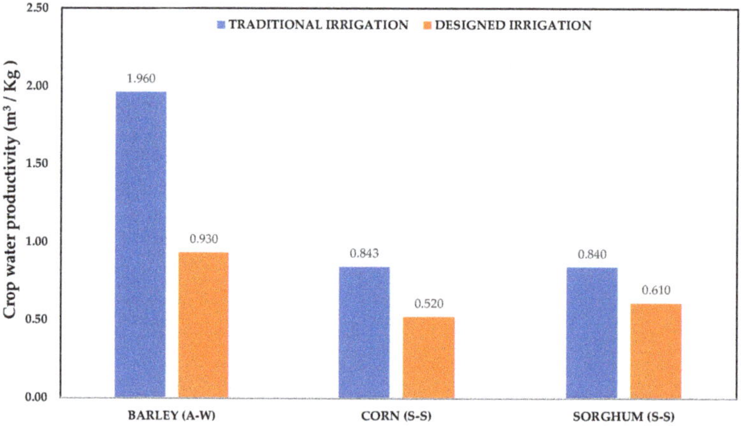

Figure 7. The water productivity obtained by cycle.

4. Conclusions

The reduction in irrigation time per hectare had a considerable impact on the irrigation depth applied to crops. In general, it can be seen that the minimum saving is of the order of 450 m^3 in clay loam, silty, and silty clay soils. The plots where there are more savings are those where they correspond to the loam, sandy clay, and silty clay loam textures. Despite the fact that this program was implemented for only 5 years, the savings obtained per cycle were still significant for the farmers, since in the low rainfall season the dams do not have enough storage to provide water, and this is where the impact of this design has been reflected: with less water, they have irrigated the same irrigation area and, on occasions, as in the 2018–2019 cycle, the savings allowed to give an additional irrigation of 15 cm to 2500 ha.

Finally, it was possible to verify that, with a design for optimal flow in each border or furrow, it helped to improve the efficiency in the use of water and helped increase the productivity of the three crops that were irrigated by gravity. Although with pressurized irrigation systems (sprinkler or drip) there is a higher WUE, this design provides an opportunity to make better use of the resource, increase productivity, and improve crop yield.

Author Contributions: Conceptualization, C.C.; methodology, C.C., I.L.-J.; B.E.-A., J.A.L.-H. and E.B.-F., software, C.C., and J.T.-A.; validation, C.C., I.L.-J.; B.E.-A., J.A.L.-H. and E.B.-F.; formal analysis, C.C., and J.T.-A.; investigation, C.C.; resources, C.C.; data curation, C.C., I.L.-J., B.E.-A., J.A.L.-H., E.B.-F. and J.T.-A.; writing—original draft preparation, C.C. and J.T.-A.; writing—review and editing, C.C.; supervision, C.C., B.E.-A. and J.A.L.-H.; project administration, C.C.; funding acquisition, C.C. All authors have read and agreed to the published version of the manuscript.

Funding: This research was supported as part of a collaboration between the National Water Commission (CONAGUA, according to its Spanish acronym); The Water Users Association of Second Unit of Module Two, Irrigation District No. 023; the Irrigation District 023, San Juan del Río, Querétaro; and the Autonomous University of Queretaro, under the program RIGRAT 2015–2019.

Acknowledgments: We would like to thank the editor and the expert reviewers for their detailed comments and suggestion for the manuscript. These were very useful to hopefully improve the quality of the manuscript.

Conflicts of Interest: The authors declare no conflict of interest.

References

1. Comisión Nacional del Agua (CONAGUA). *Estadísticas del agua en México*; CONAGUA: Coyoacán, México, 2018; p. 306.
2. Chávez, C.; Fuentes, C. Optimization of furrow irrigation by an analytical formula and its impact on reduction of the water applied. *Agrociencia* **2018**, *52*, 483–496.
3. Chávez, C.; Fuentes, C. Design and evaluation of surface irrigation systems applying an analytical formula in the irrigation district 085, La Begoña, Mexico. *Agric. Water Manag.* **2019**, *221*, 279–285. [CrossRef]
4. Hoogeveen, J.; Faures, J.M.; Peiser, L.; Burke, J.; van de Giesen, N. GlobWat—a global water balance model to assess water usage in irrigated agriculture. *Hydrol. Earth Syst. Sci.* **2015**, *19*, 3829–3844. [CrossRef]
5. Unver, O.; Bhaduri, A.; Hoogeveen, J. Water-use efficiency and productivity improvements towards a sustainable pathway for meeting future water demand. *Water Secur.* **2017**, *1*, 21–27. [CrossRef]
6. Koech, R.; Langat, P. Improving Irrigation Water Use Efficiency: A Review of Advances, Challenges and Opportunities in the Australian Context. *Water* **2018**, *10*, 1771. [CrossRef]
7. Bhaduri, A.; Manna, U. Impacts of water supply uncertainty and storage on efficient irrigation technology adoption. *Nat. Resour. Model.* **2014**, *27*, 1–24. [CrossRef]
8. Pereira, L.S.; Cordery, I.; Iacovides, I. Improved indicators of water use performance and productivity for sustainable water conservation and saving. *Agric. Water Manag.* **2012**, *108*, 39–51. [CrossRef]
9. Keen, B.; Slavich, P. Comparison of irrigation scheduling strategies for achieving water use efficiency in highbush blueberry. *N. Z. J. Crop. Hort. Sci.* **2012**, *40*, 3–20. [CrossRef]
10. Du, T.; Kang, S.; Zhang, J.; Davies, W.J. Deficit irrigation and sustainable water-resource strategies in agriculture for China's food security. *J. Exp. Bot.* **2015**, *66*, 2253–2269. [CrossRef]
11. Koech, R.; Smith, R.; Gillies, M. Trends in the use of surface irrigation in Australian irrigated agriculture: An investigation into the role surface irrigation will play in future Australian agriculture. *Water J. Aust. Water Assoc.* **2015**, *42*, 84.
12. Fuentes, C.; Chávez, C. Analytic Representation of the Optimal Flow for Gravity Irrigation. *Water* **2020**, *12*, 2710. [CrossRef]
13. Smith, R.J.; Uddin, M.J.; Gillies, M.H. Estimating irrigation duration for high performance furrow irrigation on cracking clay soils. *Agric. Water Manag.* **2018**, *206*, 78–85. [CrossRef]
14. Salahou, M.K.; Jiao, X.; Lü, H. Border irrigation performance with distance-based cut-off. *Agric. Water Manag.* **2018**, *201*, 27–37. [CrossRef]
15. Nie, W.B.; Li, Y.B.; Zhang, F.; Ma, X.Y. Optimal discharge for closed end border irrigation under soil infiltration variability. *Agric. Water Manag.* **2019**, *221*, 58–65. [CrossRef]
16. Saucedo, H.; Fuentes, C.; Zavala, M. The Saint-Venant and Richards equation system in surface irrigation: (2) Numerical coupling for the advance phase in border irrigation. *Ing. Hidraul. Mex.* **2005**, *20*, 109–119.
17. Saucedo, H.; Zavala, M.; Fuentes, C. Complete hydrodynamic model for border irrigation. *Water Technol. Sci.* **2011**, *2*, 23–38.

18. Liu, K.H.; Jiao, X.Y.; Li, J.; An, Y.H.; Guo, W.H.; Salahou, M.K.; Sang, H. Performance of a zero-inertia model for irrigation with rapidly varied inflow discharges. *Int. J. Agric. Biol. Eng.* **2020**, *13*, 175–181. [CrossRef]
19. Fuentes, S.; Trejo-Alonso, J.; Quevedo, A.; Fuentes, C.; Chávez, C. Modeling Soil Water Redistribution under Gravity Irrigation with the Richards Equation. *Mathematics* **2020**, *8*, 1581. [CrossRef]
20. Green, W.H.; Ampt, G.A. Studies in soil physics, I: The flow of air and water through soils. *J. Agric. Sci.* **1911**, *4*, 1–24.
21. Moré, J.J. The Levenberg-Marquardt algorithm: Implementation and theory. In *Numerical Analysis*; Watson, G.A., Ed.; Springer: Berlin/Heidelberg, Germany, 1978; pp. 105–116.
22. Allen, R.G.; Pereira, L.S.; Raes, D.; Smith, M. Crop evapotranspiration: Guide-lines for computing crop water requirements. In *FAO Irrigation and Drainage Paper No. 56*; FAO: Rome, Italy, 1998; p. 300.
23. Doorenbos, J.; Pruitt, W.O. *Guidelines for Predicting Crop Water Requirements; Irrigation and Drainage Paper No. 24*; Review; FAO-ONU: Roma, Italy, 1977; p. 144.
24. Han, X.; Wei, Z.; Zhang, B.; Han, C.; Song, J. Effects of crop planting structure adjustment on water use efficiency in the irrigation area of Hei River Basin. *Water* **2018**, *10*, 1305. [CrossRef]
25. Car, N.J.; Christen, E.W.; Hornbuckle, J.W.; Moore, G.A. Using a mobile phone Short Messaging Service (SMS) for irrigation scheduling in Australia-Farmers' participation and utility evaluation. *Comput. Electron. Agric.* **2012**, *84*, 132–143. [CrossRef]

Publisher's Note: MDPI stays neutral with regard to jurisdictional claims in published maps and institutional affiliations.

Article

Response of Chosen American *Asparagus officinalis* L. Cultivars to Drip Irrigation on the Sandy Soil in Central Europe: Growth, Yield, and Water Productivity

Roman Rolbiecki [1], Stanisław Rolbiecki [1], Anna Figas [2], Barbara Jagosz [3], Piotr Prus [4], Piotr Stachowski [5], Maciej J. Kazula [6], Małgorzata Szczepanek [7,*], Wiesław Ptach [8], Ferenc Pal-Fam [9], Hicran A. Sadan [1] and Daniel Liberacki [5]

[1] Department of Agrometeorology, Plant Irrigation and Horticulture, Faculty of Agriculture and Biotechnology, UTP University of Science and Technology in Bydgoszcz, 85-029 Bydgoszcz, Poland; rolbr@utp.edu.pl (R.R.); rolbs@utp.edu.pl (S.R.); hicran_sadan_76@hotmail.com (H.A.S.)

[2] Department of Agricultural Biotechnology, Faculty of Agriculture and Biotechnology, UTP University of Science and Technology in Bydgoszcz, 85-029 Bydgoszcz, Poland; figasanna@utp.edu.pl

[3] Department of Plant Biology and Biotechnology, Faculty of Biotechnology and Horticulture, University of Agriculture in Krakow, 31-425 Kraków, Poland; Barbara.Jagosz@urk.edu.pl

[4] Laboratory of Economics and Agribusiness Advisory, Department of Agronomy, Faculty of Agriculture and Biotechnology, UTP University of Science and Technology in Bydgoszcz, 85-029 Bydgoszcz, Poland; piotr.prus@utp.edu.pl

[5] Department of Land Improvement, Environmental Development and Spatial Management, Faculty of Environmental Engineering and Mechanical Engineering, Poznań University of Life Sciences, 60-649 Poznań, Poland; piotr.stachowski@up.poznan.pl (P.S.); daniel.liberacki@up.poznan.pl (D.L.)

[6] Department of Agronomy and Plant Genetics, University of Minnesota, St. Paul, MN 55108-6024, USA; maciek.kazula@gmail.com

[7] Department of Agronomy, Faculty of Agriculture and Biotechnology, UTP University of Science and Technology in Bydgoszcz, 85-029 Bydgoszcz, Poland

[8] Department of Remote Sensing and Environmental Research, Institute of Environmental Engineering, Warsaw University of Life Sciences, 02-776 Warszawa, Poland; wieslaw_ptach@sggw.pl

[9] Institute of Plant Production, Hungarian University of Agriculture and Life Sciences (MATE), Kaposvár Campus, H-7400 Kaposvár, Hungary; Pal-Fam.Ferenc.Istvan@uni-mate.hu

* Correspondence: Malgorzata.Szczepanek@utp.edu.pl

Citation: Rolbiecki, R.; Rolbiecki, S.; Figas, A.; Jagosz, B.; Prus, P.; Stachowski, P.; Kazula, M.J.; Szczepanek, M.; Ptach, W.; Pal-Fam, F.; et al. Response of Chosen American *Asparagus officinalis* L. Cultivars to Drip Irrigation on the Sandy Soil in Central Europe: Growth, Yield, and Water Productivity. *Agronomy* **2021**, *11*, 864. https://doi.org/10.3390/agronomy11050864

Academic Editor: Aliasghar Montazar

Received: 31 March 2021
Accepted: 25 April 2021
Published: 28 April 2021

Publisher's Note: MDPI stays neutral with regard to jurisdictional claims in published maps and institutional affiliations.

Abstract: The aim of this study was to verify the response of 13 American asparagus cultivars cultivated for green spear on surface postharvest drip irrigation. Irrigation, used to compensate for periodic deficiencies in precipitation, allows for high- and good-quality crops for many species. The field experiment was carried out in 2006–2008 on a very light sandy soil in central Europe (Poland). Irrigation treatments were applied using the tensiometer indications. Water requirements of asparagus were calculated on the base of reference evapotranspiration and crop coefficients. The following evaluations were made: Height, diameter, and number of summer stalks, as well marketable yield, weight, and number of consumption green spears. Drip irrigation applied for 2 years (2006–2007) in the postharvest period had a positive effect on all studied traits in both summer stalks and green spears in 2007–2008. A significant increase in the height, number, and diameter of summer stalks, as well an increase in the marketable yield, weight, and number of green spears was observed for most of the cultivars. In general, postharvest drip irrigation of asparagus cultivated in very light sandy soil significantly contributes to the increase in productivity of American cultivars of this species.

Keywords: *Asparagus officinalis* L.; cultivars; spears yield; sandy soil; water requirements; IWUE

1. Introduction

Asparagus is a perennial vegetable species. Therefore, choosing the most advantageous cultivars for cultivation is a very important factor in yielding. Thanks to intensive breeding work carried out in many countries around the world, new cultivars of asparagus are quickly emerging. New cultivars of asparagus are usually very fertile, with relatively high soil and water requirements [1,2]. Therefore, to obtain maximum marketable yields of a given asparagus cultivar, it is recommended to create optimal growth and development conditions during the growing season. Maximum asparagus production possibilities can be achieved by applying organic and mineral fertilization adapted to species nutritional needs and ensuring optimal humidity, with the use of irrigation supplementing deficiency in precipitation. Due to the specific method of cultivation, i.e., harvesting of spears in early spring, the height and quality of asparagus sprouts depend on the amount of ingredients stored in asparagus rootstocks during the growing season of the previous year [1–3].

In recent years in Poland, asparagus (*Asparagus officinalis* L.) has been observed as a vegetable gaining increasing economic significance. On the one hand, this phenomenon is related to the increase in exports of asparagus spears to European Union countries (mainly Germany), and on the other hand, with an increase in demand for this valuable vegetable among domestic consumers, changing their eating habits noticeably. Basic "heavy" species of vegetables, primarily root vegetables such as potatoes, carrots, parsley or red beet, are replaced with low-caloric species with high biological and flavor values [1,2].

Asparagus is a plant grown primarily in light soils with low water content, i.e., limited retention capacity [1,4]. On the one hand, due to a deep-reaching and well-developed root system, asparagus is relatively resistant to water deficiency in soil [5–7]. On the other hand, asparagus, as a light soil plant, reacts very positively to irrigation treatments, which are used in the postharvest period, usually from June to August, in the climatic and soil conditions of Central Europe. Postharvest irrigation significantly increases the yield of asparagus spears in the following year [3,4,8–16].

One of the elements of sustainable plant production, which has the task of protecting the soil and plant raw materials, is melioration treatments, among which drip irrigation is of great importance in commercial crops. The aim of the study was to verify the response of 13 chosen American asparagus cultivars grown for green spear production to surface drip irrigation on sandy soil in the region of central Europe (Poland).

2. Materials and Methods

2.1. Plant Material and Location of the Experiment

The field experiment was carried out in the years 2006–2008 at Kruszyn Krajenski near Bydgoszcz (central Poland) on a sandy soil (Figure 1).

The soil was classified to Typic Hapludolls. The clay content in the topsoil was 7% and, in the subsoil, the clay content ranged from 3% to 5%. The average organic matter content was 1.19%. The water reserve to 1 m depth of soil at field capacity was 87 mm and the available water was 68 mm. The field experiment was conducted in a randomized block design of a 2-factorial "split-plot" system with 4 replications. The first factor was irrigation used in 2 variants: O–non-irrigated plots (control) and D–drip-irrigated plots. The second factor was 13 American's cultivars asparagus (*Asparagus officinalis* L.) including Jersey Giant, Jersey Knight, Jersey Supreme, Jersey Deluxe, Jersey King, Atlas, Grande, Apollo, Purple Passion, UC 157, NJ 953, UC 115, and JWC 1. The asparagus crowns were planted 10th of April 2003.

Terms of single irrigation treatments of asparagus were determined on the basis of tensiometer indications according to Horticultural Institut in Geisenheim (Germany) [17]. During the irrigation season, the soil water potential was not less than -50 kPa. The surface drip irrigation of asparagus plants performed done using 16 mm diameter linear drip line T-Tape, with a 20 cm distance between the emitters. The flow rate was $5\,l\,m^{-1}\,h^{-1}$.

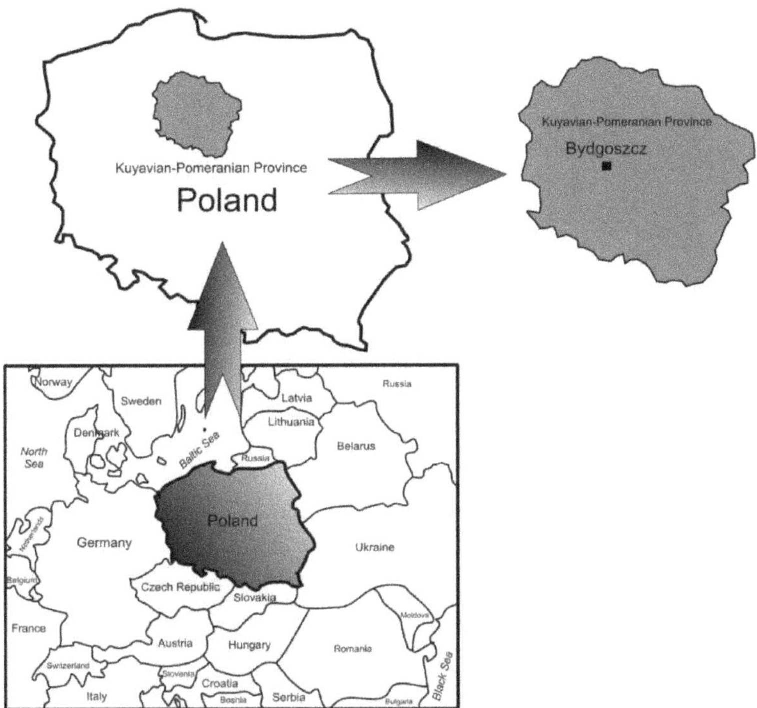

Figure 1. Location of the Bydgoszcz region—within the Kuyavian-Pomeranian Province—in Poland and in Europe.

The standard growing techniques as recommended for asparagus under Polish conditions according to Knaflewski [1] were applied. The asparagus was cultivated for green spears. The plot area for harvest was 15.12 m^2 (24 pcs × 35 cm × 180 cm). Green spears were daily collected for 9 (2007) or 10 weeks (2008) depending on the year of harvest. The observation included both summer stalks in the years 2006–2007 and green consuming spears in the years 2007–2008. The following evaluations were made: Height (cm), number, and diameter (mm) of summer stalks, as well as marketable yield (t·ha^{-1}), weight (g), and number of green spears. The irrigation water use efficiency (IWUE) was also calculated, which is the quotient of the increase in yield obtained during irrigation and the seasonal dose of water used during irrigation. Irrigation water use efficiency (kg·ha^{-1}·mm^{-1}), which presents the effectiveness of water use, was calculated for a marketable yield of green spears using the following Equation (1):

$$\text{IWUE} = \frac{(y - a)}{x},\qquad (1)$$

where

y = yield after irrigation (kg),

a = yield without irrigation (kg),

x = seasonal dose of water used in irrigation (mm).

The experimental data height, number, and diameter of summer stalks, as well marketable yield, weight, and number of green spears were statistically processed by variation analysis. Mean values were verified with Tukey's test at a 5% level.

2.2. Assessment of Water and Irrigation Needs

Reference evapotranspiration (ETo) was determined using Hargreaves' model [18] modified by Droogers and Allen [19] and expressed in Equation (2):

$$ETo = HC \times Ra \times (Tmax - Tmin)^{HE} \times \left(\frac{Tmax + Tmin}{2} + HT \right), \ (mm) \qquad (2)$$

where

HC = empirical Hargreaves coefficient = 0.0025,
Ra = extraterrestrial radiation (mm day^{-1}),
Tmax = maximum air temperature (°C),
Tmin = minimum air temperature (°C),
HE = Hargreaves exponent = 0.5,
HT = Hargreaves temperature coefficient = 16.8.

Potencial evapotranspiration (ETp) was determined by the method of crop coefficients [20] using crop coefficients for asparagus as determined by Hargreaveas' model modified by Droogers and Allen [14]. To take into account the specificity of drip irrigation (limited wetted area of soil), Equation (3) was applied [21]:

$$ETp = ETo \times kc \times kr, \ (mm) \qquad (3)$$

where

ETo = reference evapotranspiration (mm),
kc = crop coefficients,
kr = reduction coefficients.

Reduction coefficients were determined according to Freeman and Garzoli's formula [22] based on the percentage of surface coverage of asparagus in the postharvest period [14]. The results are presented in Table 1.

Table 1. Values of the correction factor k_r according to Freeman and Garzoli.

Area Covering [%]	Values of Factor k_r
10	0.10
20	0.20
30	0.30
40	0.40
50	0.75
60	0.80
70	0.85
80	0.90
90	0.95
100	1

Drought index (D) according to Stenz [23] is expressed in Equation (4):

$$D = \frac{ETo}{P}, \qquad (4)$$

where

ETo = reference evapotranspiration (mm),
P = precipitation in a given period (mm).

To describe the precipitation conditions during the experiment, the worldwide recommended Standardized Precipitation Index (SPI) was used [24]. Equation (5) was used to calculate SPI:

$$SPI = \frac{f(P) - \mu}{\sigma}, \qquad (5)$$

where

f(P) = transformed totals of precipitation in a given period,
μ = mean value of the normalized historical precipitation series,
σ = mean standard deviation of the normalized historical precipitation series.

2.3. Meteorological Conditions

Precipitation conditions in individual decades of the growing and irrigation seasons for asparagus were described according to the season classification, based on the Standardized Precipitation Index, as presented in Table 2.

Table 2. Classification of the period as dependent on Standardized Precipitation Index values [24].

Period	Standardized Precipitation Index
Extreme drought	≤ -2.00
Severe drought	from -1.99 to -1.50
Moderate drought	from -1.49 to -1.00
Normal	from -0.99 to 0.99
Moderate wet	from 1.00 to 1.49
Severe wet	from 1.50 to 1.99
Extreme wet	≥ 2.00

Numerical data from measurements taken in the meteorological station of the then Department (now the Laboratory) of Land Reclamation and Agrometeorology at the University of Science and Technology located in Mochełek near Bydgoszcz were used in the calculations.

Average air temperature values during the growing season (April–September) in the research years (2006–2007) were higher than in the long-term (1971–2000) by 0.8 °C and 0.5 °C, respectively (Table 3). In the first year of the study, high air temperature values were found in July (22.4 °C), with the value exceeding the long-term mean by as much as 4.2 °C. In the second year of the study, high air temperature values were found in June (18.2 °C, i.e., 1.9 °C above the long-term mean).

Table 3. Air temperature (°C) in the 2006–2007 vegetation season.

Year	10-Day Period	Apr	May	June	July	Aug	Sept	Mean
2006	1	5.3	12.9	11.8	22.7	17.6	15.2	
	2	7.3	13.1	18.9	21.8	17.4	15.7	
	3	8.7	11.4	19.7	22.7	15.0	14.6	
	Mean	7.1	12.5	16.8	22.4	16.6	15.2	15.1
2007	1	5.9	9.3	18.8	15.7	18.6	12.6	
	2	9.3	12.7	19.5	21.1	18.6	11.3	
	3	10.2	19.0	16.2	17.3	16.4	13.2	
	Mean	8.5	13.8	18.2	18.0	17.8	12.4	14.8
Mean for 2006–2007		7.8	13.1	17.5	20.2	17.2	13.8	14.9
Mean for 1971–2000		7.6	13.1	16.3	18.0	17.7	13.1	14.3
Difference (+/−)		+0.2	0.0	+1.3	+2.2	−0.5	+0.7	+0.6

Total precipitation of the growing season in the research years, in relation to the long-term mean, was higher by 54 mm (i.e., by 19%) (Table 4). Higher precipitation totals were recorded in the year 2007 (367 mm, i.e., 131% of the long-term mean). Particularly heavy rainfall occurred in June and July 2007, amounting to 103 mm and 112 mm, respectively, which constituted 172% or 167% of the long-term mean. In the first year of the study, the highest rainfall was recorded in August (115 mm, i.e., 225% of the long-term mean).

Table 4. Rainfalls (mm) in the 2006–2007 vegetation season.

Year	10-Day Period	Month						Mean
		Apr	May	June	July	Aug	Sept	
2006	1	0	10	7	0	75	37	
	2	0	20	15	26	23	0	
	3	45	34	0	5	17	4	
	Mean	45	64	22	31	115	42	317
2007	1	5	21	43	79	3	18	
	2	0	23	24	4	11	5	
	3	3	5	36	29	46	13	
	Mean	8	49	103	112	60	36	367
Mean for 2006–2007		26	56	62	71	87	39	342
Mean for 1971–2000		25	43	60	67	51	42	288
Difference (+/−)		+1	+13	+2	+4	+36	−3	+54

Air temperature during the harvest of asparagus spears (from the third decade of April to the third decade of June) is a very important element of asparagus cultivation [1,14].

2.4. Irrigation Needs

Irrigation was carried out after the harvest period, during the summer months, as recommended [25]. In 2006, irrigation began on July 4 and ended on August 31. The irrigation season lasted 59 days. During the time, 10 single doses were used, amounting to 92 mm. In 2007, the irrigation season was shorter and lasted 37 days (from 15 July to 20 August). A total of 54 mm water was used in 6 doses.

The values of Stenz index, presented in Table 5, indicate the irrigation needs of asparagus cultivars in July and August of the years 2006 and 2007.

Table 5. Values of the Stenz index for the 2006–2007 irrigation period.

Year	Month	
	July	August
2006	6.2	0.9
2007	1.2	2.1

The values of the Standardized Precipitation Index, summarized in Table 6, reflect the actual precipitation conditions occurring in individual decades of the irrigation season. The first decade of July in 2006 was extremely dry, while the second decade of July in 2007 was moderately dry.

Table 6. Values of the Stenz index for the 2006–2007 irrigation period.

Month	10-Days Period	Year	
		2006	2007
July	1	−2.17	2.00
	2	0.08	−1.49
	3	−0.63	0.68
August	1	1.95	−0.69
	2	0.69	0.06
	3	0.31	1.17

The smallest daily water needs under drip irrigation (below 1 mm) were found in the third decade of June when summer stalks were just beginning to grow (Figure 2). However, irrigation needs were much higher in the next 2 months of the irrigation season (from 3.2 mm to 4.4 mm). Depending on weather conditions (particularly temperature), asparagus irrigation needs variation in individual years. For comparison, in 9 years of trials with European cultivars (Gijnlim, Ramos, Vulkan) conducted in central Poland (region of Bydgoszcz), the average daily value of field water consumption at drip irrigated plots in June, July, and August, was 1.7 mm, 3.5 mm, and 4.4 mm, respectively [14]. In the lysimeter studies by Paschold, et al. [12] daily water consumption under drip irrigation was, on average, 1.6 mm in June, 2 mm in July, and 3 mm in August.

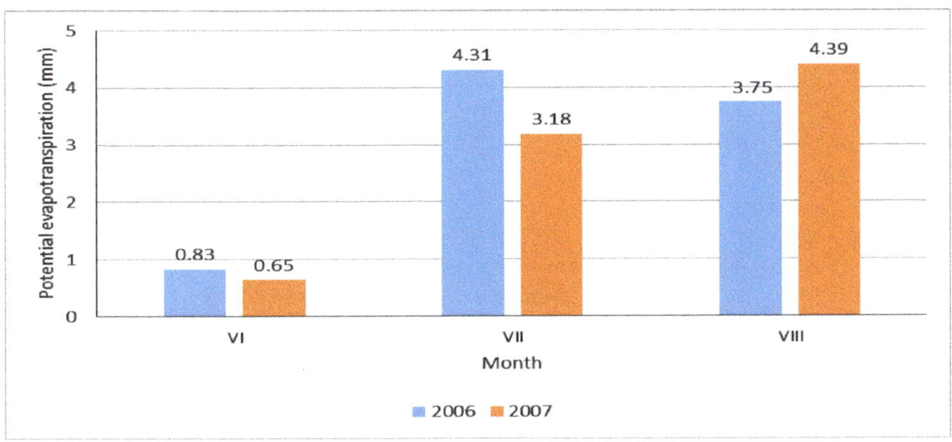

Figure 2. Daily values of the potential evapotranspiration (mm) of the asparagus plants under drip irrigation conditions in particular months of the irrigation period.

The total water needs of asparagus in the postharvest period were higher in the first year of the study (258 mm) than in the second year of the study (241 mm) (Figure 3). For comparison, in the Rolbiecki study [14], the water needs of asparagus plants under drip irrigation ranged from 205 mm to 272 mm in individual years. The results of tests carried out by Paschold and Kunzelmann [26] in Germany's climatic conditions indicated that the water needs of asparagus plants from June 20 to September 1 ranged between 150–241 mm. Moreover, in prior studies carried out in Germany, Hartmann [8] estimated that, depending on precipitation conditions, the water needs of asparagus plants ranged between 179 mm and 240 mm. In turn, Pardo et al. [27] recorded the seasonal water consumption of asparagus under lysimeter conditions ranging from 274 mm to 294 mm. In the latest research on asparagus water needs in various regions of Poland reported by Rolbiecki et al. [28], the irrigation needs of asparagus plants were found to be, on average, 228 mm in the period from July 1 to August 31 in the central-north-west region of Poland, which also covers the area considered in the present study.

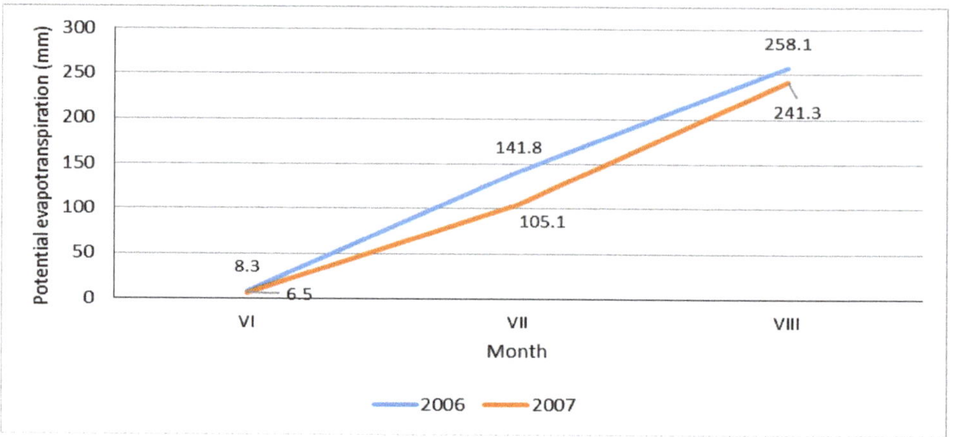

Figure 3. Cumulative potential evapotranspiration (mm) of the asparagus plants under drip irrigation conditions in particular months of the irrigation period.

3. Results and Discussion

3.1. Height, Number, and Diameter of Asparagus Summer Stalks

Drip irrigation, used after the green spears harvest in the growing season preceding the next asparagus harvest, significantly increased the height of asparagus summer stalks for the studied cultivars and years, on average, from 152 cm to 172 cm (Table 7). Stalk height due to drip irrigation increased, on average, by 20 cm (13%). The highest increase in stalk height (above 180 cm) under drip irrigation, on average, for 2 years of research, was obtained for the cultivars Apollo (196 cm), Grande (185 cm), Atlas (182 cm), Jersey King, and Purple Passion (181 cm). The best response to drip irrigation, with reference to this characteristic, was noted in Atlas cultivar. Drip irrigation increased the height of its summer stalks by 40 cm (i.e., by 28%). Some researchers have also shown a significant impact of irrigation on the height of summer stalks in the year preceding the harvest. For example, in the third and fourth year of cultivation, Drost and Wilcox-Lee [29], Hartmann [8,9], Pardo et al. [27], Paschold et al. [12], and Sterrett et al. [30] obtained an average increase in summer stalk height of 14 cm in comparison to non-irrigated control variants.

Table 7. Height of the asparagus summer stalks (cm) as dependent on cultivar and irrigation.

Cultivar	Control (Without Irrigation)			Drip Irrigation			Increase of the Height of Summer Stalks		
	2006	2007	Mean	2006	2007	Mean	2006	2007	Mean
Jersey Giant	133	139	136	164	168	166	31	29	30
Jersey Knight	158	161	159	166	175	170	8	14	11
Jersey Supreme	149	151	150	165	169	167	16	18	17
Jersey Deluxe	151	155	153	161	180	170	10	25	17
Jersey King	156	162	159	179	183	181	23	21	22
Atlas	139	145	142	179	185	182	40	40	40
Grande	160	165	162	181	190	185	21	25	23
Apollo	169	172	170	192	201	196	23	29	26
Purple Passion	150	155	152	178	185	181	28	30	29
UC 157	141	146	143	166	176	171	25	30	27
NJ 953	155	161	158	158	170	164	3	9	6
UC 115	159	162	160	159	171	165	0	9	4
JWC 1	122	149	135	130	157	143	8	8	8
Mean	149	156	152	165	178	172	18	22	20

2006: LSD$_{0.05}$ for: Irrigation = 6.343; Cultivars = 16.591; Interaction: Cultivars/Irrigation = 25.831; Irrigation/Cultivars = 13.599. 2007: LSD$_{0.05}$ for: Irrigation = 7.225; Cultivars = 14.739; Interaction: Cultivars/Irrigation = 22.341; Irrigation/Cultivars = 12.664. 2006–2007: LSD$_{0.05}$ for: Irrigation = 2.663; Cultivars = 11.936; Interaction: Cultivars/Irrigation = 16.879; Irrigation/Cultivars = 9.603.

Postharvest drip irrigation, applied in the growing season preceding the next asparagus harvest, significantly increased the number of asparagus summer stalks per plant for the studied cultivars and years, on average, from 7.9 to 11.1 (Table 8). Stalk number per plant due to drip irrigation increased, on average, by 3.2 pcs (40%). The highest increase in stalk number per plant (12 and more) under drip irrigation, on average, for 2 years of research, was observed for Jersey King (13.3), Apollo (12.6), Jersey Supreme, Atlas, and UC 157 (12) cultivars. The best response to drip irrigation, with reference to this characteristic, was noted in UC 157 and Grande. Drip irrigation increased the number of stalks per plant of these cultivars by 5.0 and 5.1 pcs (i.e., by 72%), respectively. The positive impact of drip irrigation on the increase in the number of summer stalks was also presented for Germany's weather conditions, which are similar to Polish weather conditions, by Hartmann [9], Hartmann et al. [10], and Paschold et al. [12]. Drost [31] reported that under dry climatic conditions, lowering sandy soil moisture below −50 kPa reduced the number and height of stalks in comparison to drip-irrigated variants. Sterrett et al. [30] obtained an increase in stalk number due to irrigation, on average, of four stalks per plant, and Battilani [32] obtained an increase of one stalk per plant. The diverse response of asparagus cultivars to drip irrigation has been confirmed, i.a., by the results of Dutch [33] and Polish [14] studies.

Table 8. Number of the asparagus summer stalks (pcs) as dependent on cultivar and irrigation.

Cultivar	Control (Without Irrigation)			Drip Irrigation			Increase of the Number of Summer Stalks		
	2006	2007	Mean	2006	2007	Mean	2006	2007	Mean
Jersey Giant	8.1	7.6	7.8	10.6	10.2	10.4	2.5	2.6	2.5
Jersey Knight	6.7	7.0	6.8	10.5	10.7	10.6	3.8	3.7	3.7
Jersey Supreme	8.0	8.2	8.1	12.1	11.9	12.0	4.1	3.7	3.9
Jersey Deluxe	8.5	7.9	8.2	9.7	10.0	9.8	1.2	2.1	1.6
Jersey King	10.5	10.3	10.4	13.2	13.5	13.3	2.7	3.2	2.9
Atlas	9.2	8.8	9.0	12.1	12.0	12.0	2.9	3.2	3.0
Grande	6.8	7.1	6.9	12.0	11.9	11.9	5.2	4.8	5.0
Apollo	8.7	9.1	8.9	12.3	12.9	12.6	3.6	3.8	3.7
Purple Passion	8.3	7.9	8.1	8.8	9.1	8.9	0.5	1.2	0.8
UC 157	7.1	6.8	6.9	11.8	12.2	12.0	4.7	5.4	5.0
NJ 953	8.5	8.1	8.3	10.7	11.2	10.9	2.2	3.1	2.6
UC 115	6.8	7.3	7.0	10.6	11.5	11.0	3.8	4.2	4.0
JWC 1	7.2	6.9	7.0	9.1	9.0	9.0	1.9	2.1	2.0
Mean	8.0	7.9	7.9	11.0	11.2	11.1	3.0	3.3	3.2

2006: $LSD_{0.05}$ for: Irrigation = 0.870; Cultivars = 2.190; Interaction: Cultivars/Irrigation = 3.655; Irrigation/Cultivars = 2.413. 2007: $LSD_{0.05}$ for: Irrigation = 0.987; Cultivars = 1.839; Interaction: Cultivars/Irrigation = 3.196; Irrigation/Cultivars = 2.417. 2006–2007: $LSD_{0.05}$ for: Irrigation = 1.160; Cultivars = 0.719; Interaction: Cultivars/Irrigation = 1.017; Irrigation/Cultivars = 0.578.

Drip irrigation, used after the harvest in the growing season preceding the next asparagus harvest, significantly increased the diameter of asparagus summer stalks for the studied cultivars and years, on average, from 11.2 mm to 12.6 mm (Table 9). Stalk diameter due to drip irrigation increased, on average, by 1.3 mm (12%). The highest increase in stalk diameter under drip irrigation, on average, for 2 years of research, was observed for Apollo (15.3 mm) and Grande (14.6 mm) cultivars. The best response to drip irrigation, with reference to this characteristic, was also noted in Apollo and Grande. Drip irrigation increased the diameter of their summer stalks by 3.6 mm and 3.2 mm, respectively (i.e., by 31% and 29%). Researchers such as, i.a., Sterrett et al. [30], Contador [34], Rubio [35], and Battilani [32], also observed a significant impact of drip irrigation on stalk diameter. As numerous authors have stated, irrigation significantly affects the growth of summer stalks and, consequently, the yield of asparagus spears in the next year [4,36–41]. Based on the results of a cultivar trial in 2003 (carried out as part of the third IACT) in which 31 asparagus cultivars were compared, Rolbiecki and Rolbiecki [2] found a significant impact of drip irrigation on the formation of summer stalks and marketable yield in the first harvest season. The impact of irrigation was manifested by the increased accumulation of carbohydrates in asparagus rootstocks of the plants growing in optimal humidity conditions, used in spring of the next harvest year [30,31,42–48].

Table 9. Diameter of the asparagus summer stalks (mm) as dependent on cultivar and irrigation.

Cultivar	Control (Without Irrigation)			Drip Irrigation			Increase of the Diameter of Summer Stalks		
	2006	2007	Mean	2006	2007	Mean	2006	2007	Mean
Jersey Giant	10.6	10.3	10.4	10.8	11.2	11.0	0.2	0.9	0.5
Jersey Knight	12.6	12.3	12.4	12.0	13.5	12.7	−0.6	1.2	0.3
Jersey Supreme	10.0	9.1	9.5	11.9	12.2	12.0	1.9	3.1	2.5
Jersey Deluxe	11.6	11.2	11.4	11.4	12.3	11.8	−0.2	1.1	0.4
Jersey King	13.2	12.4	12.8	13.0	13.6	13.3	−0.2	1.2	0.5
Atlas	14.0	11.2	12.6	11.0	13.8	12.4	−3.0	2.6	−0.2
Grande	11.4	11.4	11.4	14.5	14.8	14.6	3.1	3.4	3.2
Apollo	11.8	11.5	11.6	15.0	15.6	15.3	3.2	4.1	3.6
Purple Passion	12.3	11.7	12.0	10.4	13.5	11.9	−1.9	1.8	0.0
UC 157	11.8	11.6	11.7	10.4	12.9	11.6	−1.4	1.3	0.0
NJ 953	10.5	10.1	10.3	13.2	13.7	13.4	2.7	3.6	3.1
UC 115	9.5	9.8	9.6	11.3	11.9	11.6	1.8	2.1	1.9
JWC 1	10.0	10.2	10.1	10.8	12.6	11.7	0.8	2.4	1.6
Mean	11.5	11.0	11.2	12.0	13.2	12.6	0.5	2.2	1.3

2006: LSD$_{0.05}$ for: Irrigation = ns; Cultivars = 3.293; Interaction: Cultivars/Irrigation = 4.657; Irrigation/Cultivars = 2.677. 2007: LSD$_{0.05}$ for: Irrigation = 0.143; Cultivars = 1.532; Interaction: Cultivars/Irrigation = 3.221; Irrigation/Cultivars = 2.774. 2006–2007: LSD$_{0.05}$ for: Irrigation = 0.499; Cultivars = 2.234; Interaction: Cultivars/Irrigation = 3.159; Irrigation/Cultivars = 1.797.

3.2. Marketable Yield, Weight and Number of Asparagus Green Spears

Postharvest drip irrigation treatment of asparagus plants used in the growing season preceding the asparagus harvest significantly increased green spear yields from 4.21 t·ha^{-1} to 6.23 t·ha^{-1} in the next growing season (Table 10). The marketable yield increase in green spears due to drip irrigation for the studied cultivars and 2 years of research was, on average, 2.02 t·ha^{-1} (48%). The highest marketable yield increase under drip irrigation—above 7 t·ha^{-1}, on average, for 2 years of research—was obtained for Apollo, NJ 953, Jersey Deluxe, and Grande cultivars. High marketable yields above 6 t·ha^{-1} were obtained for UC 157, Jersey King, UC 115, Jersey Supreme, and Jersey Giant cultivars. Marketable yields above 5 t·ha^{-1} were obtained for Purple Passion and Jersey Knight. The lowest yields under drip irrigation (above 4 t·ha^{-1}) were collected from JWC 1 and Atlas. The best response to drip irrigation was noted in Jersey Deluxe and Jersey Giant. Drip irrigation increased their marketable yields by 4.37 t·ha^{-1} (157%) and 3.91 t·ha^{-1} (142%), respectively. No significant response to drip irrigation was found in Purple Passion, Jersey King, and Atlas. The response to drip irrigation in other tested cultivars was positive, and the observed marketable yield increases ranged from 1.71 t·ha^{-1} to 2.78 t·ha^{-1} (35–79%). The results obtained are consistent with the results of studies conducted by other scientists. Hartmann [8,9], in weather conditions characteristic for Germany, obtained yield increases of 50% in his trials on sandy soil and 25% on loamy sand due to drip irrigation. Moreover, Paschold et al. [4,11,13], carrying out a cultivar trial using drip irrigation, recorded the highest marketable yields for Gijnlim cultivar. In Poland, Rolbiecki [14] also obtained the highest marketable yields for both control (no irrigation) (6.5 t·ha^{-1}) and drip irrigated plots (10.9 t·ha^{-1}) for Gijnlim. Pinkau and Grutz [49] obtained significantly lower marketable yield increases in spears, on average, of 12.4%, observing the highest yield increases in dry years. Mulder and Lavrijsen [33], in an asparagus cultivar trial under sprinkler irrigation conducted in the Netherlands, obtained the highest yields for Gijnlim, amounting to 4.5 t·ha^{-1} in the first harvest year and 12.8 t·ha^{-1} in the second year. Numerous other authors have also presented a significant impact of irrigation on yield increases in asparagus spears grown in different than Polish climate zones [15,16,29,30,47].

Table 10. Marketable yield of the asparagus green spears (t·ha^{-1}) as dependent on cultivar and irrigation.

Cultivar	Control (Without Irrigation)			Drip Irrigation			Increase of the Marketable Yield of Green Spears		
	2007	2008	Mean	2007	2008	Mean	2007	2008	Mean
Jersey Giant	2.55	2.95	2.75	6.46	6.87	6.66	3.91	3.92	3.91
Jersey Knight	3.43	3.51	3.47	5.64	5.81	5.72	2.21	2.30	2.25
Jersey Supreme	4.18	4.32	4.25	6.16	6.70	6.43	1.98	2.38	2.18
Jersey Deluxe	2.66	2.91	2.78	6.96	7.34	7.15	4.30	4.43	4.37
Jersey King	6.15	5.80	5.97	6.04	6.50	6.27	−0.11	0.70	0.30
Atlas	4.57	4.41	4.49	4.67	4.92	4.79	0.10	0.51	0.30
Grande	5.42	4.80	5.11	6.91	7.25	7.08	1.49	2.45	1.97
Apollo	5.77	5.40	5.58	7.21	7.84	7.52	1.44	2.44	1.94
Purple Passion	5.75	5.01	5.38	4.96	5.64	5.30	−0.79	0.63	−0.08
UC 157	4.16	3.98	4.07	5.72	6.58	6.15	1.56	2.60	2.08
NJ 953	5.13	4.22	4.67	6.93	7.51	7.22	1.80	3.29	2.55
UC 115	3.11	3.96	3.53	5.75	6.87	6.31	2.64	2.91	2.78
JWC 1	2.51	2.89	2.70	3.91	4.91	4.41	1.40	2.02	1.71
Mean	4.26	4.17	4.21	5.95	6.52	6.23	1.69	2.35	2.02

2007: LSD$_{0.05}$ for: Irrigation = 0.473; Cultivars = 1.572; Interaction: Cultivars/Irrigation = 2.223; Irrigation/Cultivars = 1.274. 2008: LSD$_{0.05}$ for: Irrigation = 0.581; Cultivars = 1.822; Interaction: Cultivars/Irrigation = 2.451; Irrigation/Cultivars = 1.422. 2007–2008: LSD$_{0.05}$ for: Irrigation = 0.211; Cultivars = 0.946; Interaction: Cultivars/Irrigation = 1.337; Irrigation/Cultivars = 0.761.

Drip irrigation, used after the harvest in the next growing season preceding the asparagus harvest, significantly increased the mean spear weight from 35.33 g to 40.35 g (Table 11). The mean spear weight due to drip irrigation increased for the studied cultivars and 2 years of research, on average, by 5.02 g (14%). The greatest mean spear weight under drip irrigation—above 40 g, on average, for 2 years of research—was obtained for the cultivars Grande (48.79 g), Jersey Knight (43.32 g), Apollo (43.03 g), Purple Passion (42.99 g), Atlas (42.71 g), and UC 115 (41.24 g). The best response to drip irrigation, with reference to this characteristic, was noted in Atlas and UC 115 cultivars. Drip irrigation increased the mean spear weight by 24% and 32%, respectively. For comparison, the spear weight for Gijnlim cultivar in asparagus trials under irrigation conducted by Mulder and Lavrijsen [33] was approximately 40 g. Paschold et al. [13] obtained values at the level of 35 g, and Sterret et al. [30] obtained values at the level of 25 g.

Table 11. Weight of the asparagus green spears (g) as dependent on cultivar and irrigation.

Cultivar	Control (Without Irrigation)			Drip Irrigation			Increase of the Weight of Green Spears		
	2007	2008	Mean	2007	2008	Mean	2007	2008	Mean
Jersey Giant	25.02	26.02	25.52	34.51	36.31	35.41	9.49	10.29	9.89
Jersey Knight	36.77	38.22	37.49	42.33	44.31	43.32	5.56	6.09	5.83
Jersey Supreme	37.63	36.61	37.12	35.95	39.45	37.70	−1.68	2.84	0.58
Jersey Deluxe	32.87	31.81	32.34	35.39	38.40	36.89	2.52	6.59	4.55
Jersey King	42.86	34.76	38.81	35.19	40.19	37.69	−7.67	5.43	−1.12
Atlas	34.80	34.22	34.51	41.70	43.73	42.71	6.90	9.51	8.20
Grande	43.55	42.55	43.05	48.29	49.30	48.79	4.74	6.75	5.74
Apollo	36.95	38.55	37.75	42.05	44.01	43.03	5.10	5.46	5.28
Purple Passion	39.46	39.46	39.46	40.99	44.99	42.99	1.53	5.53	3.53
UC 157	35.97	32.97	34.47	37.49	39.50	38.49	1.52	6.53	4.02
NJ 953	37.52	36.52	37.02	36.78	39.88	38.33	−0.74	3.36	1.31
UC 115	30.23	32.33	31.28	40.24	42.24	41.24	10.01	9.91	9.96
JWC 1	29.95	31.05	30.50	37.44	38.44	37.94	7.44	7.39	7.44
Mean	35.66	35.01	35.33	39.10	41.60	40.35	3.44	6.59	5.02

2007: LSD$_{0.05}$ for: Irrigation = ns; Cultivars = 8.822; Interaction: Cultivars/Irrigation = 11.316; Irrigation/Cultivars = 7.066. 2008: LSD$_{0.05}$ for: Irrigation = 6.343; Cultivars = 6.022; Interaction: Cultivars/Irrigation = 9.416; Irrigation/Cultivars = 6.116. 2007–2008: LSD$_{0.05}$ for: Irrigation = 1.031; Cultivars = 4.622; Interaction: Cultivars/Irrigation = 6.536; Irrigation/Cultivars = 3.718.

Drip irrigation, applied after the harvest in the growing season preceding the next asparagus harvest, significantly increased the number of green spears per plant from 6.50

to 9.49 (Table 12). Green spears number per plant increased due to drip irrigation for the studied cultivars and 2 years of research, on average, by 2.98 pcs (46%). The highest increase in green spears number per plant (10 or more) under drip irrigation conditions, on average, for 2 years of research, was observed for the cultivars Jersey Deluxe (11.10), Jersey King (10.90), Jersey Supreme (10.78), Jersey Giant (10.72), NJ 953 (10.46), and Apollo (10.00). The best response to drip irrigation, with reference to this characteristic, was noted in Jersey Deluxe cultivar. Drip irrigation increased the number of green spears per plant in this cultivar by more than six spears (i.e., by 123%). Hartmann [9], Paschold et al. [11], Rolbiecki and Rolbiecki [2], and Rolbiecki [14] also obtained an increase in the number of asparagus spears due to drip irrigation.

Table 12. Number of the asparagus green spears (pcs) as dependent on cultivar and irrigation.

Cultivar	Control (Without Irrigation)			Drip Irrigation			Increase of the Number of Green Spears		
	2007	2008	Mean	2007	2008	Mean	2007	2008	Mean
Jersey Giant	5.20	6.30	6.25	10.17	11.27	10.72	4.97	4.97	4.97
Jersey Knight	5.44	5.50	5.47	7.64	8.02	7.83	2.20	2.52	2.36
Jersey Supreme	6.60	7.60	7.10	10.27	11.30	10.78	3.67	3.7	3.68
Jersey Deluxe	4.94	5.02	4.98	11.65	10.55	11.10	6.71	5.53	6.12
Jersey King	8.41	8.44	8.42	10.35	11.45	10.90	1.94	3.01	2.47
Atlas	6.49	6.22	6.35	7.53	8.53	8.03	1.04	2.31	1.67
Grande	6.47	6.01	6.24	8.20	8.65	8.42	1.73	2.64	2.18
Apollo	7.10	6.92	7.01	9.69	10.32	10.00	2.59	3.40	2.99
Purple Passion	8.66	9.66	9.16	7.32	10.32	8.82	−1.34	0.66	−0.34
UC 157	6.77	5.61	6.19	9.07	9.84	9.45	2.30	4.23	3.26
NJ 953	7.81	7.01	7.41	10.87	10.05	10.46	3.06	3.04	3.05
UC 115	4.46	5.21	4.83	8.56	9.22	8.89	4.10	4.01	4.05
JWC 1	5.17	6.20	5.68	7.15	8.83	7.99	1.98	2.63	2.30
Mean	6.42	6.59	6.50	9.11	9.87	9.49	2.69	3.28	2.98

2007: $LSD_{0.05}$ for: Irrigation = 0.747; Cultivars = 2.116; Interaction: Cultivars/Irrigation = 2.993; Irrigation/Cultivars = 1.732. 2008: $LSD_{0.05}$ for: Irrigation = 0.824; Cultivars = 1.811; Interaction: Cultivars/Irrigation = 2.459; Irrigation/Cultivars = 1.562. 2007–2008: $LSD_{0.05}$ for: Irrigation = 0.372; Cultivars = 1.666; Interaction: Cultivars/Irrigation = 2.356; Irrigation/Cultivars = 1.340.

3.3. Irrigation Water Use Efficiency

The average value of irrigation water use efficiency (IWUE) for the harvest years and cultivars was 31 $kg \cdot ha^{-1} \cdot mm^{-1}$ (Table 13). Irrigation water use efficiency in the first year of research (2007) was lower (19 $kg \cdot ha^{-1} \cdot mm^{-1}$) than in the second year of study (2008) (44 $kg \cdot ha^{-1} \cdot mm^{-1}$). This was due to the fact that, for the irrigation season in 2006, a higher seasonal irrigation norm was used than in 2007. Obviously, asparagus yields harvested at drip irrigated plots in 2007 were the result of irrigation carried out in 2006, and, respectively, postharvest irrigation carried out in 2007 had an impact on asparagus yields harvested in 2008. Jersey Deluxe and Jersey Giant were characterized by the highest IWUE, which, on average, for 2 years of research, was 64 $kg \cdot ha^{-1} \cdot mm^{-1}$ and 57 $kg \cdot ha^{-1} \cdot mm^{-1}$, respectively. The lowest IWUE (4 $kg \cdot ha^{-1} \cdot mm^{-1}$–6 $kg \cdot ha^{-1} \cdot mm^{-1}$) was found in Atlas, Jersey King, and Purple Passion cultivars. The other tested cultivars, on average, for 2 years of research, were described by IWUE as ranging from 26 $kg \cdot ha^{-1} \cdot mm^{-1}$ to 41 $kg \cdot ha^{-1} \cdot mm^{-1}$. Diverse irrigation water use efficiency in the cultivation of various asparagus cultivars has been confirmed by the results of previous trials conducted in the central Poland (region of Bydgoszcz) with other cultivars of asparagus [2,14] and vegetable species [50,51] or berry plants [52–55].

Table 13. Irrigation water use efficiency ($kg \cdot ha^{-1} \cdot mm^{-1}$) as dependent on cultivar and year studied.

Cultivar	Irrigation Water Use Efficiency		
	2007	2008	Mean
Jersey Giant	42	73	57
Jersey Knight	24	43	33
Jersey Supreme	22	44	33
Jersey Deluxe	47	82	64
Jersey King	-	13	6
Atlas	-	9	4
Grande	16	45	30
Apollo	16	45	30
Purple Passion	-	12	6
UC 157	17	48	32
NJ 953	20	61	40
UC 115	29	54	41
JWC 1	15	37	26
Mean	19	44	31

4. Conclusions

Drip irrigation of 13 cultivars of asparagus, applied for 2 years in the postharvest period (2006–2007), had a positive effect on all studied characteristics in both summer stalks and consumption green spears of this vegetable. On average, in 2006–2007, a significant increase in the height, number, and diameter of summer stalks was visible in 13, 12, and 9 of 13 studied asparagus cultivars, respectively. On average, in 2007–2008, the increase in marketable yield, weight, and number of green spears was significant for 12, 11, and 12 of 13 tested asparagus cultivars, respectively. The highest marketable yield increase of green spears under drip irrigation—above 7 $t \cdot ha^{-1}$, on average, for 2 years of research—was obtained for Jersey Deluxe, Grande, Apollo, and NJ 953 cultivars. The best response to drip irrigation was noted in Jersey Deluxe and Jersey Giant. Drip irrigation increased the marketable yields of these cultivars by 157% and 142%, respectively. Jersey Deluxe and Jersey Giant cultivars were characterized by the highest irrigation water use efficiency (calculated for the marketable yield of green spears), which, on average, for 2 years of research, was 64 $kg \cdot ha^{-1} \cdot mm^{-1}$ and 57 $kg \cdot ha^{-1} \cdot mm^{-1}$, respectively. High irrigation water use efficiency, over 30 $kg \cdot ha^{-1} \cdot mm^{-1}$, was observed in the case of 9 of 13 tested cultivars. In summary, postharvest irrigation of asparagus cultivated on a very light sandy soil significantly improved both the summer stalk characteristics, as well as the marketable yield features of consumption green spears of this vegetable species. Drip irrigation of asparagus, applied after the harvest in the growing season preceding the next green spears harvest, makes it possible to significantly increase the productivity of this vegetable, which contributes to the sustainable crop production.

Author Contributions: Conceptualization—R.R., S.R., and A.F.; methodology—R.R., S.R., and A.F.; software—R.R., S.R., and A.F.; validation—R.R., P.S., and F.P.-F.; formal analysis—R.R., P.S., D.L., and F.P.-F.; investigation—R.R., S.R., and A.F.; resources—R.R. and S.R.; data curation—R.R. and S.R.; writing—original draft preparation—R.R., S.R., A.F., B.J., P.S., W.P., P.P., M.J.K., and H.A.S.; writing—review and editing—R.R., S.R., A.F., B.J., P.S., W.P., P.P., M.S., and M.J.K.; visualization—R.R., S.R., B.J., P.P., F.P.-F., M.J.K., and H.A.S.; supervision—R.R., S.R., B.J., P.P., F.P.-F., M.S., and M.J.K. All authors have read and agreed to the published version of the manuscript.

Funding: This research received no external funding.

Institutional Review Board Statement: Not applicable.

Informed Consent Statement: Not applicable.

Conflicts of Interest: The authors declare no conflict of interest.

References

1. Knaflewski, M. *Szparag [Asparagus]*; Ha-Ka: Komorniki, Poland, 1995; pp. 3–92.
2. Rolbiecki, R.; Rolbiecki, S. Effect of surface drip irrigation on asparagus cultivars in central Poland. *Acta Hortic.* **2008**, *776*, 45–50. [CrossRef]
3. Wichrowska, D.; Rolbiecki, R.; Rolbiecki, S.; Jagosz, B.; Ptach, W.; Kazula, M.; Figas, A. Concentrations of some chemical components in white asparagus spears depending on the cultivar and post-harvest irrigation treatments. *Folia Hort.* **2018**, *30*, 147–154. [CrossRef]
4. Paschold, P.J.; Hermann, G.; Artelt, B. Asparagus varieties tested over 6 years. *Gemüse* **1999**, *35*, 261–266.
5. Kaniszewski, S. *Nawadnianie Warzyw Polowych [Irrigation of Field Vegetables]*; PlantPress: Kraków, Poland, 2005; pp. 3–115.
6. Kaniszewski, S. Technologia nawadniania warzyw. In Proceedings of the Nawadnianie warzyw w uprawach polowych, Skierniewice, Poland, 12–13 July 2005; pp. 5–19.
7. Kaniszewski, S. Nawadnianie Warzyw [Irrigation of Vegetables]. In *Nawadnianie Roślin [Plant Irrigation]*; Nowak, L., Karczmarczyk, S., Eds.; PWRiL: Poznań, Poland, 2006; pp. 295–332.
8. Hartmann, H.D. Die Bewasserung bei Spargel und ihre Auswikung auf die Pflanze. *Arch. Gart.* **1981**, *29*, 167–175.
9. Hartmann, H.D. The influence of irrigation on the development and yield of asparagus. *Acta Hortic.* **1981**, *119*, 309–316. [CrossRef]
10. Hartmann, H.D.; Hermann, G.; Kirchner-Ness, R. Effect of weather during the vegetative period on the yield of asparagus in the following year. *Gartenbauwissenschaft* **1990**, *55*, 30–34.
11. Paschold, P.J.; Hermann, G.; Artelt, B. Comparison of international asparagus cultivars under Rhine-Valley conditions in Germany. *Acta Hortic.* **1996**, *415*, 257–262. [CrossRef]
12. Paschold, P.J.; Artelt, B.; Hermann, G. The water need of asparagus (*Asparagus officinalis* L.) determined in a lysimeter station. *Acta Hortic.* **2004**, *664*, 529–536. [CrossRef]
13. Paschold, P.J.; Artelt, B.; Hermann, G. Comparison of white asparagus cultivars (*Asparagus officinalis* L.) in Germany. *Acta Hortic.* **2008**, *776*, 379–385. [CrossRef]
14. Rolbiecki, R. *Ocena Potrzeb i Efektów Mikronawodnień Szparaga (Asparagus officinalis L.) na Obszarze Szczególnie Deficytowym w Wodę [Assessment of the Needs and Effects of Micro-Irrigation of Asparagus (Asparagus officinalis L.) in an Area Particularly Water-Deficient]*; UTP: Bydgoszcz, Poland, 2013; pp. 1–103.
15. Brainard, D.C.; Był, B.; Hayden, Z.D.; Noyes, D.C.; Bakker, J.; Werling, B. Managing drought risk in changing climate: Irrigation and cultivar impact on Michigan asparagus. *Agr. Water Manag.* **2019**, *213*, 773–781. [CrossRef]
16. Campi, P.; Mastrorilli, M.; Stellacci, A.M.; Modugno, F.; Palumbo, A.D. Increasing the effective use of water in green asparagus through deficit irrigation strategies. *Agr. Water Manag.* **2019**, *217*, 119–130. [CrossRef]
17. Paschold, P.J.; Weithaler, A. Eignung von Sensoren zum Steuern der Bewässerung bei Freilandgemüse. *Z. Bewässerungswirtschaft* **2000**, *35*, 51–62.
18. Hargreaves, G.H.; Samani, Z.A. Reference crop evapotranspiration from temperature. *Appl. Eng. Agric.* **1985**, *1*, 96–99. [CrossRef]
19. Treder, W.; Wójcik, K.; Żarski, J. Wstępna ocena możliwości szacowania potrzeb wodnych roślin na podstawie prostych pomiarów meteorologicznych [Initial assessment of the possibility of estimating the water needs of plants on the basis of simple meteorological measurements]. *Zesz. Nauk. Inst. Sad. Kw. Skiern.* **2010**, *18*, 143–153.
20. Allen, R.G.; Pereira, L.S.; Raes, D.; Smith, M. *Crop Evapotranspiration—Guidelines for Computing Crop Water Requirements*; FAO Irrigation and Drainage Paper 56; FAO: Rome, Italy, 1998; p. 300.
21. Vermeiren, L.; Jobling, G.A. *Localized Irrigation*; FAO Irrigation and Drainage Paper 36; FAO: Rome, Italy, 1984; p. 198.
22. Smith, M. *CROPWAT a Computer Program for Irrigation Planning and Management*; FAO Irrigation and Drainage Paper 46; FAO: Rome, Italy, 1992; p. 132.
23. Łabędzki, L. Susze rolnicze. Zarys problematyki oraz metody monitorowania i klasyfikacji [Agricultural droughts]. *Woda. Środowisko. Obsz. Wiejskie. Rozpr. Nauk. Monogr.* **2006**, *17*, 1–107.
24. McKee, T.B.; Doesken, J.N.; Kleist, J. The Relationship of Drought Frequency and Duration to Time Scales. In Proceedings of the 8th Conference on Applied Climatology, Anaheim, CA, USA, 17–22 January 1993; American Meteorological Society: Boston, MA, USA, 1993; Volume 17, pp. 179–183.
25. Paschold, P.J.; Kleber, J.; Mayer, N. Geisenheimer Bewässerungssteuerung 2002. *Z. Bewässerungswirtschaft* **2002**, *37*, 5–15.
26. Paschold, P.J.; Kunzelmann, G. *Umweltschonender und Effizienter Bewässerungseinsatz im Gemüseanbau. Praktische Empfehlungen zur Bewässerungss-Teuerung und Bewässerungstechnik im Hessischen Ried*; Hessisches Ministerium für Umwelt, Landwirtschaft und Forsten: Wiesbaden, Germany, 2002; pp. 1–23.
27. Pardo, A.; Arbizou, J.; Suso, M.L. Evapotranspiration and crop coefficients in white asparagus. *Acta Hortic.* **1997**, *449*, 187–192. [CrossRef]
28. Rolbiecki, S.; Rolbiecki, R.; Jagosz, B.; Ptach, W.; Stachowski, P.; Kazula, M. Water needs of asparagus plants in the different regions of Poland. *Annu. Set Environ. Prot.* **2019**, *21*, 1227–1237.
29. Drost, D.T.; Wilcox-Lee, D. Soil water deficits and asparagus: Bud size and subsequent spear growth. *Sci. Hortic.* **1997**, *70*, 145–153. [CrossRef]
30. Sterrett, S.B.; Ross, B.B.; Savage, C.P., Jr. Establishment and yield of asparagus as influenced by planting and irrigation method. *J. Am. Soc. Hortic. Sci.* **1990**, *115*, 29–33. [CrossRef]
31. Drost, D.T. Soil water deficits reduce growth and yield of asparagus. *Acta Hortic.* **1999**, *479*, 383–390. [CrossRef]

32. Battilani, A. Response of asparagus (*Asparagus officinalis* L.) to post-harvesting irrigation. *Acta Hortic.* **1997**, *449*, 181–187. [CrossRef]
33. Mulder, J.H.; Lavrijsen, P. First results of the "Third International Asparagus Cultivar Trial" planted in Horst, the Netherlands. *Acta Hortic.* **2008**, *776*, 367–372. [CrossRef]
34. Contador, F.F. Asparagus (*Asparagus officinalis* L.) Response to Different Irrigation Regimes and Nitrogen Fertilization Levels. Second Year after Planting. Master's Thesis, University of Chile, Santiago, Chile, 1991; p. 92.
35. Rubio, F.A.H. Irrigation and Nitrogenous Fertilization during the First Establishment Year of an Asparagus Plantation. Master's Thesis, University of Chile, Santiago, Chile, 1992; p. 105.
36. Kaufmann, F. Intensivierung der Spargelproduktion durch Bewasserrung. *Gartenbau* **1977**, *24*, 73–74.
37. Bussell, W.T. Asparagus—Irrigation and establishment methods. *Asparagus Res. Newsl.* **1985**, *3*, 1–21.
38. Hartmann, H.D. Possibility of predicting asparagus yields in central Europe. In Proceedings of the 6th International Asparagus Symposium, Guelph, ON, Canada, 5–9 August 1985; pp. 318–323.
39. Jerez, B.J.A. Differencial Irrigation in Two-Year Old Asparagus (*Asparagus officinalis* L.). Master's Thesis, Univ. de Concepcion, Chillan, Chile, 1990; p. 115.
40. Kaufmann, F. Principles of plant density for green asparagus harvested by different methods. *Acta Hortic.* **1990**, *271*, 227–233. [CrossRef]
41. Ferreyra, E.R.; Peralta, A.J.M.; van Selles, S.G.; Fritsch, F.N.; Contador, F.F. Asparagus crop (*Asparagus officinalis* L.) response to different water regimes during the first two establishment seasons. *Agric. Tec.* **1995**, *55*, 1–8.
42. Dowtown, W.J.S.; Torokvalvy, W. Photosynnthesis in developing asparagus plants. *Aust. J. Plant Physiol.* **1975**, *2*, 367–375.
43. Wilcox-Lee, D. Soil matrics potential, plant water relations and growth in asparagus. *Hort. Sci.* **1987**, *22*, 22–24.
44. Pressman, E.; Schaffer, A.A.; Compton, D.; Zamski, E. The effect of low temperature and drought on carbohydrate content of asparagus. *J. Plant Physiol.* **1989**, *134*, 209–213. [CrossRef]
45. Pressman, E.; Schaffer, A.A.; Compton, D.; Zamski, E. Carbohydrate content of young asparagus plants as affected by temperature regimes. *J. Plant Physiol.* **1994**, *143*, 621–624. [CrossRef]
46. Martin, S.; Hartmann, H.D. The content and distribution of carbohydrates in asparagus. *Acta Hortic.* **1990**, *271*, 443–446. [CrossRef]
47. Roth, R.L.; Gardner, B.R. Asparagus yield response to water and nitrogen. *Trans. Am. Soc. Agric. Eng.* **1989**, *32*, 105–112. [CrossRef]
48. Drost, D.T. Irrigation budget and plant growth of asparagus. *Acta Hortic.* **1996**, *415*, 343–350. [CrossRef]
49. Pinkau, H.; Grutz, E.M. Effect of nitrogen, sprinkler irrigation and length of harvesting period on the yield and quality of green asparagus. *Arch. Gart.* **1985**, *33*, 177–189.
50. Rolbiecki, R.; Rolbiecki, S. Effects of micro-irrigation systems on lettuce and radish production. *Acta Hortic.* **2007**, *729*, 331–335. [CrossRef]
51. Rolbiecki, R.; Rolbiecki, S.; Figas, A.; Jagosz, B.; Stachowski, P.; Sadan, H.A.; Prus, P.; Pal-Fam, F. Requirements and effects of surface drip irrigation of mid-early potato cultivar Courage on a very light soil in Central Poland. *Agronomy* **2021**, *11*, 1–13. [CrossRef]
52. Rolbiecki, S.; Rzekanowski, C. Influence of sprinkler and drip irrigation on the growth and yield of strawberries grown on sandy soils. *Acta Hortic.* **1997**, *439*, 669–672. [CrossRef]
53. Rolbiecki, S.; Rolbiecki, R.; Rzekanowski, C. Response of black currant (*Ribes nigrum* L.) cv. 'Titania' to micro-irrigation under loose sandy soil conditions. *Acta Hortic.* **2002**, *585*, 649–652. [CrossRef]
54. Rolbiecki, S.; Rolbiecki, R.; Rzekanowski, C. Effect of micro-irrigation on the growth and yield of raspberry (*Rubus idaeus* L.) cv. 'Polana' grown in very light soil. *Acta Hortic.* **2002**, *585*, 653–657. [CrossRef]
55. Rolbiecki, S.; Rolbiecki, R.; Rzekanowski, C.; Derkacz, M. Effect of different irrigation regimes on growth and yield of 'Elsanta' strawberries planted on loose sandy soil. *Acta Hortic.* **2004**, *646*, 163–166. [CrossRef]

Article

Evaluating Irrigation and Farming Systems with Solar MajiPump in Ethiopia

Tewodros T. Assefa [1,*], Temesgen F. Adametie [2], Abdu Y. Yimam [1], Sisay A. Belay [1], Yonas M. Degu [3], Solomon T. Hailemeskel [3], Seifu A. Tilahun [1], Manuel R. Reyes [4] and P. V. Vara Prasad [4]

[1] Faculty of Civil and Water Resource Engineering, Bahir Dar Institute of Technology, Bahir Dar University, Bahir Dar 26, Ethiopia; abdukemer62@gmail.com (A.Y.Y.); sisayasress@gmail.com (S.A.B.); sat86@cornell.edu (S.A.T.)
[2] Department of Irrigation and Drainage, Pawe Agricultural Research Center, Ethiopian Institute of Agricultural Research, Pawe 25, Ethiopia; temesgenfentahun09@gmail.com
[3] Faculty of Mechanical and Industrial Engineering, Bahir Dar Institute of Technology, Bahir Dar University, Bahir Dar 26, Ethiopia; yonasmitiku2000@gmail.com (Y.M.D.); soltektata@gmail.com (S.T.H.)
[4] Sustainable Intensification Innovation Lab and Department of Agronomy, Kansas State University, Manhattan, KS 66506, USA; mannyreyes@ksu.edu (M.R.R.); vara@ksu.edu (P.V.V.P.)
[*] Correspondence: ttaffese@gmail.com; Tel.: +251-91-210-0610

Citation: Assefa, T.T.; Adametie, T.F.; Yimam, A.Y.; Belay, S.A.; Degu, Y.M.; Hailemeskel, S.T.; Tilahun, S.A.; Reyes, M.R.; Prasad, P.V.V. Evaluating Irrigation and Farming Systems with Solar MajiPump in Ethiopia. *Agronomy* **2021**, *11*, 17. https://dx.doi.org/10.3390/agronomy11010017

Received: 25 November 2020
Accepted: 21 December 2020
Published: 23 December 2020

Publisher's Note: MDPI stays neutral with regard to jurisdictional claims in published maps and institutional affiliations.

Abstract: Small-scale irrigation in Ethiopia is a key strategy to improve and sustain the food production system. Besides the use of surface water for irrigation, it is essential to unlock the groundwater potential. It is equally important to use soil management and water-saving systems to overcome the declining soil fertility and the temporal water scarcity in the region. In this study, the solar MajiPump was introduced to enable dry season crop production in Ethiopia using shallow groundwater sources. The capacity of the MajiPumps (MP400 and MP200) was tested for the discharge head and discharge using three types of solar panels (150 W and 200 W rigid, and 200 W flexible). Besides, drip irrigation and conservation agriculture (CA) farming systems were evaluated in terms of water productivity and crop yield in comparison to the farmers' practice (overhead irrigation and tilled farming system). Results indicated that the maximum discharge head capacity of the MajiPumps was 18 m, 14 m, 10 m when using MP400 with 200 W rigid, MP400 with 200 W flexible, and MP200 with 150 W rigid solar panels, respectively. The corresponding MajiPump flow rates ranged from 7.8 L/min to 24.6 L/min, 3 L/min to 25 L/min, and 3.6 L/min to 22.2 L/min, respectively. Compared to farmer's practice, water productivity was significantly improved under the CA farming and the drip irrigation systems for both irrigated vegetables (garlic, onion, cabbage, potato) and rainfed maize production. The water productivity of garlic, cabbage, potato, and maize was increased by 256%, 43%, 53%, and 9%, respectively, under CA as compared to conventional tillage (CT) even under overhead irrigation. Thus, farmers can obtain a significant water-saving benefit from CA regardless of water application systems. However, water and crop productivity could be further improved in the combined use of MajiPump with CA and drip irrigation (i.e., 38% and 33% water productivity and 43% and 36% crop productivity improvements were observed for potato and onion, respectively). Similarly, compared to CT, the use of CA significantly increased garlic, cabbage, potato, and maize yield by 170%, 42%, 43%, and 15%, respectively under the MajiPump water-lifting system. Overall, the solar-powered drip irrigation and CA farming system were found to be efficient to expand small-scale irrigation and improve productivity and livelihoods of smallholder farmers in Ethiopia.

Keywords: solar MajiPump; water and crop productivity; small-scale irrigation; conservation agriculture; Ethiopia

1. Introduction

Agriculture has been practiced for centuries and is regarded as the main source of food and income for the rural communities of Ethiopia [1], which accounts for more than

80% of the total population [2]. However, rainfed agriculture has frequently suffered from uneven distribution of rainfall and frequent drought shocks, leading to food insecurity of the poor rural communities [3,4]. In response to such recurrent challenges, small-scale irrigation has been considered as one of the main strategies to alleviate food and income shortages [5,6] and enhance the livelihoods of farmers in Ethiopia [7–12]. Small-scale irrigation often refers to distributed irrigation, small private irrigation, smallholder irrigation, or farmer-led irrigation [13]. In recent years, there is a keen interest in small-scale irrigation due to its cost-effectiveness [14] and sustainable management as compared to large-scale irrigation [13]. It is believed that Ethiopia has more than 6 million hectares of land that is appropriate for small-scale irrigation use [15] and ample water resources suitable for irrigation [15–18]. Nevertheless, irrigated agriculture comprises only 3% of the national food production, using less than 5% of the cultivated land for irrigation [4,19] due to various constraints. Xie et al. [13] depicted that Ethiopia has the potential to add about 1 million ha of land irrigated by small-scale irrigation systems by 2030.

Despite the considerably large potential for irrigation, there are several challenges for the wider adoption of small-scale irrigation in Ethiopia. Some of these challenges include temporal water scarcity [10], poor management of soil and water [7,11,20], lack of water storage facilities, limited opportunities for gravity-fed irrigation, lack of access to irrigation technology, high initial and operation cost of irrigation technologies, and limited capital investments [15,19,21]. On the other hand, limited rainfall and prolonged dry spells entail the need for the efficient use of both surface and groundwater sources, conservation agricultural (CA) practices, efficient water distribution and application systems [22–25]. It is evidenced that in Ethiopia, CA practice provides dual benefits of improved water [11,16,22,25–28] and improved soil conservation [29]. In terms of water application technology, the drip irrigation system is considered the most efficient and water-saving system [30,31]. CA in this study refers to the minimum soil disturbance with no-till practice, year-round organic mulch cover with grass, and diverse cropping in rotation, whereas CT refers to the traditional tillage with no-organic mulch cover and diverse cropping in rotation.

Groundwater is believed to be stable in the face of climate change as compared to surface water and would serve as a source of irrigation [15,32,33]. The role of efficient, labor-saving, and cost-effective water-lifting technologies is vital in unlocking groundwater potential for smallholder farmers [34]. Treadle pumps, rope and washer, pulley, and bucket have been used by smallholders in Ethiopia as a means of water-lifting technologies. However, these technologies are labor-intensive and only just used as a means of water-lifting beyond domestic use (e.g., drinking and cooking), and not for irrigation. Motor pumps (diesel or petrol) have been used by some farmers for irrigation but constrained due to high energy demand, limited access to fuel, and the alarming increase in the cost of fuel, and thus leading to increased risks in irrigated crop production [34]. In some urban areas, electric motor pumps might be feasible and used for urban agriculture. However, electricity access is rare for the rural community of Ethiopia [35]. In response to such challenges, several researchers suggested the use of solar pumps due to their high labor productivity, environmental sustainability, and use of renewable energy sources [36–38]. Ethiopia, as a tropical region, has ample solar energy [39,40] that can be captured for water lifting and pumping systems.

The MajiPump is a solar-powered water-lifting technology that was introduced in Ethiopia in 2017 by the Appropriate Scale Mechanization Consortium of the Feed the Future Sustainable Intensification Innovation Lab (SIIL). The solar MajiPump, a submersible pump, uses solar energy to extract water from wells and surface ponds. A solar panel is connected to the MajiPump by an electric cable driven by the direct current (DC). However, the discharge head and discharge capacity of the MajiPump is not known beyond the company specification. Evaluating the use of these pumps under field conditions and their impact on crop yields is critical for scaling and adoption of these technologies. Widescale use of efficient water applications in combination with improved crop and soil water management technologies are vital for income generation and increase resilience in the face of climate change, and to reverse the decline of soil fertility. Such systems need to be tested for both vegetable production systems, which are becoming more popular due to demand for vegetables from urban markets and for high-value grain crops such as maize (*Zea mays* L.). Smallholder vegetable production is considered as a strategic approach to minimize children's death and stunting caused by malnutrition, which is a serious challenge in Ethiopia [41], by providing healthy and nutritious diets. Thus, the objectives of this study were to evaluate small-scale irrigation package: (1) MajiPumps (MP400 and MP200) for its discharge head and discharge capacity with different solar panels; (2) drip water application system with the common farmer's overhead irrigation practice using hose; and (3) CA with farmer's conventional tillage (CT) practice; in terms of water productivity and yields of key crops [garlic (*Allium sativium* L.), onion (*Allium cepa* L.), cabbage (*Brassica oleracea* L. var. captata), potato (*Solanum tuberosum* L.), and maize]. The results from this study would assist decision-makers and other stakeholders in scaling small-scale irrigation technologies and exploring groundwater potential in Ethiopia.

2. Materials and Methods

2.1. Study Area

This study was conducted in the central Ethiopian highlands at two experimental sites: Affesa in Dangila district and Alefa in Bure district (Figure 1). Affesa (36.83° N, 11.25° E) and Alefa (37.06° N, 10.62° E) sites are located about 80 km and 150 km southwest of Bahir Dar, respectively. The elevation of Affesa site ranges from 2132 to 2219 m above MSL, whereas Alefa site elevation ranges from 1983 to 2033 m above MSL. Both Dangila and Bure districts are categorized under moist sub-tropical regions. Dangila has an average annual rainfall of 1578 mm and a mean annual temperature of 17 °C [42], whereas, the mean annual rainfall and temperature in Bure ranges from 1386 mm to 1757 mm and 14 °C to 24 °C, respectively [43]. Based on soil laboratory analysis, clay soil is the dominant soil texture in Affesa (46% clay and 36% silt) and the soil type in Alefa is dominated by loam soil (44% sand and 29% silt). The dominant rainfed crops in both Dangila and Bure districts include maize, millets (sorghum: *Sorghum bicolor* L., or pearl millet: *Panicum sp.*), barley (*Hordeum vulgare* L.), teff [*Eragrostis tef* (Zucc.) Trotter], and wheat (*Triticum aestivum* L.), [44]. While, onion, potato, cabbage, pepper, tomato, and garlic are the dominant irrigated vegetables in both Dangila and Bure district [22,45]. In the Affesa site, farmers used to practice irrigation using river sources, whereas groundwater use was limited to domestic purposes in both sites due to lack of access to affordable water-lifting technology.

2.2. Experimental Design

The experiment was laid as a paired t-design to compare the effects of conservation practices and irrigation systems. The paired t-test is mathematically powerful in comparing two-paired measurements that have intrinsic relationships and allows good control of individual differences without necessarily having a large sample size [46]. De Winter [47] proved the applicability of paired t-test as low as two replicates. Several studies, including Yimam et al. [22], Belay et al. [26], and Assefa et al. [11] have used paired-t design for similar purposes. The experimental design and setup were described for each site (i.e., Alefa and Affesa) separately as shown below.

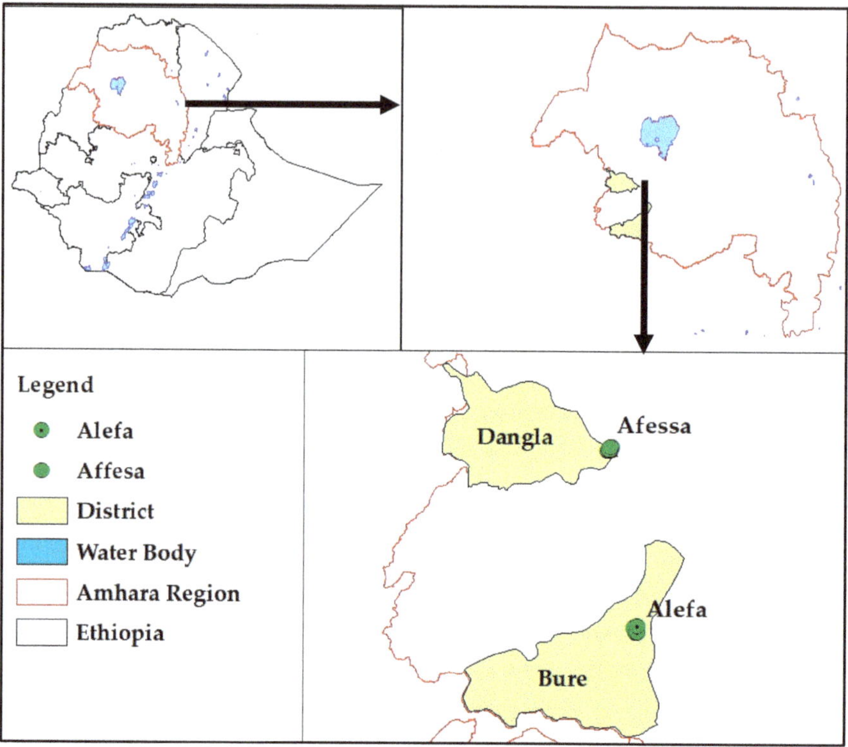

Figure 1. Location map of experimental sites in the northern part of Ethiopia.

Alefa site

CA was the treatment and CT was the control in which each participant was involved in both (CA and CT) practice with overhead irrigation (hose). A total of 10 replicates (i.e., farmers) were used. The size of each plot was 100 m^2 (Figure 2a), which then equally divided; 50 m^2 for each management (i.e., randomly assigned to CA and CT). The 100 m^2 plot has 10 beds with 30 cm furrows in between (i.e., 70 cm by 10 m bed size and 30 cm by 10 m furrow size). The experimental plots are not evenly distributed, approximately the distance between the plots ranges from 300 m to more than 1 km. CA was evaluated against CT for their impacts on water productivity, crop growth characteristics, and yield.

Affesa site

Drip irrigation was the treatment, and overhead irrigation (hose) was the control in which each participant was involved in both drip and overhead irrigation under CA practice (Figure 2b). A total of 10 replicates (i.e., farmers) were used each having 100 m^2 plot size. The size of each plot was equally divided; 50 m^2 for each management (i.e., randomly assigned to drip and overhead irrigation). Similar to Alefa, the 100 m^2 plot has 10 beds with 30 cm furrows in between. In this case, drip irrigation was evaluated against overhead irrigation for their impacts on water productivity, crop growth characteristics, and yield. Besides, maize-forage vetch (*Vicia* sp.) inter-cropping (500 m^2 in size as shown in Figure 2c) was introduced during the rainy season (i.e., 250 m^2 with CA and another 250 m^2 with CT practice at random) to provide an alternative source of mulch for conservation practice and simultaneously evaluate the effect of CA on rainfed maize productivity.

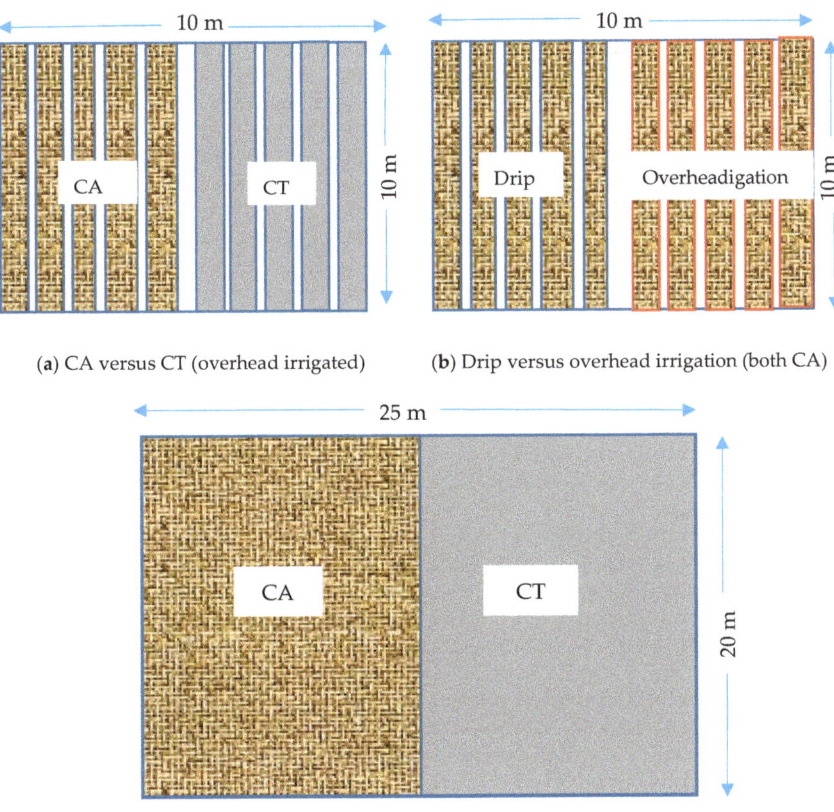

(**a**) CA versus CT (overhead irrigated)   (**b**) Drip versus overhead irrigation (both CA)

(**c**) CA versus CT under (rainfed)

Figure 2. Experimental design: (**a**) conservative agriculture (CA) versus conventional tillage (CT) both under overhead irrigation using hose, (**b**) drip versus overhead irrigation using hose both under CA, (**c**) CA versus CT intercropped with forage vetch under rainfed maize production

A total of 20 farmers participated in this on-farm experiment: 10 farmers in Alefa and 10 farmers in Affesa. A series of discussions were conducted with the local government and community leaders to select potential farmers for this research. Farmer's willingness to participate in this research was confirmed through a focus group discussion. The availability of shallow groundwater, a 100 m^2 plot for vegetable production, and a 500 m^2 plot for maize-forage production that was close to the household or at a walking distance was considered as additional criteria as a home gardening principle [48] to identify potential farmers. On the 100 m^2 vegetable plot, a total of 10 beds (70 cm by 10 m) were prepared with 30 cm furrows in between. Farmers produced various vegetables (garlic, onion, cabbage, and potato) in the dry season (2018 to 2020) with irrigation. Each farmer used a solar MajiPump to extract water from shallow groundwater well to an elevated (about 1.5 m high from the ground) water storage tank (1000 L in size). Water was then applied to the plots from the water storage tanks through gravity using the drip system or overhead using a hose, depending on the experimental design. In the rainy season, farmers grew maize, and then inter-crop forage vetch after the maize reaches the maturity stage. The vetch forage production uses partly rainfall and then residual moisture from the rainy season. The variety of seeds for the vegetables, maize, and forage vetch was the same for all farmers.

2.3. Soil Physico-Chemical Properties

Soil samples were taken from both Alefa and Affesa sites before the intervention of our treatments mainly to observe if there was variability between experimental plots for a paired t-design. Five sampling plots were randomly selected from a total of 10 plots at each site. Considering the maximum root depth of the various crops grown in the study sites (i.e., onion, garlic, potato, cabbage, and maize) in Yimam et al. [22], Iwama [49], and Gao, et al. [50], and the soil layer classification in Westerveld, et al. [51] and Hsu, et al. [52], soil samples were collected from three depths (i.e., 0–30 cm, 30–60 cm, and 60–90 cm). A total of 30 soil samples were collected: 15 samples from each experimental site. Soil laboratory analysis was conducted at Amhara Design and Supervision Works Enterprise (ADSWE) to determine the various physio-chemical properties (i.e., field capacity, permanent wilting point, soil texture, available organic matter, pH, total N, available P, and available K). A representative soil mass of about 1 kg was sampled from each plot. Soil texture was measured using hydrometer, while field capacity and wilting point were determined using pressure (porous) plate apparatus. The details of each laboratory analysis and approach used by ADSWE can be found from Tesema et al. [53]. Coefficient of variation was calculated from the 15 soil samples at each experimental site (i.e., Alefa and Affesa) for each soil physio-chemical properties.

2.4. Climate Data

Climatic data (rainfall, maximum and minimum temperature, wind speed, sunshine hour, relative humidity) were collected from the nearby meteorological station, Dangila for Affesa site, and Bure for Alefa site. The CROPWAT 8 model was used to estimate the reference evapotranspiration (ETo) using the Penman-Monteith method [54]. Mean monthly rainfall reaches its maximum value in July (401 mm in Affesa and 301 mm in Alefa), and mean month evapotranspiration reaches its maximum value in April (112 mm in Affesa and 128 mm in Alefa) as shown in Figure 3. Spatial and temporal rainfall variability is high demanding irrigation to prevent crop failure and increase the cycle of crop production [11,55]. Irrigation use for dry season vegetable production in the study sites ranges from October to June.

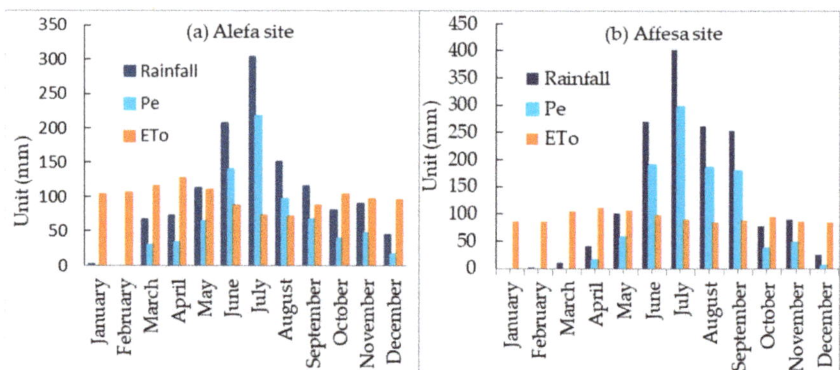

Figure 3. Climatic characteristics (rainfall; Pe, effective rainfall; ETo, reference crop transpiration) at the research sites of (**a**) Alefa and (**b**) Affesa (2018–2020).

Effective rainfall (*Pe*) was determined using the United States Department of Agriculture Soil Conservation Service (USDA-SCS) method [56,57] as shown in (Equation (1)). The effective rainfall was used later to determine water productivity for each crop under both irrigated and rainfed systems.

$$\left(\begin{array}{ll} Pe = 0.8P - 25 & P > 75 \; mm/month \\ Pe = 0.6P - 10 & P < 75 \; mm/month \end{array} \right) \tag{1}$$

where *Pe* and *P* are effective rainfall and precipitation in mm/month, respectively.

2.5. Agronomic Data

Garlic, cabbage, potato, onion, and maize crops were grown in the study sites. Farmers at each experimental site (i.e., Alefa or Affesa) grew the same crop during each cropping cycle. Activities including planting, mulch application, fertilizer, and pesticide application and crop harvest information were monitored for each crop during each cropping cycle (Table 1). On average, 1 kg m^{-2} of dried grass mulch was applied for CA plots. Farmers applied Urea (46-0-0: N-P-K) fertilizer at a rate of 100 kg ha^{-1} for irrigated onion and potato. For rainfed maize, 240 kg ha^{-1} of DAP (18-46-0: N-P-K), 500 kg ha^{-1} of Urea fertilizer, and 5 L ha^{-1} of Diazinon 60% chemical were applied. Crop characteristics such as plant height, potato tuber diameter, garlic and onion bulb diameter, cabbage head diameter, and crop/vegetable yield were recorded. Plant height was monitored every week. Measuring tape and Caliper were used to measure plant height and diameter, respectively. The digital balance was used to determine weight crop yield. A paired t-test was used at a 5% significance level to determine the effects of management practices on all crop characteristics.

2.6. Solar MajiPump and Its Applicability

In this study, two types (i.e., MP400 and MP200) of submersible brushless DC motor MajiPumps were used. Both MP400 (1.8 kg of mass) and MP200 (1.5 kg of mass) pumps have similar looks from the outside (Figure 4a). The basic difference between the two pumps is the water-lifting capacity; MP400 could lift to 25 m discharge head using a 160 W (24 VDC) panel whereas the MP200 pump could lift to 10 m using an 80 W (12 VDC) panel. Both pumps could deliver 34 L/min flow rate for open flow. Two types of solar panels were used for the MP400 pump (monocrystalline 200 W rigid panel and two flexible panels of 200 W connected in series), and a monocrystalline 150 W rigid panel was used for the MP200 pump. The monocrystalline 200 W rigid panel is 1.58 m by 0.85 m in size with 12.7 kg total mass (Figure 4b). Whereas the monocrystalline 150 W rigid panel is 1.48 m by 0.67 m in size with 10.22 kg total mass (Figure 4d). The thin film (amorous) flexible 100 W solar panel (Figure 4d) is 1.05 m by 0.54 m size and 1.4 kg of mass each.

Table 1. Crop rotation and management activities at Affesa and Alefa experimental sites.

Site	Vegetable	Management Activity	Date
Alefa	Irrigated Garlic (1st cycle)	Plot preparation	5 January 2019
		Mulch application [2]	14 January 2019
		Planting	16 January 2019
		Harvest	11 May 2019
	Irrigated Cabbage (2nd cycle)	Tillage [1]	6 October 2019
		Mulch application [2]	19 October 2019
		Transplanting	20 October 2019
		Harvest	25 February 2020
	Irrigated Potato (3rd cycle)	Tillage [1]	28 February 2020
		Mulch application [2]	3 March 2020
		Planting	5 March 2020
		Harvest	30 June 2020
Affesa	Irrigated Potato (1st cycle)	Plot preparation	5 February 2019
		Mulch application [2]	10 February 2019
		Planting	11 February 2019
		UREA[3] application	13 March 2019
		Harvest	20 May 2019
	Irrigated Onion (2nd cycle)	Tillage [1]	10 December 2019
		Mulch application [2]	14 December 2019
		Planting	15 December 2019
		URAE [3] application	26 January 2020
		Harvest	25 March 2020
	Rainfed Maize	Plot preparation	27 March 2019
		Mulch application [2]	15 May 2019
		Planting	23 May 2019
		DAP [3] application	23 May 2019
		UREA [3] application	14 June 2019
		Diazinon 60% [4] application	7 July 2019
		Forage inter-cropping	21 September 2019
		Maize harvest	18 October 2019
		Forage harvest	21 November 2019

Note [1] Only for CT plots; [2] only for CA plots; [3] Fertilizer; [4] Pesticide.

Figure 4. Pictures; (**a**) MajiPump, (**b**) 200 W rigid solar panel, (**c**) 150 W rigid solar panel, (**d**) 2 × 100 W flexible solar panel, and (**e**) solar meter. The 200 W rigid and 2 × 100 W flexible panels were connected in series with MP400 pumps, whereas the 150 W panel was connected with an MP200 pump.

The MajiPumps were tested using the different panels for their discharge head and flow rate capacity. Digital Tenmars TM-207 (Figure 4e) solar power meter (0.1 W/m^2 resolution) was used to measure solar intensity. The MP400 and MP200 pumps were immersed in the water after a $\frac{3}{4}$ inch high-density polyethylene (HDPE) pipe was fitted and a digital stopwatch was used to record the time taken for a specific volume of water. The analog water meter was used to measure flow volume at different discharge head. A tape measure was used to measure water level depth (head) where the pump is installed. When the solar panel and pump setup, a connection cable was used to connect the pump with the panel and supply power. At this instant, where the pump starts to run, both the water meter and stopwatch start recording. The pump runs until the amount of water yield reaches 100L and the flow rate was determined (L/S). The maximum discharge head and flow rate were considered as the capacity of the pumps (MP400 and MP200). At the experimental sites (Alefa and Affesa), groundwater depth from the surface was monitored throughout the year to compare it with the MajiPumps discharge head capacity and determine the applicability of pumps in the study area.

2.7. Water and Crop Productivity Data Analysis

Farmers could decide the irrigation interval and amount based on their field observation on soil moisture. Farmer's water application practice (i.e., application dates and amounts) were recorded from each plot after every irrigation based on the availability of water in the fixed water storage tanks (1000 L) for the dry season production. Crop yield (Y) was measured as weight during harvest separately for each soil and water management (i.e., CA, CT, and overhead and drip irrigation). Water productivity (WP), the amount of yield per unit volume of water [58], was computed as a quotient of crop yield and amount of water applied (irrigation and effective rainfall) as shown in Equation (2). The effects of management practices on irrigation water use, crop yield, and water productivity were analyzed using a paired t-test at a 5% significance level. Besides, the variability of forage production among participant farmers due to effort and commitment was analyzed using the coefficient of variation.

$$\left(WP = \frac{Y}{I + Pe} \right) \tag{2}$$

where WP, Y, I, Pe are water productivity (kg/m^3), yield (kg), irrigated water (m^3), and effective rainfall (m^3), respectively.

3. Results

3.1. Soil Properties Across Experimental Plots

Several soil physio-chemical properties were analyzed at various soil depths (i.e., 0–30 cm, 30–60 cm, and 60–90 cm) to check variability across experimental plots (Table 2). Gallardo [59] explained that variability of soil properties can be best described using the coefficient of variation (CV); high variability when the CV is greater than 91% and low variability if otherwise. Based on CV analysis, the variability of soil properties was low across experimental plots at both sites (i.e., Alefa and Affesa), satisfying precondition for paired management comparisons. The soil class in Alefa site is clay loam (0 to 30 cm) and clay (30 to 90 cm), whereas it is clay (0 to 60 cm) and silt clay (60 to 90 cm) in Affesa site. Soil salinity was generally low in the highlands of Ethiopia where the experimental sites are located [60].

Table 2. Mean value of soil physio-chemical properties across experimental plots for Alefa and Affesa.

Site	Soil Properties	No.	Soil Depth							
			0–30 cm				30–60 cm		60–90 cm	
			Mean	Max.	Min.	CV	Mean	CV	Mean	CV
Alefa	pH	H_2O	5.7	6.4	5.1	9.7	5.7	3.9	5.2	2.3
	Texture	% sand	43.8	61	27	28.3	19.0	33.3	23.8	45.5
		% silt	28.6	31	27	5.9	23.8	14.1	22.6	40.3
		% clay	27.6	44	12	43.6	57.2	16.3	53.6	36.5
	OC	%	3.1	4.5	2.1	29.2	1.7	22.6	1.5	37.2
	OM	%	5.4	5.4	7.7	29.1	2.9	22.7	2.6	37.2
	TN	%	0.3	0.4	0.2	30.2	0.1	21.4	0.1	39.5
	Av. P	ppm	36.9	71.8	11.1	71.7	11.1	31.5	4.6	112
	Av. K	ppm	127.0	163	100	18.4	96.3	28.0	83.0	43.7
	FC	%	30.2	31	29	3.7	29.7	4.4	35.9	6.8
	PWP	%	16.5	21.5	11.1	25.7	13.7	7.0	21.4	4.8
Affesa	pH	H_2O	4.6	5	4.2	7.8	4.55	4.9	4.6	3
	Texture	% sand	19	22	14	18.2	19.5	30.6	13.0	19.9
		% silt	35.5	50	26	31.0	18.5	29.8	20.0	21.6
		% clay	45.5	55	36	19.5	62.0	7.4	67.0	9.9
	OC	%	2.5	2.7	2.3	7.8	1.8	35.7	2.1	24.8
	OM	%	4.3	4.6	3.9	7.8	3.1	35.9	3.6	25.0
	TN	%	0.2	0.2	0.18	20.7	0.1	31.1	0.2	21.6
	Av. P	ppm	8.6	13	3.6	44.1	3.7	11.5	4.7	25.5
	Av. K	ppm	41.6	56	28.5	27.3	38.4	30.2	33.8	30.5
	FC	%	27.6	30	27	1.5	27.6	1.0	26.8	3.4
	PWP	%	17.2	18	16	3.7	16.9	2.1	17.0	4.3

Note: CV, Max., Min., OC, OM, TN, Av. P, Av. K, FC, PWP are coefficient of variation, maximum, minimum, organic carbon, organic matter, total nitrogen, available phosphorous, available potassium, field capacity, and permanent wilting point, respectively. Maximum and minimum values of soil properties were provided for the topsoil (0–30 cm).

3.2. MajiPumps Capacity and Groundwater Depth in the Study Area

The MajiPumps (MP400 and MP200) discharge head and discharge relations of the different solar panels (i.e., 200 W flexible, 200 W, and 150 rigid) best fitted ($R^2 > 0.8$) with the natural logarithmic function (Figure 5). The MajiPumps were tested at the various solar intensity in the study region ranging from 1026 W/m^2 to 1230 W/m^2. The maximum discharge heads for MP400 with 200W rigid panel (Figure 5a), MP400 with 200 W flexible panel (Figure 5b), and MP200 with rigid panel (Figure 5c) were observed to be 18 m, 14 m, and 10 m, respectively. The minimum water yields capacity of the pumps from the shallow groundwater wells at the point of maximum discharge heads (i.e., 18 m, 14 m, and 10 m) were found 0.13 L/S, 0.05 L/S, and 0.06 L/S, for MP400 with 200W rigid panel, MP400 with 200 W flexible panel, and MP200 with 150 W rigid panels, respectively. The maximum pumping discharges for MP400 with 200 rigid panels, MP400 with 200 W flexible panel, and MP200 with 150 W rigid panel were found 0.41 L/S, 0.25 L/S, and 0.37 L/S, respectively.

Groundwater depth at Alefa and Affesa sites were monitored (Figure 6) throughout the year to explore potentials for irrigation using the solar MajiPump water-lifting technology. Water levels were measured from the second week of March 2019 to the first week of August 2020 for Alefa and from the last week of February 2019 to the first week of July 2020. The depth of groundwater level in the shallow well ranges from 1.6 m (in October) to 9 m (last week of April) at Alefa site, whereas it ranges from 2 m (first week of August) to 12 m (first week of April) at Affesa site.

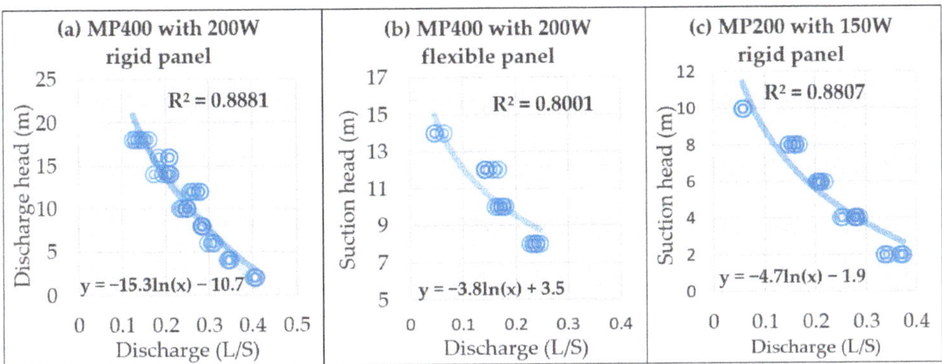

Figure 5. Discharge head and discharge curves of MajiPumps for solar intensity ranges in the study region (1026 to 1230 W/m²).

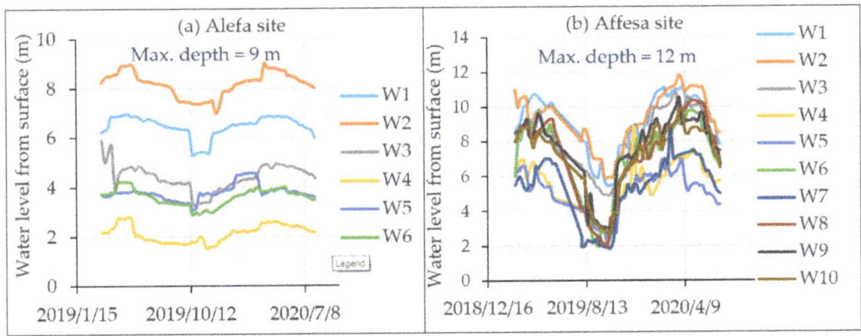

Figure 6. Groundwater depth from surface at Alefa (**a**) and Affesa (**b**) sites.

3.3. Effects of Farming and IrrigationSystems on Crop Growth Characteristics

A one-tailed paired t-test was conducted to analyze plant height for garlic, cabbage, potato, onion, and maize (Table 3). The mean plant height was significantly improved under CA for garlic and potato, whereas the improvements for cabbage and maize height was not statistically significant. The mean plant height of garlic and potato were improved under CA, respectively, by 17% and 7% when compared with the CT. Drip irrigation significantly improved potato height by 8% when compared with the overhead water application system, whereas the improvement of onion height was not statistically significant.

Similarly, a one-tailed paired t-test was used to analyze potato tuber diameter, cabbage head diameter, garlic, and onion bulb diameter under the different soil management practices and irrigation systems (Table 4). Both garlic bulb diameter and cabbage head diameter were significantly improved under CA when compared with CT, which were increased by 35% and 26%, respectively, under CA. Potato tuber diameter was significantly improved under the drip water application when compared with the overhead water application system, whereas the improvement for the onion bulb diameter was not statistically significant. The mean potato tuber diameter was improved by 23% under drip irrigation when compared with under the overhead system.

Table 3. Mean value of plant height under different tillage (CA versus CT) and water management (drip versus overhead water application) systems.

Management	Statistics	Plant Height (cm)			
		Garlic	Cabbage	Potato	Maize
Sample size	**N**	**120**	**90**	**90**	**90**
CA	Mean	44.8	30.6	44	293.4
	Max.	55	56	52	327
	Min.	13	12	12	118
CT	Mean	38.4	29.3	41.3	292.3
	Max.	50	37	47	325
	Min.	8	13	9	116
CA ǀ CT	SEM±	1.95 ǀ 2.2	0.7 ǀ 0.7	3 ǀ 2.3	16.9 ǀ 17.3
	p-value	0.0005 ***	0.3	0.02 **	0.19
		Potato	Onion		
Sample size	**N**	105	105		
Drip	Mean	34.1	49		
	Max.	46.4	67		
	Min.	17.5	5		
Overhead	Mean	31.7	48		
	Max.	43.9	65		
	Min.	15.9	4		
Drip ǀ Overhead	SEM±	4.8 ǀ 4.9	0.63 ǀ 0.93		
	p-value	0.00003 ***	0.09		

Note: N, SEM, Max., Min., **, *** are sample size, standard error of the mean, maximum, minimum, and significance at $p < 0.05$ and $p < 0.001\%$, respectively.

Table 4. Mean value of bulb, tuber, and cabbage head diameter different tillage (CA versus CT) and water management (drip versus overhead water application) systems.

Management	Statistics	Diameter (cm)	
		Garlic Bulb	Cabbage Head
Sample size	**N**	**120**	**90**
CA	Mean	3.5	9.3
	Max.	5	11.6
	Min.	1.9	7.2
CT	Mean	2.6	7.4
	Max.	4	9.5
	Min.	1.5	5.6
CA ǀ CT	SEM±	0.18 ǀ 0.17	0.6 ǀ 0.5
	p-value	0.001 ***	0.0004 ***
		Potato tuber	Onion bulb
Sample size	**N**	105	105
Drip	Mean	3.8	4.0
	Max.	5.2	6
	Min.	1.95	2
Overhead	Mean	3.1	3.8
	Max.	5.34	6
	Min.	1.9	2
Drip ǀ Overhead	SEM±	0.18 ǀ 0.09	0.13 ǀ 0.26
	p-value	0.002 **	0.2

Note: N, SEM, Max., Min. **, *** are sample size, standard error of the mean, maximum, minimum, and significance at $p < 0.01$ and $p < 0.001$, respectively.

3.4. Direct Effects of Farming and Irrigation Systems on Water Use and Crop Yield

A one-tailed paired t-test was used to analyze the impacts of soil and water management practices on the amount of water applied (irrigation plus effective rainfall), Table 5. The mean water use was significantly reduced ($\alpha = 5\%$) under CA for all vegetables (Table 5). The mean water uses of garlic, cabbage, and potato mean water uses were reduced, respectively, by 18%, 8%, and 9% under CA when compared to CT. However, the water use difference was not statistically significant ($\alpha = 5\%$) between drip and overhead water application systems, both under CA practice.

Table 5. The mean value of crops totals water uses under different tillage (CA versus CT) and water management (drip versus overhead water application) systems.

Management	Statistics	Water Use (mm)		
		Garlic	Cabbage	Potato
	N	8	6	6
	Mean	316	380	294
CA	Max.	425	497	304
	Min.	267	294	281
	Mean	386	414	323
CT	Max.	435	301	331
	Min.	308	549	313
CA I CT	SEM±	18.7 I 16.3	33.4 I 38.7	4.5 I 3.6
	p-value	0.0007 ***	0. 0002 ***	0.0003 ***
		Potato	Onion	
	N	7	7	
	Mean	341	247	
Drip	Max.	374	265	
	Min.	295	204	
	Mean	350	246	
Overhead	Max.	398	270	
	Min.	304	207	
Drip I Overhead	SEM±	10.3 I 11.7	5.9 I 6.0	
	p-value	0.21	0.5	

Note: N, SEM, Max., Min., *** are sample size, standard error of the mean, maximum, minimum, and significance at $p < 0.001$, respectively.

Crop yields were significantly increased under CA and drip irrigation systems (Table 6) for all crops (i.e., irrigated vegetables and rainfed maize production). The mean crop yields of the garlic bulb, fresh cabbage, potato tuber, and maize grain were increased by 170%, 42%, 43%, and 15% under CA when compared with CT, respectively, though water applications were significantly reduced under CA practice. Similarly, the mean crop yields of potato tuber and onion bulb were significantly increased under the drip irrigation system when compared with the overhead water application using a hose, though both irrigation systems were under CA practice. Potato and onion yields were increased by 43% and 36%, respectively, under drip water application when compared to overhead water application using a hose. On the other hand, farmers were able to harvest from about 5 t ha^{-1} to 12.5 t ha^{-1} of forage biomass beside 7.2 t ha^{-1} of maize production without using irrigation

3.5. Effects of CA and Irrigation System on Water Productivity

Water productivity was found significantly increased under CA and drip irrigation systems for all crops during both irrigated and dry season production (Table 7). The mean water productivity of garlic, cabbage, potato, and maize was increased, respectively, by 256%, 43%, 53%, and 9% under CA when compared with CT. Similarly, water productivity of potato and onion was significantly increased under a drip irrigation system when compared with an overhead system, both irrigation systems were under CA practice. The mean water productivity of potato and onion was increased by 38% and 33% under, respectively, under the drip water application system when compared with overhead water application using a hose.

Table 6. The mean value of crop yields under different tillage (CA versus CT) and water management (drip versus overhead water application) systems.

Management	Statistics	Crop Yield (t ha^{-1})			
		Garlic Bulb	Cabbage Fresh	Potato Tuber	Maize Grain
Sample size	N	8	6	6	6
CA	Mean	10	78	38.6	8.3
	Max.	17	95	50	10
	Min.	3.55	62.5	30	6.3
CT	Mean	3.7	55	27	7.2
	Max.	6.5	72.5	35	9.5
	Min.	1	45	20	5
CA | CT	SEM±	1.5 | 0.77	4.5 | 4.3	2.8 | 2.3	0.54 | 0.63
	p-value	0.0003 ***	0. 007 ***	0.00008 ***	0.001 ***
		Potato Tuber	Onion Bulb		
Sample size	N	7	7		
Drip	Mean	38.6	9.1		
	Max.	43.9	11.2		
	Min.	21.9	6		
Overhead	Mean	27	6.7		
	Max.	35.3	9		
	Min.	18	3.6		
Drip | Overhead	SEM±	3.5 | 2.1	0.58 | 0.61		
	p-value	0.016 *	0.0001 ***		

Note: N, SEM, Max., Min., *, *** are sample size, standard error of the mean, maximum, minimum, and significance at $p < 0.05$, and $p < 0.001$%, respectively.

Table 7. Mean value of water productivity under different tillage (CA versus CT) and water management (drip versus overhead water application) systems.

Management	Statistics	Water Productivity (kg m^{-3})			
		Garlic Bulb	Cabbage Fresh	Potato Tuber	Maize Grain
Sample size	N	8	6	6	6
CA	Mean	3.2	20.4	13.3	1.2
	Max.	4.6	24.9	16.4	1.5
	Min.	1.3	16	10	0.9
CT	Mean	1	14.3	8.7	1.1
	Max.	1.7	20.8	10.6	1.5
	Min.	0.3	9.8	6	0.7
CA | CT	SEM±	0.42 | 0.18	1.6 | 1.8	3.3 | 1.8	0.08 | 0.09
	p-value	0.0001 ***	0. 003 **	0.002 **	0.001 ***

Table 7. *Cont.*

Management	Statistics	Water Productivity (kg m^{-3})			
		Garlic Bulb	Cabbage Fresh	Potato Tuber	Maize Grain
		Potato Tuber	Onion Bulb		
Sample size	N	7	7		
Drip	Mean	9.8	3.6		
	Max.	13.9	4.7		
	Min.	5.8	2.5		
Overhead	Mean	7.1	2.7		
	Max.	9.5	3.9		
	Min.	3.8	1.6		
Drip	Overhead	SEM±	3.5	2.1	0.58	0.61		
	p-value	0.016 *	0.0001 ***		

Note: N, SEM, Max., Min., *, **, *** are sample size, standard error of the mean, maximum, minimum, and significance at $p < 0.05$, $p < 0.01$ and $p < 0.001$, respectively.

4. Discussion

4.1. Evaluation of Solar MajiPumps

On the specification of the MajiPumps, MP400 and MP200 would lift water up to 25 m (using 160 W 24VDC solar panel) and 10 m (using 80 W 12 VDC solar panel), respectively, both having 34 L/min open flow rate. During our field experiment, we observed that MP400 and MP200 pumps would lift water to a maximum of 18 m (using 200 W 24 VDC solar panel) and 10 m (using 150 W 12 VDS solar panels). The pipe maximum flow (3/4 inch in size) was found 24.6 L/min for MP400 with 200 W rigid panel and 22.4 L/min with a 150 W rigid panel. Various factors including solar intensity, panel types, and surrounding temperature could affect pump discharge head and discharge.

The minimum discharges from the shallow groundwater wells at the point of maximum discharge heads (i.e., 18 m, 14 m, and 10 m) were 7.8 L/min, 3 L/min, and 23.6 L/min, for MP400 with 200 W rigid panel, MP400 with 200 W flexible panel, and MP200 with 150 W rigid panel, respectively. This minimum pump discharges could fill the 1000 L water storage tanks that farmers used for this experiment in 2.1 hr., 5.6 hr., and 4.6 hr., respectively. When the solar MajiPumps (i.e., MP400 with 200 W rigid panel, MP400 with 200 W flexible panel and MP200 with 150 W rigid panel) provides the maximum water yield capacity (i.e., 24.61 L/min, 15 L/min, and 22.2 L/min), they could fill the 1000 L water storage tanks in 0.68 hr., 1.11 hr., and 0.75 hr., respectively. Considering 8-h effective solar intensity in a day, MP400 and MP200 pumps would lift a maximum of about 11,764 L/day and 10,666 L/day, respectively. This would help to provide irrigation between half and one hectare of land depending on crop types, farming systems, water application systems, and cropping season.

4.2. Effects of CA on Water Productivity, and Crop Yields

The water-saving capacity of CA was found significantly higher when compared to CT for the various irrigated vegetables and rainfed maize production. This was mainly due to a reduction of water loss from soil evaporation associated with the grass mulch cover in CA. Consequently, soil moisture would be maintained and available for crop use in the CA practice. Assefa et al. [28] found up to 49% reduction of evapotranspiration and up to 40% increment of soil moisture in CA practice for various vegetables, supporting the claim that reduction of water loss is mainly from reduced soil evaporation. Significant improvement of water productivity associated with CA practice was observed in the Ethiopian highlands [13,27,28]. CA was tested with drip irrigation previously [11,16,22, 25–28] showed a significant increase in water productivity. Similarly, in our study CA under both a drip irrigation system and overhead irrigation significantly increased water productivity. Thus, farmers who could not afford to buy a drip irrigation system would still get a significant water-saving benefit from the use of CA, even with overhead irrigation.

Crop yield was found significantly higher in CA when compared to CT for the various irrigated vegetables and rainfed maize production. This was mainly due to an improvement in soil quality (nutrients) and water use efficiency in CA. Assefa, Jha, Reyes, Worqlul, Doro, and Tilahun [29] found more than 6% and 4% increment of soil organic C and total N, respectively, under CA when compared with CT. Besides, CA decreased nutrient loss due to either runoff or percolation. Belay et al. [26] found a significant decrease of NO_3-N (up to 44%) and PO_4-P (up to 50%) in runoff and leachate under CA as compared to CT. This provides more readily available nutrients in CA for plant growth, leading to improved crop yields. Crop yield improvements in CA for this study (15–170%) were found to be consistent with Assefa et al. [11] and Belay et al. [26], which showed 9% to more than 100% yield improvements in CA with the drip irrigation system. This indicates, CA would still provide a significant improvement of soil quality and crop yield regardless of the irrigation practice (overhead or drip system).

4.3. Effects of Drip Irrigation on Water Productivity and Crop Yield

A significant water saving was observed under the drip system (38% for potato and 33% for onion), mainly due to the capability of the drip system in delivering water uniformly to the root of crops and minimizing water losses. Drip irrigation systems reduce water application to open ground or soil spaces that are not directly used by crops, which would rather facilitate weed growth. The result suggests that the combined use of drip irrigation and CA provides a significant water use efficiency as compared to the combination of CA and overhead irrigation. Assefa et al. [16] found nearly a threefold water saving capacity when combined with CA as compared to overhead irrigation with the tilled system. This will help minimize the overexploitation of shallow groundwater wells and maximize irrigated crop production. A similar result was reported by Kigalu, et al. [61] which found a quadratic response of water productivity for the drip system as compared to overhead irrigation in Tanzania.

The effect of the drip irrigation system was significant in improving crop productivity as compared to overhead irrigation. Potato and onion yield was increased by 43% and 36%, respectively under drip irrigation. Dawit, et al. [62] reported similar results, improved crop yields for drip irrigation in the eastern part of Ethiopia. The uniform water application and minimum soil nutrient loss associated with the drip system would be the main reason for improved crop yield. Fandika, et al. [63] result indicated a higher tomato yield response associated with the uniform water application in the drip system. Whereas, Elhindi, et al. [64] and Mirjat, et al. [65] observed minimum loss of soil minerals, and fertilizers when using drip irrigation.

4.4. Comparison of MajiPump with Previous Pulley Studies on Water Productivity

Water productivity was significantly improved under MajiPump water-lifting system when compared with a pulley system [11,22,29] for the same crop types (Figure 7). The water productivity values of garlic, onion, and cabbage under the MajiPump with CA were 3.2 kg m^{-3}, 9 kg m^{-3}, and 20 kg m^{-3}, respectively, while the corresponding values were 1.1 kg m^{-3}, 2.6 kg m^{-3}, and 9.2 kg m^{-3} for the pulley system with CA (i.e., 190%, 246%, and 117% improvements, respectively). Besides, the MajiPump showed a significant water productivity improvement in the conventional tilled (CT) system. The water productivity of garlic, onion, and cabbage under MajiPump with CT was 1 kg m^{-3}, 6.7 kg m^{-3}, and 14.4 kg m^{-3}, respectively, and 0.6 kg m^{-3}, 0.2 kg m^{-3}, and 6.9 kg m^{-3} under the pulley with CT system (i.e., 67% to 325% higher than the pulley with CT system). Moreover, the water productivity under the MajiPump with CT was higher than the pulley system with CA. The water productivity under the MajiPump with CT was 6.7 kg m^{-3} and 14.4 kg m^{-3}, while it was 2.6 kg m^{-3} and 9.2 kg m^{-3} under the pulley system with CA, respectively, for onion and cabbage vegetables (i.e., 158% and 56% improvements, respectively).

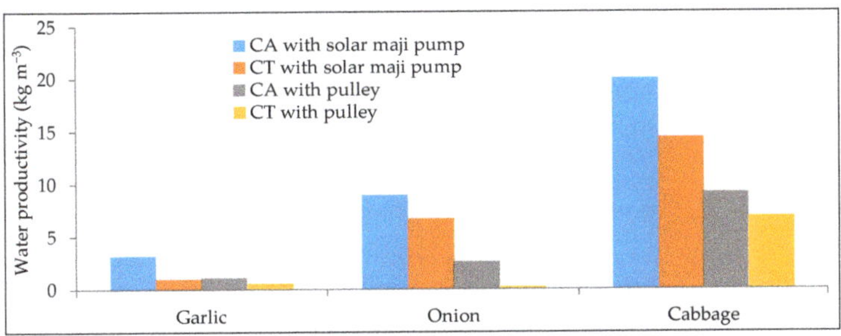

Figure 7. Comparison of MajiPump and pulley system on water productivity.

In general, the highest water productivity benefit could be gained through the combined use of MajiPump with CA practice. The highest water productivity in the MajiPump system was attributed to its additional advantage to increasing labor productivity. Farmers in the study area explained that water-lifting using the MajiPump took 5 to 10 min while a pulley system took 1.5 to 2 h to fill a 1.5 m height elevated 500 L water storage tank. Besides, filling a water storage tank with the pulley system requires two persons at a time when the labor time in the MajiPump system is only to connect the pump and solar panel. The minimum labor demand in using the MajiPump initiated smallholder farmers to provide enough irrigation water for vegetables, and thus increasing their water and labor productivity.

5. Summary and Conclusions

This research showed the potential benefits of the solar-powered water-lifting system (MajiPump) and CA technologies on water productivity and crop yields under on-farm conditions of smallholder farmers in Ethiopia. The capacity of two MajiPumps used in this study (MP400 and MP200) were found to extract water up to a maximum depth of 10 m using MP200 with 150 W rigid panel, 14 m using MP400 with 200 W flexible panel, and to 18 m using MP400 with 200 W rigid panel from shallow groundwater wells. The corresponding flow rate discharge capacity for these pumps and panel sizes were in the range of 7.8 L/min to 24.6 L/min, 3 L/min to 15 L/min, and 3.6 L/min, to 22.2 L/min, respectively.

Water and crop productivity were significantly increased under the CA farming system when compared with CT, both using farmers' common overhead irrigation (hose system). Water productivity was improved by 9% to 256%, and crop productivity was improved by 15% to 170% depending on the types of crops, and seasons of production (i.e., dry irrigated and rainfed). This shows the CA farming system has increased significant benefits (water-saving and crop yield increment) to farmers even using traditional overhead irrigation. However, the use of drip irrigation with the CA system further improved water and crop productivity as compared to the combination of the CA system with overhead irrigation. Besides, a significant increase in water productivity was observed in the combined use of MajiPump and CA when compared with the pulley water-lifting system. We conclude that the solar MajiPump with CA and drip irrigation is a promising approach to expand small-scale irrigation that can improve some key vegetable and grain crops of smallholder farmers in Ethiopia [7].

Author Contributions: T.T.A. contributed to the experimental design, data analysis and interpretation, and drafted the manuscript; T.F.A. contributed to the experimental design, data acquisition and analysis; A.Y.Y. contributed to experimental design, data collection and analysis; S.A.B. contributed to the data analysis and interpretation, revised the manuscript; Y.M.D. contributed to data collection and revising the manuscript; S.T.H. contributed to data collection and revising the manuscript, S.A.T. contributed to the data analysis and revising the manuscript; M.R.R. contributed to revising the manuscript; P.V.V.P. contributed to revising the manuscript. All authors have read and agreed to the published version of the manuscript.

Funding: This research was funded by the Appropriate Scale Mechanization Consortium of Feed the Future Innovation Lab for Collaborative Research on Sustainable Intensification (Cooperative Agreement No. AID-OAA-L-14-00006, Kansas State University) funded by United States Agency for International Development (USAID). The opinions expressed herein are those of the author(s) and do not necessarily reflect the views of the USAID or Kansas State University.

Data Availability Statement: The data presented in this study are available on request from the corresponding author.

Acknowledgments: We would like to acknowledge the Ethiopian National Meteorological Agency (ENMA) for providing quality data for this research. Contribution no. 21-122-J from Kansas Agricultural Experiment Station.

Conflicts of Interest: The authors declare no conflict of interest.

References

1.   Diao, X.; Nin Pratt, A. Growth options and poverty reduction in Ethiopia—An economy-wide model analysis. *Food Policy* **2007**, *32*, 205–228. [CrossRef]
2.   Weldearegawi, B.; Ashebir, Y.; Gebeye, E.; Gebregziabiher, T.; Yohannes, M.; Mussa, S.; Berhe, H.; Abebe, Z. Emerging chronic non-communicable diseases in rural communities of Northern Ethiopia: Evidence using population-based verbal au-topsy method in Kilite Awlaelo surveillance site. *Health Policy Plan.* **2013**, *28*, 891–898. [CrossRef] [PubMed]
3.   Belachew, T.; Hadley, C.; Lindstrom, D.; Mariam, A.G.; Lachat, C.; Kolsteren, P. Food insecurity, school absenteeism and educational attainment of adolescents in Jimma Zone Southwest Ethiopia: A longitudinal study. *Nutr. J.* **2011**, *10*, 29. [CrossRef] [PubMed]
4.   Awulachew, S.B.; Yilma, A.D.; Loulseged, M.; Loiskandl, W.; Ayana, M.; Alamirew, T. *Water Resources and Irrigation Development in Ethiopia*; IWMI: Colombo, Siri Lanka, 2007; Volume 123.
5.   Tesfaye, A.; Bogale, A.; Namara, R.E.; Bacha, D. The impact of small-scale irrigation on household food security: The case of Filtino and Godino irrigation schemes in Ethiopia. *Irrig. Drain. Syst.* **2008**, *22*, 145–158. [CrossRef]
6.   Haile, G.G.; Kasa, A.K. Irrigation in Ethiopia: A review. *Acad. J. Agric. Res.* **2015**, *3*, 264–269.
7.   Aseyehegu, K.; Yirga, C.; Rajan, S. Effect of Small-Scale Irrigation on the Income of Rural Farm Households: The Case of Laelay Maichew District, Central Tigray, Ethiopia. *J. Stored Prod. Postharvest Res.* **2011**, *2*, 208–215. [CrossRef]
8.   Mengistie, D.; Asmamaw, D.K. Assessment of the Impact of Small-Scale Irrigation on Household Livelihood Improvement at Gubalafto District, North Wollo, Ethiopia. *Agriculture* **2016**, *6*, 27. [CrossRef]
9.   Kuma, A.N.Y.; Alemu, A.A.T.; Nigussie, A.; Adisu, A.; Desalegn, K. Onion Production for Income Generation in Small Scale Irrigation Users Agropastoral Households of Ethiopia. *J. Hortic.* **2015**, *2*, 1–5. [CrossRef]
10. Gebrehiwot, N.T.; Mesfin, K.A.; Nyssen, J. Small-scale irrigation: The driver for promoting agricultural production and food security (the case of Tigray Regional State, Northern Ethiopia). *Irrig. Drain. Syst. Eng.* **2015**, *4*, 1000141. [CrossRef]
11. Assefa, T.; Jha, M.; Reyes, M.; Tilahun, S.; Worqlul, A.W. Experimental evaluation of conservation agriculture with drip irri-gation for water productivity in Sub-Saharan Africa. *Water* **2019**, *11*, 530. [CrossRef]
12. Berg, M.V.D.; Ruben, R. Small-Scale irrigation and income distribution in Ethiopia. *J. Dev. Stud.* **2006**, *42*, 868–880. [CrossRef]
13. Xie, H.; You, L.; Dile, Y.T.; Worqlul, A.W.; Bizimana, J.-C.; Srinivasan, R.; Richardson, J.W.; Gerik, T.; Clark, N. Mapping development potential of dry-season small-scale irrigation in Sub-Saharan African countries under joint biophysical and economic constraints—An agent-based modeling approach with an application to Ethiopia. *Agric. Syst.* **2021**, *186*, 102987. [CrossRef]
14. You, L.; Ringler, C.; Wood-Sichra, U.; Robertson, R.; Wood, S.; Zhu, T.; Nelson, G.; Guo, Z.; Sun, Y. What is the irrigation potential for Africa? A combined biophysical and socioeconomic approach. *Food Policy* **2011**, *36*, 770–782. [CrossRef]
15. Worqlul, A.W.; Jeong, J.; Dile, Y.T.; Osorio, J.; Schmitter, P.; Gerik, T.; Srinivasan, R.; Clark, N. Assessing potential land suita-ble for surface irrigation using groundwater in Ethiopia. *Appl. Geogr.* **2017**, *85*, 1–13. [CrossRef]
16. Assefa, T.; Jha, M.K.; Worqlul, A.W.; Reyes, M.R.; Tilahun, S.A. Scaling-Up Conservation Agriculture Production System with Drip Irrigation by Integrating MCE Technique and the APEX Model. *Water* **2019**, *11*, 2007. [CrossRef]
17. Awulachew, S.B. Irrigation potential in Ethiopia: Constraints and opportunities for enhancing the system. *Gates Open Re-search* **2019**, *3*. [CrossRef]
18. Worqlul, A.W.; Collick, A.S.; Rossiter, D.G.; Langan, S.; Steenhuis, T.S. Assessment of surface water irrigation potential in the Ethiopian highlands: The Lake Tana Basin. *Catena* **2015**, *129*, 76–85. [CrossRef]
19. Bacha, D.; Namara, R.E.; Bogale, A.; Tesfaye, A. Impact of small-scale irrigation on household poverty: Empirical evidence from the Ambo district in Ethiopia. *Irrig. Drain.* **2011**, *60*, 1–10. [CrossRef]
20. Derib, S.D.; Descheemaeker, K.; Haileslassie, A.; Amede, T. Irrigation water productivity as affected by water management in a small-scale irrigation scheme in the blue nile basin, ethiopia. *Exp. Agric.* **2011**, *47*, 39–55. [CrossRef]

21. Theis, S.; Lefore, N.; Meinzen-Dick, R.; Bryan, E. What happens after technology adoption? Gendered aspects of small-scale irrigation technologies in Ethiopia, Ghana, and Tanzania. *Agric. Hum. Values* **2018**, *35*, 671–684. [CrossRef]
22. Yimam, A.Y.; Assefa, T.T.; Adane, N.F.; Tilahun, S.A.; Jha, M.K.; Reyes, M.R. Experimental Evaluation for the Impacts of Conservation Agriculture with Drip Irrigation on Crop Coefficient and Soil Properties in the Sub-Humid Ethiopian Highlands. *Water* **2020**, *12*, 947. [CrossRef]
23. Evans, A.E.; Giordano, M.; Clayton, T. *Investing in Agricultural Water Management to Benefit Smallholder Farmers in Ethio-pia*; AgWater Solutions Project Country Synthesis Report; IWMI: Colombo, Siri Lanka, 2012; Volume 152.
24. Namara, R.E.; Hope, L.; Sarpong, E.O.; De Fraiture, C.; Owusu, D. Adoption patterns and constraints pertaining to small-scale water lifting technologies in Ghana. *Agric. Water Manag.* **2014**, *131*, 194–203. [CrossRef]
25. Belay, S.A.; Schmitter, P.; Worqlul, A.W.; Steenhuis, T.S.; Reyes, M.R.; Tilahun, S.A. Conservation Agriculture Saves Irrigation Water in the Dry Monsoon Phase in the Ethiopian Highlands. *Water* **2019**, *11*, 2103. [CrossRef]
26. Belay, S.A.; Assefa, T.T.; Prasad, P.V.V.; Schmitter, P.; Worqlul, A.W.; Steenhuis, T.S.; Reyes, M.R.; Tilahun, S.A. The Response of Water and Nutrient Dynamics and of Crop Yield to Conservation Agriculture in the Ethiopian Highlands. *Sustainability* **2020**, *12*, 5989. [CrossRef]
27. Assefa, T.T. Experimental and Modeling Evaluation of Conservation Agriculture with Drip Irrigation for Small-scale Agri-culture in Sub-Saharan Africa. Ph.D. Thesis, North Carolina Agricultural and Technical State University, Greensboro, NC, USA, 2018.
28. Assefa, T.T.; Jha, M.K.; Reyes, M.R.; Worqlul, A.W. Modeling the Impacts of Conservation Agriculture with a Drip Irrigation System on the Hydrology and Water Management in Sub-Saharan Africa. *Sustainability* **2018**, *10*, 4763. [CrossRef]
29. Assefa, T.; Jha, M.; Reyes, M.; Worqlul, A.; Doro, L.; Tilahun, S. Conservation agriculture with drip irrigation: Effects on soil quality and crop yield in sub-Saharan Africa. *J. Soil Water Conserv.* **2020**, *75*, 209–217. [CrossRef]
30. Heumesser, C.; Fuss, S.; Szolgayová, J.; Strauss, F.; Schmid, E. Investment in irrigation systems under precipitation uncer-tainty. *Water Resour. Manag.* **2012**, *26*, 3113–3137. [CrossRef]
31. Burney, J.; Woltering, L.; Burke, M.; Naylor, R.; Pasternak, D. Solar-powered drip irrigation enhances food security in the Sudano–Sahel. *Proc. Natl. Acad. Sci. USA* **2010**, *107*, 1848–1853. [CrossRef]
32. Gowing, J.; Walker, D.; Parkin, G.; Forsythe, N.; Haile, A.T.; Ayenew, D.A.; Alamirew, D. Can shallow groundwater sustain small-scale irrigated agriculture in sub-Saharan Africa? Evidence from N-W Ethiopia. *Groundw. Sustain. Dev.* **2020**, *10*, 100290. [CrossRef]
33. Siebert, S.; Burke, J.; Faures, J.-M.; Frenken, K.; Hoogeveen, J.; Döll, P.; Portmann, F.T. Groundwater use for irrigation–A global inventory. *Hydrol. Earth Syst. Sci.* **2010**, *14*, 1863–1880. [CrossRef]
34. Nigussie, L.; Lefore, N.; Schmitter, P.; Nicol, A. *Gender and Water Technologies: Water Lifting for Irrigation and Multiple Purposes in Ethiopia*; International Livestock Research Institute (ILRI); East Africa and Nile Basin Office: Addis Ababa, Ethiopia, 2017.
35. Gray, C.; Mueller, V. Drought and Population Mobility in Rural Ethiopia. *World Dev.* **2012**, *40*, 134–145. [CrossRef] [PubMed]
36. Biswas, S.; Iqbal, M.T. Dynamic Modelling of a Solar Water Pumping System with Energy Storage. *J. Sol. Energy* **2018**, *2018*, 1–12. [CrossRef]
37. Kelley, L.C.; Gilbertson, E.; Sheikh, A.; Eppinger, S.D.; Dubowsky, S. On the feasibility of solar-powered irrigation. *Renew. Sustain. Energy Rev.* **2010**, *14*, 2669–2682. [CrossRef]
38. Gupta, E. The impact of solar water pumps on energy-water-food nexus: Evidence from Rajasthan, India. *Energy Policy* **2019**, *129*, 598–609. [CrossRef]
39. Chandel, S.; Naik, M.N.; Chandel, R. Review of solar photovoltaic water pumping system technology for irrigation and community drinking water supplies. *Renew. Sustain. Energy Rev.* **2015**, *49*, 1084–1099. [CrossRef]
40. Foley, G. *Photovoltaic Applications in Rural Areas of the Developing World*; World Bank: Washington, DC, USA, 1995; Volume 304.
41. Beriso, B.S. Prevalence of protein-energy malnutrition in children under five years of age admitted to pediatric wards at Asella Referral and Teaching Hospital, Arsi Zone, Oromiya, Ethiopia. *East Afr. J. Sci.* **2019**, *13*, 81–88.
42. Belay, M.; Bewket, W. Traditional Irrigation and Water Management Practices in Highland Ethiopia: Case Study IN Dangila Woreda. *Irrig. Drain.* **2013**, *62*, 435–448. [CrossRef]
43. Lemma, T.; Sehai, E.; Hoekstra, D. *Status and Capacity of Farmer Training Centers (FTCs) in the Improving Productivity and Market Success (IPMS) Pilot Learning Woredas (PLWs)*; International Livestock Research Institute (ILRI): Addis Ababa, Ethiopia, 2011.
44. Walker, D.; Parkin, G.; Schmitter, P.; Gowing, J.; Tilahun, S.A.; Haile, A.T.; Yimam, A.Y. Insights from a multi-method re-charge estimation comparison study. *Groundwater* **2019**, *57*, 245–258. [CrossRef] [PubMed]
45. Abay, A. *Market Chain Analysis of Red Pepper: The Case of Bure Woreda, West Gojjam Zone, Amhara National Regional State, Ethiopia*; Haramaya University: Harar Haramaya, Ethiopia, 2010.
46. Eng, J. Sample Size Estimation: How Many Individuals Should Be Studied? *Radiology* **2003**, *227*, 309–313. [CrossRef] [PubMed]
47. De Winter, J.C. Using the Student's t-test with extremely small sample sizes. *Pract. Assess. Res. Eval.* **2013**, *18*, 10.
48. Assefa, T.; Jha, M.; Reyes, M.; Srinivasan, R.; Worqlul, A.W. Assessment of suitable areas for home gardens for irrigation po-tential, water availability, and water-lifting technologies. *Water* **2018**, *10*, 495. [CrossRef]
49. Iwama, K. Physiology of the Potato: New Insights into Root System and Repercussions for Crop Management. *Potato Res.* **2008**, *51*, 333–353. [CrossRef]
50. Gao, Y.; Duan, A.; Qiu, X.; Liu, Z.; Sun, J.; Zhang, J.; Wang, H. Distribution of roots and root length density in a maize/soybean strip intercropping system. *Agric. Water Manag.* **2010**, *98*, 199–212. [CrossRef]

51. Westerveld, S.M.; McKeown, A.W.; McDonald, M.R. Distribution of nitrogen uptake, fibrous roots and nitrogen in the soil profile for fresh-market and processing carrot cultivars. *Can. J. Plant Sci.* **2006**, *86*, 1227–1237. [CrossRef]
52. Hsu, S.-L.; Hung, J.; Wallace, A. Soil pH Variation Within a Soil. I. pH Variation in Soil Pores Observed in a Column-Leaching Method. *Commun. Soil Sci. Plant Anal.* **2004**, *35*, 319–329. [CrossRef]
53. Tesema, M.; Schmitter, P.; Nakawuka, P.; Tilahun, S.A.; Steenhuis, T.; Langan, S. Evaluating Irrigation Technologies to Im-prove Crop and Water Productivity of Onion in Dangishta Watershed During the Dry Monsoon Phase. In Proceedings of the Fourth International Conference on the Advancement of Science and Technology in Civil and Water Resources Engineering, Bahir Dar, Ethiopia, 13–29 November 2019.
54. Zotarelli, L.; Dukes, M.D.; Romero, C.C.; Migliaccio, K.W.; Morgan, K.T. *Step by Step Calculation of the Penman-Monteith Evapotranspiration (FAO-56 Method)*; Institute of Food and Agricultural Sciences, University of Florida: Gainesville, FL, USA, 2010.
55. Burney, J.; Naylor, R.L.; Postel, S.L. The case for distributed irrigation as a development priority in sub-Saharan Africa. *Proc. Natl. Acad. Sci. USA* **2013**, *110*, 12513–12517. [CrossRef]
56. *Water Requirements for Irrigation and the Environment*; Springer Science and Business Media LLC: Berlin, Germany, 2009.
57. Allen, R.G.; Pereira, L.S.; Raes, D.; Smith, M. Crop evapotranspiration-Guidelines for computing crop water requirements-FAO Irrigation and drainage paper 56. *FAO Rome* **1998**, *300*, D05109.
58. Ali, M.; Talukder, M. Increasing water productivity in crop production—A synthesis. *Agric. Water Manag.* **2008**, *95*, 1201–1213. [CrossRef]
59. Gallardo, A. Spatial Variability of Soil Properties in a Floodplain Forest in Northwest Spain. *Ecosystems* **2003**, *6*, 564–576. [CrossRef]
60. Aredehey, G.; Libsekal, H.; Brhane, M.; Welde, K.; Giday, A.; Moral, M.T. Top-soil salinity mapping using geostatistical approach in the agricultural landscape of Timuga irrigation scheme, South Tigray, Ethiopia. *Cogent Food Agric.* **2018**, *4*, 1514959. [CrossRef]
61. Kigalu, J.M.; Kimambo, E.I.; Msite, I.; Gembe, M. Drip irrigation of tea (*Camellia sinensis* L.): 1. Yield and crop water produc-tivity responses to irrigation. *Agric. Water Manag.* **2008**, *95*, 1253–1260. [CrossRef]
62. Dawit, M.; Dinka, M.O.; Leta, O.T. Implications of Adopting Drip Irrigation System on Crop Yield and Gender-Sensitive Issues: The Case of Haramaya District, Ethiopia. *J. Open Innov. Technol. Mark. Complex.* **2020**, *6*, 96. [CrossRef]
63. Fandika, I.R.; Kadyampakeni, D.M.; Zingore, S. Performance of bucket drip irrigation powered by treadle pump on tomato and maize/bean production in Malawi. *Irrig. Sci.* **2011**, *30*, 57–68. [CrossRef]
64. Elhindi, K.M.; El-Hendawy, S.; Abdel-Salam, E.; Elgorban, A.; Ahmed, M. Impacts of fertigation via surface and subsurface drip irrigation on growth rate, yield and flower quality of *Zinnia elegans*. *Bragantia* **2015**, *75*, 96–107. [CrossRef]
65. Mirjat, M.; Jiskani, M.; Siyal, A.; Mirjat, M. Mango production and fruit quality under properly managed drip irrigation sys-tem. *Pak. J. Agric. Agric. Eng. Vet. Sci.* **2011**, *27*, 1–12.

MDPI
St. Alban-Anlage 66
4052 Basel
Switzerland
Tel. +41 61 683 77 34
Fax +41 61 302 89 18
www.mdpi.com

Agronomy Editorial Office
E-mail: agronomy@mdpi.com
www.mdpi.com/journal/agronomy